DYNAMICS OF
MARINE ECOSYSTEMS

Biological–Physical Interactions in the Oceans
Second Edition

DYNAMICS OF
MARINE ECOSYSTEMS

Biological–Physical Interactions in the Oceans
Second Edition

K. H. MANN & J. R. N. LAZIER
DEPARTMENT OF FISHERIES AND OCEANS
BEDFORD INSTITUTE OF OCEANOGRAPHY
DARTMOUTH, NOVA SCOTIA

b
**Blackwell
Science**

©1996 by Blackwell Science, Inc.

Editorial Offices:
Commerce Place, 350 Main Street
  Malden, Massachusetts 02148
  USA
Osney Mead, Oxford OX2 0EL
  England
25 John Street, London WC1N 2BL
  England
23 Ainslie Place, Edinburgh EH3 6AJ
  Scotland
54 University Street, Carlton, Victoria 3053
  Australia

Other Editorial Offices:
Blackwell Wissenschafts-Verlag GmbH
  Kurfürstendamm 57, 10707 Berlin
  Germany
Blackwell Science KK
  MG Kodenmacho Building, 7-10
  Kodenmacho Nihombashi, Chuo-ku
  Tokyo 104
  Japan

Distributors:
USA
  Blackwell Science, Inc.
  Commerce Place
  350 Main Street
  Malden, Massachusetts 02148
  (Telephone orders: 800-215-1000 or
  781-388-8250; fax orders: 781-388-8270)

Canada
  Login Brothers Book Company
  324 Saulteaux Crescent
  Winnipeg, Manitoba, R3J 3T2
  (Telephone orders: 204-837-2987)

Australia
  Blackwell Science Pty, Ltd.
  54 University Street
  Carlton, Victoria 3053
  (Telephone orders: 03-9347-0300;
  fax orders: 03-9349-3016)

Outside North America and Australia
Blackwell Science, Ltd.
c/o Marston Book Services, Ltd.
P.O. Box 269
Abingdon
Oxon OX14 4YN
England
(Telephone orders: 44-01235-465500;
fax orders: 44-01235-465555)

Acquisitions: Jane Humphreys
Production: Irene Herlihy
Manufacturing: Lisa Flanagan

Typeset by BookMasters, Inc.
Printed and bound by Edwards Brothers Inc.

Printed in the United States of America
99 00 01 02 5 4 3 2

The Blackwell Science logo is a trade mark of
Blackwell Science Ltd., registered at the United
Kingdom Trade Marks Registry

Library of Congress Cataloging-In-Publication
Data

Mann, K. H. (Kenneth Henry), 1923–
    Dynamics of marine ecosystems: biological-
physical interactions in the oceans / K. H. Mann
& J. R. N. Lazier.—2nd ed.
       p.   cm.
    Includes bibliographical references (p.   )
    and index.
    ISBN 0-86542-539-6 (alk. paper)
    1. Marine ecology. 2. Biotic communities.
I. Lazier, J. R. N.   II. Title
QH541.5.S3M25   1996
574.5'2632—dc20   96-8573
                                   CIP

Dedicated to the memory of

GORDON A. RILEY

A good friend and colleague

who was thinking and writing about

these same topics 50 years ago

Contents

Preface to Second Edition

We have been surfing a wave of interest in biological-physical interactions in the ocean. In the course of updating this text we have found that there has been an explosion of publications in integrated oceanography, particularly at the two ends of the space–time scale.

At the scale of millimeters to meters there have been important advances in our understanding of the influence of turbulence on processes in plankton. There have been numerous experimental studies on the effects of turbulence on nutrient uptake by phytoplankton, and on the encounter rate between predator and prey. A new understanding has been reached of the physical processes that cause dissolved organic matter to aggregate on surfaces and form colloids, which in turn aggregate until there are organic particles of the size range suitable for food for plankton and benthos.

At the scale of hundreds of meters, there have been numerous field experiments on the relationship between the mixed layer and the deep ocean, designed to elucidate the mechanisms governing the seasonal rise and fall of phytoplankton and zooplankton biomass. When combined with the results of remote sensing on a much larger scale, it has been possible to describe and quantify the seasonal patterns of primary production for all parts of the world ocean. This development holds enormous potential for advancing our understanding of the role of the ocean biota in global climate change.

In the past five years we have come to a much deeper understanding of the importance of the major interannual patterns of atmospheric change, such as the Southern Oscillation and the Aleutian and Icelandic low pressure systems. Local shifts in biological production and the concomitant changes in fish stocks are now seen as part of global scale processes.

A view of the global ocean and the atmosphere as a coupled system is just coming into focus for the first time and the implications for biology are exciting. They are likely to lead to a new appreciation of the role of global processes, over which we have little or no control, in determining the magnitude and distribution of the fish and shellfish stocks that are an important part of the food resources of the human population. This, in turn, should lead to a more conservative approach to the management of marine living resources.

We thank Neil Bellefontaine, Steve McPhee, Jim Elliott, Allyn Clarke, Mike Sinclair, Don Gordon, and Paul Keizer for their continued interest and support during the revision of this volume. We are especially grateful to Jane Humphreys, our editor in the Cambridge office of Blackwell Science, for her friendly advice and assistance throughout the process, and to our wives Isabel Mann and Catherine Lazier for their continuing support and encouragement.

K.H.M., J.R.N.L.

Preface to First Edition

In an earlier book by the senior author entitled *Ecology of Coastal Waters: A Systems Approach,* marine ecosystems were described in terms of their characteristic primary production, whether by phytoplankton, seaweed, mangrove, marsh grass or seagrass. Estimates were presented of the annual mean values for primary production, and pathways of energy flow were traced through the food webs. One chapter was devoted to water movement and productivity. A reviewer commented that the book was too much about mean flows and not enough about variance.

Reflecting on this, it was clear that much of the variance in marine productivity is a function of water movement. Decomposition and liberation of nutrients tend to take place in deep water or on the sea floor and water movement is needed to bring those nutrients back up into the euphotic zone for use by the primary producers. Tides give water movement a diurnal and fortnightly periodicity, while seasonal changes in solar heating impose changes in the mixed layer on scales of months to a year. Long-term climatic cycles impose their own variations on water movement and hence on biological productivity. The theme for this book began to crystallize as an expansion of the earlier chapter on water movement and productivity considered at a range of temporal and spatial scales.

While the ideas were developing, remarkable changes were occurring in oceanography. More and more, biological oceanographers were teaming up with physical and chemical oceanographers to study marine ecosystems in their totality. Physical oceanographers were increasingly able to explain to their colleagues what was going on in gyres, at fronts, on banks or in estuaries, and the biologists were developing the instrumentation needed to obtain continuous records of biological variables to supplement the spot samples that had been characteristic of biological oceanography for decades. Satellite observations of ocean colour were giving large-scale perspectives on ocean productivity undreamed of by earlier generations. The feeling emerged that marine ecology was coming of age. It was developing a new maturity based on the integration of disciplines, and in the process yielding important new understanding about ecosystem function. We therefore decided to use the theme of physical processes and productivity as a starting point for an account of recent developments in marine ecology, in which physics, chemistry and biology are inter-related aspects of the dynamics of marine ecosystems.

Formal courses in oceanography tend to separate marine biology and marine physics. Our aim is to emphasize the links between the two subjects by presenting in each chapter the relevant physical processes along with the biology. Because the reader is expected to have a more complete background in biology than in physics the two subjects are written from slightly different viewpoints. The presentation of the physics is fairly elementary and emphasizes the important physical processes, while the presentation of the biology emphasizes the recent development of the field. To assist the physical presentation we have used some mathematical symbols and equations simply because they are part of the language of the subject and provide a useful shorthand for presenting ideas. It is for example much easier and more precise to write the symbol for the derivative of a variable than to write it out every time it is used. In a further attempt to present the physics in manageable portions some of the details required for a more advanced understanding have been separated into boxes which may be skipped on first reading. The references for both the physics and biology, though numerous, are by no means exhaustive and where good reviews exist we have drawn attention to them and left the reader to find the original sources.

In view of the current concern about the role of the oceans in climate change, we believe that there will be an increasing need to understand the integrated biological-physical functioning of marine ecosystems. We therefore hope that professional researchers in the various disciplines of oceanography will find this book of value in broadening their understanding of marine ecology, as an aid to defining those research programs that will be needed if we are to anticipate the consequences of global change.

In covering such a broad field we have relied heavily on the advice and assistance of many colleagues. For biological material Glen Harrison, Steve Kerr, Alan Longhurst, Eric Mills, Trevor Platt and Mike Sinclair have been particularly helpful, while on the physics and chemistry side we have enjoyed the advice of Allyn Clarke, Fred Dobson, David Greenberg, Ross Hendry, Edward Horne, Peter Jones, Hal Sandstrom, John Loder, Neil Oakey and Stuart Smith. We thank Mark Denny, Mike Keen and Jim McCarthy for helpful comments on various parts of the manuscript, and we particularly thank our editor, Simon Rallison, for his most helpful advice and guidance at all stages of this project. We wish to thank Betty Sutherland and her library staff in the Bedford Institute of Oceanography for expert assistance with the literature and for suffering more or less continuous occupation of part of the library over an extended period. We also wish to thank Steve McPhee, Jim Elliott and Mike Sinclair of the Science Branch of the Department of Fisheries and Oceans for supporting us in our endeavours and our wives Isabel Mann and Catherine Lazier for their encouragement, enthusiasm and patience.

K.H.M., J.R.N.L.

1

Marine Ecology Comes of Age

Marine ecology, as traditionally understood, is the study of marine organisms and their relationship with other organisms and with the surrounding environment. The subject parallels similar studies of organisms on land but, while terrestrial organisms are relatively easy to observe and manipulate, marine organisms are much more inaccessible. This inaccessibility has led to a slower growth of knowledge. The physical factors leading to fertile and infertile areas are very different on land than in the ocean. The nutrients required by land plants are generated nearby from the decaying remains of previous generations, but decaying matter in the ocean tends to sink and leave the sunlit euphotic layer where plants grow. The nutrients supplied by the decay are thus unavailable for plant growth unless some physical mechanisms bring the nutrients back up to the surface. This book is largely concerned with those mechanisms and the resulting biological phenomena.

It is now possible to add an extra dimension to marine ecology. Instead of putting the organisms at the center of the picture and considering them in relationship to other organisms and the environment, it is possible to work with marine ecosystems in which physical, chemical, and biological components are equally important in defining total system properties. Those properties include production of living organisms such as fish, but flux of carbon dioxide as determined by both physical and biological processes may be more important in the context of climate change.

Interest and research activity in marine ecology are intensifying. There are many reasons for this trend, of which four may be mentioned:

1 The physical processes underlying some of the large-scale biological phenomena are now better understood. For example, strides have been made in deciphering the mechanisms underlying the major coastal upwelling systems of the world and their variations in time, the connection between variations in the Peruvian upwelling system and the southern oscillation in the atmosphere, and the physical mechanisms giving rise to the biologically important tidal fronts.

2 There have been important advances in our ability to make continuous, fine-scale biological measurements by means of automated sensors feeding into computers. It is now possible to collect biological data with a coverage and resolution comparable with the best physical data. Examples are the measurement of chlorophyll by means of satellites or with towed fluorimeters, and the electronic counting of zooplankton. The satellite image in Plate 1 shows the global distribution of chlorophyll in surface waters, and reveals a great deal about the incidence of upwelling and the exchange of gases with the atmosphere.

3 The need to understand marine ecological processes influencing the greenhouse effect and other aspects of world climate is becoming more urgent. The flux of carbon dioxide from the atmosphere into surface waters and on down into the deep ocean, as a result of biological processes, is believed to be an important part of the mechanism of climate change.

In this connection, the development of the spring bloom in temperate waters and the formation of patches of high productivity in otherwise unproductive tropical oceans are fields of active research with promising futures.

4   The increased understanding of fundamental processes is beginning to influence the management of the ocean's living resources. The way in which year-to-year changes in the ocean's characteristics affect the growth and survival of fish larvae and the effects of changing weather patterns on the distribution of fish are two topics that will receive a great deal of attention in the coming decades.

For all of these reasons, marine ecology has changed rapidly and may be said to have come of age. The dominant theme of this book is that physical processes create the conditions for many important biological processes; the biology cannot be understood in isolation. One good example of this close link is found in shelf-sea tidal fronts, which have been known for many years to be biologically important. As a result of elegant analysis the location of these fronts can now be predicted with simple calculations involving the speed of the tidal current and the depth of the water. Another example is the discovery that spring phytoplankton blooms are generated when organisms are confined to the euphotic layer by vertical stratification developed in the water by the sun's warming. In this volume, the connections between the physical and biological processes are emphasized and brought into focus more sharply than before.

The nature of the relationships between physical and biological processes is subtle and complex. Not only do the physical processes create a structure, such as a shallow mixed layer, or a front, within which biological processes may proceed, but they also influence the rates of biological processes in many indirect ways. Discussion of this relationship has most often been in terms of energy flow. Biologists often model food web relationships in terms of the flow of solar energy, captured in photosynthesis by the phytoplankton and passed from organism to organism by means of feeding transfers. The physical phenomena such as currents, turbulence, and stratification also rely on solar energy, transmitted to the water directly as heat or indirectly as momentum from the wind. These two fluxes of solar energy are in one sense quite distinct: organisms do not use the energy of water motion for their metabolic needs. In another sense, they are inter-related. Water movement alters the boundary layers around organisms, transports nutrients and waste products, assists migrations, and influences the rate of encounter between planktonic predators and their prey. Stratification causes the retention of planktonic organisms in the upper layer of the ocean, making light more available but limiting access to inorganic nutrients. Water temperature has a profound influence on the rates at which biological processes proceed, and differences in water motion, from place to place, largely determine the kinds of organisms colonizing those places. From a biological point of view, the physical energy is termed auxiliary energy, which literally means 'helping energy.'

LENGTH SCALES

In approaching the subject it is useful to have a feeling for the dimensions of the organisms and phenomena to be discussed (Fig. 1.01). Ocean basins are typically 10,000 km wide and confine the largest biological communities. The average depth of the ocean is 3800 m but the depths of the euphotic layer (≈ 100 m) and the mixed layer (≈ 100 m) are more often critical to open-ocean biological processes.

dioxide content of the atmosphere would be much higher, the earth would be much hotter, and the circulation of the oceans would be totally different. Even on the short time scale there are examples of phytoplankton altering the penetration of light and heat into the water column and hence the functioning of the ecosystem. Interactions between physics and biology are not entirely, or even mainly, in one direction.

The third question has been much discussed without any real resolution. It is a problem for ecologists generally, for we do not understand how to include bacterial processes on scales of millimeters and seconds in the same models that deal with animals that live for decades and may range over thousands of kilometers. Marine ecologists have the added difficulty that the biological events take place in a physical medium that exhibits processes on the same range of scales, thus compounding the difficulties.

We have found it useful to keep these questions in mind as we review the developments of marine ecology as an integrated physical, chemical, and biological discipline.

PART A
PROCESSES ON A SCALE OF LESS THAN 1 KILOMETER

2

Biology and Boundary Layers

2.1 INTRODUCTION

In this chapter we shall explore the intimate relationships between the small-scale processes in sea water and the lives of plants and animals. In order to do so, we shall have to shed many of the concepts that are ingrained in our way of thinking simply because we inhabit bodies of a particular size. To take one example, it seems natural for us to think that if we are in the sea and use our arms to push water backward, we shall move forward, coasting for many seconds or minutes before the viscosity of the water brings us to a halt. For a microorganism this case is not true. Viscosity is all-important. A picoplankton cell of about $1\,\mu m$ diameter swimming at about $30\ \mu m\ s^{-1}$ and then stopping would come to a halt in about $0.6\ \mu s$, having travelled only about $10^{-4}\ \mu m$ (Purcell 1977). Alternatively, a small crustacean that extended a pair of stiff limbs at right angles to the body and attempted to 'row' itself forward would rock forward and back, staying in exactly the same place. Hence, traditional ideas about the locomotion of small organisms have to be drastically modified. In order to do so we have to understand that motion through the water is a function of two key variables, momentum and viscosity, and that the relative proportions of these variables change according to the scale of events being studied.

Consider the situation of a planktonic larva that is approaching the sea bed and is about to choose a site for settlement and metamorphosis. Interesting laboratory studies have been made, showing, for example, how certain larvae respond to chemical cues. In the real world, most areas of the sea bottom are exposed to one or two daily cycles of tidal currents. As Simpson (1981) put it, 'in stress terms, these tidal streams are equivalent to hurricane-force winds in the atmosphere blowing regularly twice per day.' Careful analysis shows that the only place that larvae can find water quiet enough for them to swim about and explore the bottom is a thin layer about 100 μm deep immediately adjacent to the sea floor. In this thin layer they barely have room to maneuver, and it turns out that they use their swimming powers only to descend, sample the sea floor, and rise again if it is unsuitable. This process has been called the 'balloonist technique.' An understanding of the situation requires familiarity with the fundamental properties of boundary layers that form around objects when the water is in motion relative to the object.

We turn now to a consideration of the small-scale boundary layers of the surfaces of phytoplankton or seaweeds. Viscosity causes the average speed of the flow to decrease from its value in the open water to zero at the boundary. The size of the turbulent eddies in the water also decreases to zero at the boundary. This change creates problems for organisms that require the transport of nutrients toward their surfaces and waste substances away from them. Turbulent eddies transport nutrients efficiently in the open water but are too weak to transport nutrients through the boundary layer. Unless some special action is taken, the organism's metabolism is restricted by the lack of turbulent transport, which is replaced by very slow transport due to molecular diffusion. This restriction is often known as diffusion limitation of metabolism.

The thickness of a boundary layer is reduced in proportion to the speed of the water moving past it. For large plants such as seaweeds, thinning of the boundary layer is achieved by attaching themselves to a solid surface in a zone where tidal currents and wave action cause vigorous water movement. This technique has its dangers, for if the water movement is excessive, the drag on the plant may tear it from its attachment. Many seaweeds are capable of making changes in their shape during growth, to reflect the trade-off between the need to maximize turbulence close to the plant surface and the need to reduce drag.

The conventional understanding of a planktonic organism is that it moves passively with the water. However, the need to overcome diffusion limitation is just as real for a phytoplankton cell as it is for a seaweed. There are two main techniques available. One is to have a heavy cell wall that causes the cell to sink through the water column. This technique is employed by diatoms and works best in mixed layers where the stirring tendency of the turbulent flow counteracts the sinking of the organisms. Without the turbulence the diatoms would all end up in deep water, where there is insufficient light for photosynthesis. Investigation of the efficacy of the technique requires an understanding of the physics governing the sinking rate of particles in a fluid, as well as the physics of small-scale boundary layers.

The second technique adopted by phytoplankton is to perform locomotory movements. This solution has been adopted by the flagellates, but it is a far from simple process because of the problems associated with overcoming viscosity at small scales.

When we take all these physical aspects of life in the sea into consideration, we find that many of our existing concepts, based on experiments carried out in the laboratory in still water, are in need of drastic revision. Grappling with the physics of turbulent flow is hard labor for many biologists, but it is absolutely essential for understanding contemporary marine ecology.

2.2 PHYTOPLANKTON AND BOUNDARY LAYERS

Phytoplankton productivity in the world ocean is now a major concern, because of the role it is thought to play in modifying the carbon dioxide content of the atmosphere and hence the scenario for global climate change. One of the major themes running through this book is the need to understand how phytoplankton productivity is influenced by the physics of the ocean. This influence operates at many scales, from ocean basin circulation, through localized areas of upwelling, down to the smallest scales of turbulence that affect individual cells. Kinetic energy is imparted to the world ocean by sun, wind, and tides, and the energy of large-scale motions is transmitted progressively to smaller and smaller scales of motion until, at very small scales, the motion is resisted by the molecular viscosity of the water and is eventually dissipated as heat. In the sections that follow, we shall make a fairly long detour into the physics of turbulence, viscosity, molecular and turbulent diffusion, and the structure of boundary layers, before returning to the physiology of phytoplankton in section 2.2.10.

2.2.1 Turbulent motion

To begin, consider the hypothetical record of velocity in the ocean shown in Fig. 2.01. The signal has a mean or average, $\bar{u}$, over the record, but at most times the velocity deviates from the mean by an amount u', called the fluctuating part of the flow. The sum of the two at any instant gives the total velocity.

The fluctuations in the motion indicate the presence of turbulence. When $u' = 0$, turbulence is absent and the flow is said to be smooth, or laminar. For turbulent flow, u' is a function of time and is made up of fluctuations of many periods randomly mixed together. The most rapid fluctuations may have a period of about a second and are the smallest turbulent motions with scales of a few millimeters. The longest fluctuation in the record may represent motions that are a few meters in size with periods of tens of minutes.

Turbulent motions of these small scales are usually assumed to be three-dimensional and statistically similar in all directions; that is, the turbulence is said to be homogeneous

Fig. 2.01 A hypothetical record of water velocity in turbulent flow illustrating the difference between the average value $\bar{u}$ and the fluctuating component u'.

or isotropic. This convention arose because many major advances in the theoretical under-standing of turbulent motion associated with important practical problems became possible only by using this assumption, which allowed a great simplification to the equations gov-erning turbulent motion. This idealized state is a fairly good assumption for scales between the viscous and buoyancy scales, calculated later, but for scales as large as or larger than the depth of water, the motions lose homogeneity and become two-dimensional eddies that are sometimes called geostrophic turbulence to denote the fact that the eddies are random, large, and adjusted to the influence of the rotation of the earth.

2.2.2 Sources of turbulent energy

The energy in the turbulent eddies is extracted from the larger-scale motions via many different instability mechanisms. The most common and widely known instability is the breaking of surface waves that occurs when the waves get too steep. The breaking con-verts the regular and predictable motion of the wave into random turbulent motion. Deeper down in the ocean, internal waves propagate on and through the vertical den-sity gradients. These waves also can become unstable and break up into turbulence. In the upper layer of the ocean, the wind, besides generating waves, forces the water to move relative to the layers below. This relative movement, or shear, can also lead to un-stable motions that break the flow up into turbulent motions. Finally, the large perma-nent currents, such as the Gulf Stream, develop meanders that create the large two-dimensional eddies of the geostrophic turbulence that eventually break up into smaller scales of motion.

2.2.3 Viscosity

The energy in turbulent motion is continually being transferred from large scales of motion to small scales. The little eddies that are 5 cm across get their energy from larger eddies, which in turn get their energy from still larger ones. This process, called the energy cascade, does not change the total amount of energy in the turbulence nor does it convert the kinetic energy of the turbulent motion to another form of energy.

  With the decrease in the size of the turbulent eddies comes an increase in the velocity gradient across the eddies. When the eddies are small enough and the velocity shear is great enough, then molecular viscosity, the internal resistance of the water, acts to resist and smooth out the gradients in velocity. This smoothing of the flow by viscosity is the way the energy in the turbulence is finally converted to heat and dissipated. The stress generated by the viscous forces is discussed in Box 2.01.

2.2.4 Comparing forces, the Reynolds number

The importance of the forces due to viscosity is often quantified by calculating the Reynolds number, which is the ratio of the inertial force to the viscous force acting on the body of in-

Box 2.01 Calculating the stress due to viscosity

Viscous stresses for most oceanic phenomena are negligible, partly because the viscosity of water is so small—in fact one of the lowest found in naturally occurring liquids—and partly because only at the smallest scales are the velocity gradients large enough to make the viscous stresses significant when compared to the other forces present. A simple calculation shows the magnitude of the viscous forces in a specific situation. The molecular viscous stress τ that one layer such as A in Fig. 2.02 exerts upon layer B is

$$\tau = \rho v \, d\bar{u}/dz \qquad (2.01)$$

where $\rho \simeq 10^3$ kg m^{-3} is the density, $v \simeq 10^{-6}$ m^2 s^{-1} is the coefficient of kinematic viscosity, and $d\bar{u}/dz$ is the gradient of the average velocity perpendicular to the flow. A typical change in mean velocity in the ocean of 1.0 m s^{-1} over 1000 m gives a velocity gradient of 10^{-3} s^{-1} and leads to a minute viscous stress of 10^{-6} N m^{-2}. Where velocity changes by 0.001 m s^{-1} over 0.01 m, as it may in small eddies, the viscous stress is a significant 10^{-4} N m^{-2}.

terest, be it an animal or fluid element. The inertial force is the force that was necessary to accelerate the body to the velocity it now possesses, or to stop the body now travelling at a constant speed under its own inertia. The ratio, a dimensionless number, works out to be the velocity u times a typical dimension d divided by the kinematic viscosity v,

$$\text{Re} = ud/v \qquad (2.02)$$

The number was originally used as a criterion of whether flow in pipes would be smooth or turbulent. The typical dimension d was the pipe diameter, but it may equally be a typical dimension, such as length of width, of an organism around which flow is being considered.

The most useful feature of dimensionless numbers, such as the Reynolds number, is that they apply, universally, to all similar situations. For pipes of any size, for example, the division between smooth and turbulent motion occurs when Re $\simeq$ 2500, which suggests that

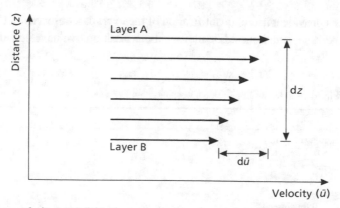

Fig. 2.02 The mean velocity $\bar{u}$, parallel to the x axis, increases by an amount $d\bar{u}$ in the distance dz, but viscosity creates a stress across the gradient that tends to retard the faster moving water at level A, and speed up the slower moving water at level B.

the viscous forces keep the flow smooth by suppressing small irregularities even though the inertial forces are 2500 times as large as the viscous forces. Since Reynolds first introduced the ratio it has been applied usefully in many quite different situations, but all for the same purpose—to compare the inertial and viscous forces.

Sometimes the ratio is applied to bodies of water. For example, if a mass of water, 1 km across, is moving with a uniform speed of 10 cm s^{-1}, Re $\simeq 10^8$ indicating a region where the viscous forces are too small to suppress the small perturbations that grow into turbulent eddies. On the other hand, if the water mass is 1 cm across and moving at 1 cm s^{-1}, Re $\simeq 10^2$ indicating a flow where viscous forces are getting to be important in suppressing small perturbations in the flow.

It is also common to see calculations of the Reynolds number of solid bodies in the water such as grains of sand and animals. A 0.1 m fish swimming at 1.0 m s^{-1}, for example, has a Reynolds number of 10^5. Obviously inertial forces dominate its life and viscosity can be ignored when considering the fish as a whole. A microscopic animal 50 μm long swimming at 10 μm s^{-1}, on the other hand, exhibits the minuscule Reynolds number of 5×10^{-4}. Inertial forces can be ignored in this animal's world, which is dominated by viscous forces. The Reynolds number is, then, a useful guide in assessing the relative strength of the inertial and viscous forces and for comparing similar situations.

The enormous range of Reynolds numbers associated with living organisms is illustrated in Table 2.1, compiled by Vogel (1981). The Reynolds number of an even wider range of organisms was calculated by Okubo (1987) using a characteristic dimension, *d*, and a typical swimming speed, *u*, for animals and bacteria, or sinking speed for phytoplankton. He performed this calculation for the whole range from bacteria to whales, then for good measure added the point for humans with a height of 2 m and a swimming speed of 1 m s^{-1}. The resulting plot, Fig. 2.03, shows that the Reynolds number increases systematically with the size of the organism according to the regression

$$\text{Re} = ud/v \simeq 1.4 \times 10^6 \times d^{1.86} \tag{2.03}$$

where *d* is measured in meters. Substituting $v = 10^{-6}$ m^2 s^{-1}, we find the relationship between the characteristic length and swimming speed,

$$u(\text{m s}^{-1}) = 1.4 \times d^{0.86} \tag{2.04}$$

These equations provide a useful quantification of the everyday observation that large animals swim faster than small ones. For additional clarification, we have added two lines to

Table 2.1 Approximations to the magnitude of the Reynolds number of various organisms. From Vogel (1981).

	Re
A large whale swimming at 10 m s^{-1}	300,000,000
A tuna swimming at the same speed	30,000,000
A duck flying at 20 m s^{-1}	300,000
A large dragonfly going 7 m s^{-1}	30,000
A copepod in a pulse of 20 cm s^{-1}	300
Flight of the smallest flying insects	30
An invertebrate larva, 0.3 mm long moving at 1 mm s^{-1}	0.3
A sea urchin sperm advancing the species at 0.2 mm s^{-1}	0.03

Fig. 2.03 Reynolds number (Re) versus organism size (d). Mammals (○), Fish (■), Amphipods (●), Zooplankton (□), Protozoa (▲), Phytoplankton (△), Bacteria (⊙), Human (◯). The heavy line is the best fit to the data. Thin lines illustrate the relationships for swimming at 1 and 10 bodylengths per second. Adapted from Okubo (1987).

Fig. 2.03 to indicate what the relationship would look like if the organisms moved at 1 and 10 bodylengths per second. Okubo's best fit line lies at an angle between these two and suggests that small organisms can move about 10 bodylengths in a second but larger animals swim at close to 1 bodylength per second. Such statements must, however, be treated with caution as the values of swimming speed are quite scattered, especially for zooplankton, and the relationship may only be accurate to within a factor of 10. It is also worth remembering that many of the data, especially those for the smallest organisms, were obtained under a microscope in a laboratory and not in the natural environment.

2.2.5 Molecular diffusion

As indicated above, turbulent energy is passed from large to small eddies and the viscosity of the water limits the size of the smallest eddies to a few millimeters diameter (see below).

Such turbulence is ineffective in transporting nutrients and wastes for organisms less than 1 mm in diameter (which make up more than half the total biomass of the oceans, Sheldon *et al.* 1972). Small organisms must rely on the flux due to molecular diffusion—that is, the slow mixing caused by the random motion of molecules. As is shown in Box 2.02, this process is very slow, requiring about 10 s to produce an effect over a distance of 100 μm.

2.2.6 Scales of turbulent structures

When dealing with small organisms living in turbulent water, it is sometimes important to estimate the distributions of velocity, temperature, and nutrients near the organisms. For example, we might want to know the sizes of the smallest turbulent eddies or the sizes of the smallest fluctuations in temperature, salinity, and nitrate.

The size of the smallest velocity fluctuation is determined by the strength of two competing forces. The force due to viscosity works to remove variations in velocity while the inertial force associated with the turbulent motions tends to create velocity fluctuations. The size of the eddies where these opposing forces are in balance is normally taken as the limiting size of the velocity fluctuations. If the viscous, or smoothing, force is represented by the kinematic viscosity ν, while the twisting or shearing force in the turbulence is represented by the rate of turbulent-energy dissipation ε W kg^{-1}, the distance across the smallest eddies, known as the viscous or the Kolmogoroff length scale, L_v, is estimated from the following equation (Gill 1982),

$$L_v = (\nu^3/\varepsilon)^{1/4} \tag{2.07}$$

Box 2.02 Fick's law and the diffusion time scale

Molecular diffusion is the slow mixing caused by the random motion of molecules. The flux of some constituent through the water due to molecular diffusion is given by Fick's first law of diffusion. This law states that if the concentration of some constituent C changes by an amount dC over a short distance dz, the flow of C down the concentration gradient of molecular motion is

$$F = -D \, dC/dz \tag{2.05}$$

If C is in kg m^{-3}, F is the flux of C in kg m^{-2} s^{-1} and D is the coefficient of molecular diffusion, which for large molecules (i.e., salt) in water is about 1.5×10^{-9} m^2 s^{-1}.

The coefficient, D, contains information related to how fast the molecules of the diffusing substance move through the fluid. To use this information, the definition of D is converted into the formula

$$D = 10^{-9} \text{ m}^2 \text{ s}^{-1} = L^2/t \tag{2.06}$$

where L and t are characteristic values of length and time. Turning the formula around yields $t = L^2/D = 10^9 L^2$, giving an estimate of the time it takes molecular diffusion to go the distance L. If we use this relation, the time it would take molecular diffusion to cause an effect over 100 μm is about $(100 \times 10^{-6})^2 \times 10^9 \approx 10$ s.

In much of the oceanographic literature, however, this length is written

$$L_v = 2\pi(v^3/\varepsilon)^{1/4} \tag{2.08}$$

where the factor of 2π has been added for mathematical convenience as it simplifies the manipulation of equations. As discussed by Lazier and Mann (1989), the factor of 2π also results in a more realistic value of the length scale. Measurements have shown that there is virtually no energy in the turbulent eddies at the scale defined by Eq. 2.07.

The variable, ε, representing the dissipation of turbulent energy, is estimated from measurements of the finest scales of the velocity gradient. By doing this, Oakey and Elliott (1980) calculate that ε in the top 50 m layer over the Scotian shelf off Nova Scotia varies from 10^{-6} W kg^{-1} when the wind is 15 m s^{-1} to 10^{-8} W kg^{-1} when the wind is less than 5 m s^{-1}. In these situations the smallest scales L_v, using Eq. 2.08, would vary roughly from 6 to 20 mm. At greater depths or when the wind is light, the level of turbulent energy decreases to roughly 10^{-9} W kg^{-1}, corresponding to a smallest eddy size of 35 mm (Osborn 1978). The smallest fluctuations of variables such as temperature and salinity are smaller than they are for velocity as is shown in Box 2.03.

At the other end of the size range are the largest turbulent eddies, which are important because they determine the vertical excursion of the small passive organisms being moved about by the turbulent flow. Near the surface these large eddies determine how much time the organism spends in the euphotic zone. The largest eddies occur when the inertial forces associated with the turbulence, which tend to stir the water, are about equal to the buoyancy forces, which tend to keep the water stratified. This size is estimated from the turbulent-energy dissipation rate ε and the buoyancy or Brunt–Väisälä frequency N (section 3.2.4), which is proportional to the density stratification. The size of the largest turbulent eddies, sometimes called the buoyancy length scale, is estimated by Gargett *et al.* (1984) to be

$$L_b = (\varepsilon/N^3)^{1/2} \tag{2.10}$$

Box 2.03 The smallest scale of different variables

We saw in section 2.2.6 that viscosity limits the size of the smallest turbulent eddies to about 6 mm in a highly energetic environment. Viscosity may be thought of as the molecular diffusion of momentum with a value of $\simeq 10^{-6}$ m^2 s^{-1} while the molecular diffusivity of heat is about 1.5×10^{-7} m^2 s^{-1} or about one-tenth the value for viscosity. The diffusivities of salt and nitrate are two orders of magnitude lower again, 1.5×10^{-9} m^2 s^{-1}. The effect of these lower diffusivities is to permit smaller fluctuations to persist longer, before being smoothed out by diffusion.

The length scale of the smallest fluctuation of any property of diffusion constant D is given by

$$L_d = 2\pi(vD^2/\varepsilon)^{1/4} \tag{2.09}$$

which is called the Batchelor scale. If ε varies from 10^{-6} to 10^{-9} as suggested earlier, the smallest scale for temperature fluctuations is 2–13 mm and the smallest scale for salt or nitrate fluctuations is 0.2–1.0 mm.

In the mixed layer if $\varepsilon \simeq 10^{-7}$ W kg^{-1} and $N \simeq 10^{-3}$ rad s^{-1}, then $L_b \simeq 10$ m. In the deep ocean, where ε is small, or in stratified regions, where N is high, the buoyancy length scale works out to about one-tenth of the value found in the mixed layer.

2.2.7 Turbulent or eddy diffusion

Across distances greater than a few millimeters the eddies of the turbulence mix the water much more effectively than does molecular motion. The eddy-caused diffusion works the same way as molecular diffusion except that the random movement of the eddies is much larger than the molecular motion. Eddy diffusion is also different from molecular diffusion in that it is the same for all properties—that is, heat, salt, and nitrate will have the same eddy diffusion constant. A typical eddy diffusivity for horizontal diffusion in the ocean is $\simeq 500$ m^2 s^{-1}, which is about 10^9 times the molecular diffusivity for heat. More details about turbulent diffusion are contained in Box 2.04.

2.2.8 Boundary layers

It was mentioned in the introduction that solid boundaries, such as the surfaces of organisms, have associated with them a boundary layer in which water movement is reduced. Since all organisms in the sea have a need to exchange molecules of O_2, CO_2, NH_3, and so

---

Box 2.04 Eddy fluxes and time scales

Ideally the turbulent flux of a constituent, C, can be determined by measuring the turbulent velocity fluctuations along with the fluctuations in the concentration of C. Such measurements are, however, difficult to make and are done only in special situations. Usually it is assumed, by analogy with the molecular case, that the flux is dependent on the gradient of C and a diffusion constant, except in this case the diffusion constant is a diffusion due to eddies rather than a molecular diffusion. Thus, the flux F of the constituent C is written

$$F = K_e \, dC/dz \qquad (2.11)$$

where K_e is the eddy diffusivity and dC/dz is the gradient in C. The value of K_e varies greatly throughout the ocean, depending partly on the level of turbulence and partly on the stratification. Also, because of stratification, eddy diffusion is not the same vertically as horizontally. One recent estimate quoted by Gargett *et al.* (1984) gives, deep in the ocean, a horizontal eddy diffusivity $K_h \simeq 500$ m^2 s^{-1} and a vertical eddy diffusivity $K_v \simeq 0.6 \times 10^{-4}$ m^2 s^{-1}.

As with the molecular case (Eq. 2.06), the eddy diffusivity contains information on the rate of diffusion. For the case of turbulent diffusion, the approximate length of time the diffusion takes over a distance L is calculated from L^2/K_e and if L is 1 km, horizontal eddy diffusion will show an effect in $L^2/K_e = 10^6/500 \simeq \frac{1}{2}$ h. In the deep stratified ocean, eddy diffusion will transport an effect vertically through 10 m in $100/0.6 \times 10^{-4} \simeq 20$ days. In the mixed layer where the eddy diffusion rate is high in all directions, the time taken to mix properties is obviously less than the time taken in the deep ocean.

forth, with the surrounding medium, the boundary layer is liable to reduce that rate of exchange. A boundary layer is also associated with the water above the sea floor, which can affect the process of exchange of essential substances between the benthic community of organisms and the overlying water. This section examines in detail the properties of boundary layers.

The fundamental property to be considered is the 'no-slip condition.' Water molecules in contact with a solid surface stick to that surface and are, therefore, stationary with respect to it. If we plot the average velocity near the boundary as in Fig. 2.04(a), we see that it increases as we move away from the boundary until we come to water moving with the 'free-stream velocity'—i.e., the velocity that it would have if the solid surface were not present. However, Fig. 2.04(a) is simplified, as if the water in a particular layer is moving at a constant velocity and always in the same direction. In real life flows are turbulent and exhibit fluctuations perpendicular to the direction of the mean flow as illustrated in Fig. 2.04(b). The amplitude of these fluctuations decreases toward the boundary as there can be no flow into or out of the boundary.

The decrease in the magnitude of the mean flow is caused by the stress, τ, or drag exerted on the water by the solid boundary. The transmission of this stress across the layer of water is accomplished, in turbulent flow, by eddy diffusion. As the boundary is approached the flow decreases until a level is reached where viscosity smooths out almost all turbulence and the stress between the water and the solid is transmitted by viscous stresses.

It is therefore possible to recognize two distinct layers within the boundary layer, the part in which turbulent fluctuations transmit the stress (Reynolds stresses) and the viscous sublayer adjacent to the boundary. If the velocity profile in Fig. 2.04 is plotted on a semi-logarithmic scale, as in Fig. 2.05 (lower line), the different layers are easily distinguished. From 10 cm down to about 0.2 cm from the boundary the profile is represented by a straight line,

Fig. 2.04 (a) A vertical profile of mean water velocity through the boundary layer above a smooth surface showing the linear sublayer where viscous stresses dominate the stress between the water and the surface and the logarithmic layer where turbulent or Reynolds stresses dominate. (b) The same profile as in (a) including time series measurements of velocity at three levels to illustrate the increase in the size of the turbulent fluctuations with height above the boundary.

Fig. 2.05 Logarithmic profiles of mean velocity through a boundary layer for the cases of smooth ($z_o = 0.03$ cm) and slightly rough ($z_o = 0.1$ cm) flow. Both profiles have $u_* = 0.25$ cm s^{-1}. The linear sublayer where viscous forces are important is a prominent feature of only the smooth flow profile.

commonly called a 'log-linear' relationship. Adjacent to the boundary, in the viscous sublayer, the profile is no longer log-linear. In fact, in this thin, viscous sublayer the velocity decreases linearly, which appears as a curve on the logarithmic plot. The convention is to call the outer region the logarithmic layer, or 'log-layer,' and the region adjacent to the boundary the linear or viscous sublayer. The equations for these layers are presented in Box 2.05.

Since boundary layers are regions of reduced turbulent transport, which may reduce the exchange of essential substances between an organism and the water, it is important to know the thickness of the layers. As we move from the surface of an organism out into the water the speed of flow increases as the effects of the no-slip condition decrease, but just where the speed reaches the free-stream value is not easily defined. The equation introduced in Box 2.05 to describe the mean flow in the boundary layer (Eq. 2.13) predicts an ever increasing velocity with distance from the boundary. This situation occurs because the equation describes only the logarithmic layer without any information about the free-stream velocity. In this situation, the practical solution is to define the thickness of the boundary layer as that distance from the boundary at which the velocity is 99% of the free-stream velocity.

The thickness of the boundary layer surrounding an object moving through water increases with increasing distance from the leading edge. A familiar analogy is that of an aircraft wing. At the front of the wing the boundary layer is very thin but increases in thickness with distance from the leading edge according to the expression (Prandtl 1969):

$$\delta = (xv/u)^{1/2} \tag{2.14}$$

where x is the distance from the leading edge, v is the viscosity of the medium, and u is the speed of the flow. When applied to an organism moving through water, two conclusions

Box 2.05 Equations for the boundary layers

Two equations are required to describe these two layers (Chriss and Caldwell 1984). For the viscous boundary layer, as we have already seen (Eq. 2.01), the viscous stress between adjacent layers moving at different velocities is given by

$$\tau = \rho v \, du/dz$$

Because τ and ρ are constant through the boundary layer, the ratio $(\tau/\rho)^{1/2}$ is also constant, and is conventionally replaced by u_*, which is called the 'friction velocity.' After integrating and introducing u_*, the velocity profile becomes

$$u(z) = u_*^2 z/v \tag{2.12}$$

From this equation, we see that in the viscous boundary layer the velocity, u, is a linear function of z, the distance from the boundary. Hence we have the straight line close to the boundary in Fig. 2.04.

For the log layer, the mean velocity is given by

$$u(z) = (u_*/k) \ln(z/z_0) \tag{2.13}$$

where k is the Von Karman constant, ≈ 0.4. In this layer the velocity increases logarithmically with distance from the boundary.

In these examples we have assumed that the boundary is a smooth surface. In the real world few surfaces are perfectly smooth and the irregularities, or roughness elements, in the surface are liable to generate turbulent wakes in the sublayer. If these wakes are not smoothed out by viscosity, the whole sublayer becomes turbulent and the flow is called 'rough.' This development has two consequences. First, the boundary exerts more stress on the water because there is more friction. The velocity at any given distance from the boundary is, therefore, less than before as indicated by the upper line in Fig. 2.05. Second, the relationship between velocity and distance from the boundary becomes logarithmic throughout the boundary layer and the linear sublayer controlled by viscosity disappears.

As indicated in Fig. 2.05, the height, z_0, where the mean velocity is predicted to go to zero is greater in rough flow than in smooth flow. This height can be used as an estimate of the size of the roughness of the surface and is called the roughness length. In practice it is about one-tenth the height of the roughness elements. The logarithmic profile is reasonably accurate above the level of the roughness elements but among the roughness elements the flow is a complicated function of position and cannot be predicted by the equations.

The variable z_0, representing the size of the roughness elements, and the friction velocity u_* are sometimes used as the typical length and velocity scales, respectively, in the Reynolds number (Eq. 2.02). In this case, it is called the roughness Reynolds number and is denoted by Re_*.

may be reached. The first is that the larger the organism, the thicker the boundary layer, since there are more places where the value of x is large. Second, the boundary layer becomes thinner as the velocity of the water relative to the organism increases. For sedentary organisms, rapidly moving water will cause a thinning of the boundary layer, and can be expected to assist exchange of substances. For motile organisms, rapid swimming can be expected to have the same effect.

2.2.9 Drag

When a machine or an animal moves through water, two forces arise to oppose the motion. The first of these drag forces originates in the no-slip condition in association with viscosity. Because of the no-slip condition the moving object is stuck to the immediately adjacent layer of fluid, which in turn pulls along, because of viscosity, some of the layers of fluid further out. The resultant shear in the water is maintained by a stress, or force per unit area (Eq. 2.01), which must be supplied by the moving object if it is to maintain its motion. Because the magnitude of this drag force is directly dependent on the surface area of the object, it is often called skin friction instead of viscous drag. In cases of low Reynolds number flow, as encountered with small plants or animals falling or swimming slowly, this drag force is the only appreciable one encountered and is an important feature of the feeding strategies as will be explored in section 2.3.2.

The other drag force, called form or pressure drag, arises because the water, which has mass, must be pushed out of the way of the moving object, and then move back into place behind. Discussion of this force is thoroughly presented by Vogel (1981), who shows that it may be represented by the formula

$$F_D \simeq \frac{1}{2} C_D \rho u^2 A \qquad (2.15)$$

where ρ is the density of the water, u is the speed of the object, and A is some measure of the cross-sectional area of the object, such as the area projected onto the plane perpendicular to the direction of motion. The term C_D, the drag coefficient, is a complicated function of velocity that must be determined empirically for each object. C_D is so complicated because of the way water flows around the moving object, especially around the stern or back end. Objects that are designed to have as little drag as possible, like airplanes and fish, are shaped in a streamlined way so that the fluid moves in behind in a smooth and orderly way. If the object is not streamlined, the motion creates vortices and turbulent wakes at the back that lead to a reduction in pressure on the back side. Lower pressure at the back than at the front generates the pressure drag force in the direction opposite to the motion. The speed at which a particular object creates the turbulent wake or the place on the object where the wake will be produced cannot be deduced *a priori,* and the form or pressure drag can be determined only by experiment for each item. For some simple shapes, such as spheres and cylinders, values of C_D have been determined once and for all as functions of Reynolds number; these values are available in the reference literature. Plants, however, change shape as the speed of the flow changes and it is not yet possible to determine the form drag of such objects without direct measurements.

2.2.10 The problem for phytoplankton

Previous sections briefly referred to the problem of boundary layers around phytoplankton cells. The generally accepted story, first worked out quantitatively in a classic paper by Munk and Riley (1952), is that phytoplankton cells suspended motionless in the water tend to use up the nutrients in the water around them. Their nutrient-uptake rate is then limited to the rate at which the nutrients can diffuse toward them. One way of overcoming this limitation is to generate movement relative to the water, so that the zone of nutrient-depleted water is periodically renewed. The cells may accomplish this goal either by sinking passively or by generating their own locomotion. As discussed earlier, there is inevitably a region around the organism in which the velocity of the water relative to the cell is less than the absolute rate of movement of the cell through the water. This region is created by the no-slip condition, which states that molecules in contact with the cell surface adhere to that surface and cannot slip past it. In the boundary layer around the cell, the water velocity relative to the cell changes from zero at the cell surface to the 'free-stream velocity' beyond the boundary layer. A sinking or swimming phytoplankton cell drags its boundary layer with it, causing fluid in the depleted region to be sheared away. New fluid in which the nutrients have not been depleted takes its place. This replacement decreases the effective size of the depleted region and increases the rate of uptake by the cell. The magnitude of the increase depends on the velocity of the cell relative to the water.

2.2.11 Sinking of phytoplankton

Most phytoplankton cells are more dense than water. The density of sea water varies from about 1.021 to 1.028 g cm^{-3}, but the density of cytoplasm ranges from 1.03 to 1.10 g cm^{-3}. Diatoms have a cell wall of hydrated silicon dioxide with a density of about 2.6 g cm^{-3} and coccolithophorids have plates of calcite, aragonite, or valerite with a density of 2.70–2.95 g cm^{-3}. Thus, plenty of evidence suggests that many phytoplankton cells are substantially more dense than water, which means that they will sink except where an upward movement of the water prevents it. The rate at which the cells fall can be estimated (Box 2.06) if the Reynolds number is $\ll 1.0$ and if the cells are approximately spherical.

Box 2.06 Falling speed of small spheres

The formula used to calculate the terminal velocity of small spheres at low Reynolds number was devised by Sir George Stokes in 1850. He first found that fluid slows the falling sphere by exerting a resisting force equal to $6\pi\rho\nu Vr$, where ρ and ν represent the density and the kinematic viscosity of the fluid, respectively, and V and r represent the velocity and radius of the sphere, respectively. This force must be equal to the weight of the sphere less the upward push due to buoyancy. The weight of a sphere is equal to its mass times the acceleration due to gravity, g, and the mass is the volume times the density, ρ', thus the weight is $\frac{4}{3}\pi r^3 \times \rho' \times g$. The upward force of buoyancy is equal to the weight of the water displaced by the sphere, $\frac{4}{3}\pi r^3 \times \rho \times g$. Equating the resisting force to the net weight gives

$$6\pi\rho\nu Vr = \frac{4}{3}\pi r^3 g(\rho' - \rho) \qquad (2.16)$$

and after rearranging, the rate of fall is given by

$$V = \frac{2}{9}\, gr^2(\rho' - \rho)/\rho\nu \qquad (2.17)$$

Fig. 2.06 Average sinking rate of phytoplankton cells grown in culture and measured in the laboratory and line predicted by Stokes' law. After Smayda (1970).

For a given excess density and viscosity, a sphere will increase its sinking speed in proportion to the square of the radius (Eq. 2.17 presented in Box 2.06). Few phytoplankton cells are spherical; nevertheless for Re ≤ 0.5, Stokes' law may be applied to nonspherical bodies without important errors (Hutchinson 1967). Roughly speaking, phytoplankton up to 500 μm in diameter fall within this category. Empirical evidence for increasing sinking speed with increasing size has been reviewed by Smayda (1970) (Fig. 2.06). Although the data fall within the 1–500 μm range, the slopes of the lines are considerably less than the value of 2 predicted by Stokes' law. In considering the reasons for this, shape can probably be ruled out. According to the theoretical analysis of Munk and Riley (1952), the effect of geometry is different in different size classes. In order of decreasing sinking rate, the pattern for 5μm particles is plate > cylinder > sphere. For 50 μm particles it is cylinder ≃ plate > sphere, and for 500 μm particles it is cylinder > sphere > plate. Moreover, the sinking rates of different shapes tend to equalize with increasing size. The data in Fig. 2.06 have not been corrected for changes in density and viscosity with temperature. At the natural extremes this variance is not negligible: a change from 0 °C to 25 °C results in lowered density and viscosity so that the sinking rate of 20 μm particles approximately doubles. There is no reason to think that the data displayed in Fig. 2.06 were carried out at such an extreme range of temperature difference.

It is seen from Fig. 2.06 that senescent cells sink faster than actively growing cells. It has also been found that preserved cells sink faster than living cells, and that natural blooms of phytoplankton tend to sink in the water column as they approach senescence. Factors that are thought to contribute to this effect include variation in the cell content of low density oils and fats, variation in the amount of light ions such as NH_4^+, and possible effects of excreted products on the viscosity of the fluid surrounding the cells. The effectiveness of sinking as a means of combating diffusion limitation is considered in section 2.2.13.

2.2.12 Swimming by phytoplankton

Swimming is the alternative to sinking, as a means of reducing the diffusion limitation on nutrient uptake. The locomotion of small organisms is a very different proposition from what we are familiar with from our own attempts at swimming or from watching macroscopic organisms such as fish or insects (Purcell 1977). For small organisms that perform movements relatively slowly, the Reynolds number is low and viscous forces predominate: A picoplankton organism of about 1 μm radius moving at about 30 μm s^{-1} generates inertial forces that are so small that after the swimming effort ceases it will come to a stop in about 0.6×10^{-6}s, travelling only 10^{-4} μm. It is rather like a human swimming in a pool of molasses and forbidden to move any part of the body faster than 1 cm min^{-1}. One consequence of this situation is that jet propulsion just does not work: it relies on inertial forces. A scallop-like animal at this scale would just expel the liquid as it closed its shell and take it in again on opening, getting nowhere in the process. The mechanisms that work best under these conditions are the 'flexible oar' (cilium) and the 'corkscrew' (flagellum pointing forward) (Purcell 1977). Of the two, flagella are the most commonly used by phytoplankton.

2.2.13 The effectiveness of swimming or sinking

A stationary phytoplankton cell cannot take up nutrients faster than they are transported toward it by diffusion (section 2.2.10). There are some types of phytoplankton that take up nutrients so slowly that diffusion is not a limitation on their metabolism. This condition is not true for most phytoplankters, which must move relative to the water by sinking or swimming to increase their nutrient uptake (Pasciak and Gavis 1974). The critical question is: how effective is relative movement between the cell and the water in increasing the rate of nutrient uptake?

We see in the following discussions that the conclusions of various studies are quantitatively the same. For small organisms in the 1–10 μm range, viscosity dominates their world and diffusion is faster than water movement in supplying nutrients through their boundary layer. As far as we can tell, these small organisms move to find better concentrations of nutrients in the environment, not to reduce their diffusion limitation.

On the other hand, flagellates larger than 10 μm, many of which are able to swim more than 10 times their own bodylength per second, may achieve a significant increase in nutrient uptake by swimming. The relative advantages are greater still for cells of 50 μm or more, not because they move faster but because they have a smaller ratio of surface area to volume and have a stronger nutrient limitation in the first place.

Most diatoms are unable to swim, but create water movement past their surfaces by sinking. Large cells sink faster than small cells, and it is only the larger cells that sink fast enough to gain a significant advantage from it in terms of nutrient uptake. In no case can these cells abolish nutrient limitation by sinking, but the differences between cells are thought to be great enough to influence interspecific competition. The evidence for these findings is as follows.

Pasciak and Gavis (1974) considered the question of diffusion limitation in conjunction with conventional Michaelis–Menten nutrient dynamics. They concluded that when the concentration of nutrients in the environment is not many times greater than the half-saturation constant of the cell, and when the cell is small, diffusion can be a severe limitation on cell metabolism. Gavis (1976) then introduced motion relative to the water into the calculations. Taking published rates of sinking of phytoplankton cells, he concluded that an appreciable increase in nutrient uptake rate would occur only if the cells are relatively large (100 μm) and the concentration of nutrients is low. In the optimum case, sinking would alleviate only about 30% of diffusion limitation.

A similar result was obtained using published swimming rates. The effect was most marked with large (100 μm) cells swimming at about 3 bodylengths per second. There was an appreciable increase in nutrient uptake but it was still less than one-third what it would have been in the absence of diffusion limitation. Hence, the conclusion was that swimming and sinking are not very effective methods of alleviating diffusion limitation, but that they might make enough difference to affect the outcome of competition among species.

Berg and Purcell (1977) and Purcell (1977) did not take into account the biological properties of the cells but considered the case of a spherical cell of radius r, propelled at a constant velocity u through water containing molecules 'for which the cell is a perfect sink.' They calculated the relationship between velocity and the fractional increase in the transport of nutrients to the cell. They showed that this increase was a complex function of ru/D, where D is the diffusion constant of the molecules absorbed. More recently Sommer (1988), building on the ideas of Goldman (1984), took this formulation by Berg and Purcell (1977) and plotted the results for 19 species of marine flagellates for which the size and swimming velocity were known (Fig. 2.07). Swimming speeds ranged from about 2 to a maximum of 100 bodylengths per second, with the smaller cells having the higher relative velocities. Even so, the conclusion (Fig. 2.07) was that cells of 5 μm or less increased their nutrient flux by only 5–30% when swimming. On the other hand, flagellates of 50 μm diameter had the potential to double or even treble their nutrient-uptake rate by swimming.

We may also apply this diagram to a consideration of sinking, treating it as a special case of locomotion. From Fig. 2.06 we see that a 50 μm diameter cell sinks at approximately 1 m d^{-1} or 12 μm s^{-1}. According to Fig. 2.07 this sink rate might increase the nutrient flux

Fig. 2.07 Fractional increase of nutrient flux to the cell surface as a function of swimming velocity and cell diameter, calculated for spherical cells, after Berg and Purcell (1977). The circles represent observations and the finely dotted line is a regression through the observations. From Sommer (1988).

by 30–40%. On the other hand, a cell of 10 μm would sink at only about 3 μm s^{-1}, which would give it no increase in nutrient-uptake rate.

There is yet another way of trying to understand that creating movement of water relative to a very small cell does not benefit nutrient uptake. Berg and Purcell (1977) discussed it in relation to 'stirring,' defined as the creation of water currents by means of an appendage, but the concept applies equally well to water movement induced by sinking or swimming. The time required for transporting a nutrient distance L by water movement is L/u, where L is the distance transported and u is the water velocity. On the other hand, the time required for transport by diffusion is L^2/D, where D is the diffusion constant (see section 2.2.5). The effectiveness of water movement versus diffusion, for any given distance and diffusion constant (the Sherwood number), is given by the equation

$$\frac{\text{time for transport by diffusion}}{\text{time for transport by movement}} = \frac{L^2/D}{L/u} = Lu/D$$

For scales on the order of 1 μm the ratio works out to approximately 10^{-2}. Diffusion is about 100 times faster than movement. Hence, in this world of low Reynolds numbers, nothing is gained by trying to reduce the diffusion barrier by generating turbulent advection. In this context, the only possible advantage to the organism of undertaking locomotion is that it might encounter nutrients in a higher concentration. Purcell (1977) summarized it by saying that the organism does not move like a cow that is grazing on pasture, it moves to find greener pastures.

2.2.14 The role of turbulence

In the foregoing discussion on diffusion limitation and the possible effects of swimming or sinking on the rate of uptake of nutrients, no mention was made of the effects of turbulent motion. Turbulence, we know, is a ubiquitous feature of the ocean but it is not immediately clear how the turbulent motion affects the flow around very small organisms or if that flow will alter the rate of molecular diffusion toward or away from a cell. Since the interaction between the small organisms and the turbulence cannot be observed directly in the ocean, we must rely on theoretical constructions and indirect observations.

The first discussion of the effect of turbulence on the molecular diffusion in the vicinity of small organisms is attributed to Munk and Riley (1952). They suggest that the turbulent-pressure fluctuations in the ocean create a small relative motion between the organisms and the water because the two are of slightly different densities. They conclude, however, that the effect of this relative motion on the diffusive flux near an organism would be negligible. The question was reexamined by Lazier and Mann (1989), who argued that turbulence is manifest across even the smallest distances by a linear shear whose magnitude depends on the strength of the turbulence. The shear is a property of the fluid and exists whether there are organisms present or not. The effect that this shear has on the diffusion toward or away from a perfectly absorbing sphere is calculated using the experimental results of Purcell (1978).

A summary of the increase in the diffusive flux due to both relative motion through the water and turbulence is given in Fig. 2.08. The two curves constructed from the result of Berg and Purcell (1977), B&P, show the change in the diffusive flux as a function of diameter for relative motion through the water at 1 and 10 diameters per second (d s^{-1}). The

Fig. 2.08 Percentage increase in the diffusive flux toward a perfectly absorbing sphere of diameter d caused by the relative motion or the shear created by turbulence, adapted from Lazier and Mann (1989). The curves for relative motion are for sinking or swimming at 1 and 10 diameters per second (d s^{-1}) based on the work of Munk and Riley (1952) (M&R) and Berg and Purcell (1977) (B&P). The data for the effect of turbulence are derived from the work of Purcell (1978) for dissipation rates of 10^{-6} and 10^{-9} W kg^{-1}. For interpretation, see text.

calculations of Munk and Riley (1952) for spheres are recast in the curves labelled M&R. On the right are two curves indicating the response to weak and strong turbulence.

The B&P results indicate that for cells less than ≃5μm in diameter sinking or swimming at even 10 d s^{-1} will have no effect on the rate of molecular diffusion in the neighborhood of the cell. For cells of larger diameter relative motion causes a significant increase in the diffusive flux. A 40 μm cell, for example, will increase the flux by about 100% if it moves through the water at 10 s^{-1}. Similar increases in flux are noted for relative motion at 1 d s^{-1} but larger cell diameters are required. The M&R curves are roughly the same as the B&P, which is encouraging considering the completely different methods of analysis. The curves representing the effect of turbulence show that even strong turbulence will not have an appreciable effect compared with that due to relative motion through the water. For stationary cells the turbulence will start to have an effect on cells greater than 100 μm in diameter. In weak turbulence, only cells larger than 1 mm will be affected.

Various authors have made practical investigations of the effect of turbulence on phytoplankton. Savidge (1981) created turbulence by placing an oscillating grid within a 25 liter culture. When the oscillation rate was increased by a factor of 10, nitrate uptake of the diatom *Phaeodactylum tricornutum* was enhanced by 21%, but phosphate uptake and carbon fixation decreased. Nevertheless, the growth rate increased 25–40%. For the flagellate *Brachiomonas submarina,* the higher level of turbulence resulted in a 60% increase in its growth rate.

The same type of apparatus was used by Howarth *et al.* (1993) to test the effect of turbulence on the nitrogen-fixing activity of heterocystic cyanobacteria. Although they used turbulence levels greater than those normally encountered in lakes or coastal waters (except perhaps in surf zones), no inhibition of nitrogen fixation was found. Howarth's group considered that, at the levels of turbulence used, anoxic microzones could not possibly have persisted on the surfaces of the cyanobacteria. In the past, anoxic microzones have been considered essential to nitrogen fixation by these organisms, and the inhibition of these zones by higher turbulence in coastal waters has been given as an explanation of the lower rates of N fixation as compared with lakes. This idea is no longer tenable.

Pasciak and Gavis (1975) subjected the diatom *Dytilum brightwellii* to shear by holding the culture between two cylinders, one of which was rotated. At a shear rate of 5 rad s^{-1}, nutrient uptake was enhanced about 10%.

Thomas and Gibson (1990, 1992) also used the apparatus with two concentric cylinders to generate shear and turbulence. They consistently found that turbulence inhibited growth in dinoflagellates. With *Gonyaulax polyedra*, which is responsible for 'red tides' in coastal waters, the threshold for growth inhibition was about 3 rad s^{-1}, and growth was inhibited completely above 8 rad s^{-1}. At the higher levels, the cells lost their flagella, so that active locomotion became impossible. These results were thought to explain the observation that calm weather is necessary for the development of red tides off southern California. With *Gymnodinium splendens*, which is an important food item for the first-feeding larvae of northern anchovies (see section 5.5.2), Thomas and Gibson (1992) found that growth inhibition was present at the low level of 1 rad s^{-1}, but did not change much over the range 1–40 rad s^{-1}. Moreover, the cells remained quite healthy at all levels of turbulence, and after 3–5 days appeared to acclimate and partially regain their normal growth rates. From these observations, Thomas and Gibson concluded that *G. splendens* is not particularly sensitive to turbulence, and the observed sensitivity of anchovy recruitment to turbulence (sections 2.3.6 and 5.5.2) is not explained by the direct effect of the turbulence on the food organisms.

Aguilera *et al.* (1994) studied the effect of turbulence on the freshwater microalga *Dunaliella viridis* growing in a chemostat. Turbulence was created by bubbling. By measuring the pressure of the air in the inflowing and outflowing lines, they were able to calculate the energy supplied to the system. Over the energy range 0–2.0 W m^{-2} they found that the specific growth rate peaked at 0.63 W m^{-2}, and the maximum rate of primary production occurred at 1.67 W m^{-2}. The authors concluded that the main effect of the turbulence was the breakdown of boundary layers around the cells. Interestingly, the carbon fixation rates (but not the growth rates) of cells in an unstirred medium were *higher* than those in low levels of turbulence (0.02–0.2 W m^{-2}).

The addition of 0.2 W m^{-2} of mechanical energy permitted 0.21 W m^{-2} of primary production in the cultures. These numbers have the same order of magnitude as those calculated for natural systems by Margalef (1978a) (see section 11.3.1).

In a review of plankton production processes, Wyatt and Jenkinson (1993) pointed out that many phytoplankton species form strings or networks that straddle different turbulent structures and may increase the flow of water past their constituent cells. In other species, individual cells secrete strands or nets of mucilaginous polymers, changing the relationship of the cells to turbulence structures and strongly modifying sinking rates. Thus, the relationship of phytoplankton to the fine structure of the water column may be modified by the activities of the cells themselves.

In a later section (2.2.16) we trace the mechanisms for the formation of aggregations of colloidal material with bacteria. Frequently, phytoplankton cells are also trapped in these aggregations, and the sinking of the aggregations may increase the flow of water past the phytoplankton cells.

In summary, these works suggest that when a species is subject to a tenfold increase in turbulent shear, it may increase the uptake of one nutrient while decreasing the uptake of another. This factor appears to have a dome-shaped relationship to turbulent shear that is different for each species, so that at high levels some species are damaged and have decreased metabolism, while others have enhanced nutrient uptake and growth. Thus, we see that the conclusions reached on theoretical grounds by Lazier and Mann (1989) are only partly borne out by experiment. Some of these differences may relate to the higher turbulence levels used in some of the experiments. Lazier and Mann (1989) considered the effects of the shear associated with turbulence levels commonly observed in the ocean—that is, turbulence dissipation rates between 10^{-6} and 10^{-9} W kg^{-1} equivalent to shear levels between 0.56 and 0.018 rad s^{-1}, respectively. In most of the laboratory experiments, the shears are considerably higher—in some cases, by an order of magnitude.

2.2.15 The paradox of cell growth in low-nutrient environments

Phytosynthesis has been shown to occur in waters where the concentration of nitrogenous nutrients falls below the limit of detection (McCarthy and Goldman 1979). How can phytoplankton be physiologically active and grow enough to maintain their populations when nutrients appear at such a low level? It has been suggested that the phytoplankton can take up the nutrients from small local concentrations such as might be produced by the excretion of a single zooplankton organism, or the regenerated nutrients from the bacterial decomposition of a particle of organic matter. Lehman and Scavia (1982a, 1982b) produced experimental evidence and models in support of the claim that the freshwater alga *Chlamydomonas* could derive nutrients from the excretion of the zooplankter *Daphnia pulex*. In contrast, Jackson (1980), Williams and Muir (1981), and Currie (1984) argued that the natural concentrations of organisms and the probable rates of nutrient dispersal by turbulent diffusion rendered the transfer of nutrients from zooplankton to phytoplankton insignificant.

The case for transfer of nutrients from bacteria to phytoplankton rests on the observation that small particles of organic matter, which are often formed on the surfaces of bubbles (Riley 1970), tend to form aggregates that are widely distributed in the ocean (Silver *et al.* 1978; Silver and Alldredge 1981). These aggregates typically contain concentrations of bacteria, protozoa, and phytoplankton. Goldman (1984b) put forward an 'aggregate spinning-wheel hypothesis' (Fig. 2.09) according to which phytoplankton, micrograzers, and bacteria live attached to the surfaces of the aggregates and are involved in the rapid recycling of nutrients. Evidence suggests the existence of both phytosynthesis and respiration within aggregates (Alldredge and Cohen 1987) and the presence of elevated concentrations of phosphate ammonium and nitrate (Shanks and Trent 1979). We shall return to this topic in section 2.2.17.

2.2.16 Uptake of nutrients by bacteria

(a) In the absence of turbulence

Jumars *et al.* (1993) reviewed uptake of nutrients by bacteria free-living in the plankton. Heterotrophic bacteria (i.e., those that do not perform photosynthesis) receive their energy

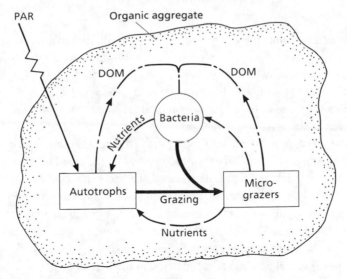

Fig. 2.09 Conceptual scheme of a microbial food chain within a discrete organic aggregate. From Goldman (1984).

supply from the uptake of organic nutrients from solution by a process known as osmotrophy. They may take the nutrients directly from the surrounding water or they may attach themselves to particles and secrete exoenzymes that convert the particulate material to soluble form. In the following discussion, we shall see that uptake from solution involves severe limitations, and that the processes that convert dissolved organic matter to particulate form are critical to the nutrition of marine bacteria in the water column.

Because the surface area of a sphere increases as the square of the radius, while volume increases as the cube of the radius, it was once thought that organisms that rely on diffusion for the transport of nutrients should maximize their surface-area-to-volume ratio by being as small as possible. Jumars *et al.* (1993) have shown that the optimum size for an osmotrophic bacterium changes with the external concentration of nutrients. In a stagnant, homogeneous ocean, the rate of flux of a nutrient to a spherical cell, J_D (mol cell^{-1} T^{-1}), by diffusion alone is given by

$$J_D = -4\pi r_0 D(C_\infty - C_0)$$

where r_0 is the radius of the cell, C_∞ is the ambient concentration of the nutrient beyond the boundary layer, and C_0 is the concentration at the cell surface. Therefore, a cell can increase the flux to itself by enhancing its active uptake system, thus lowering C_0, or by increasing the cell radius. The rate of use of the diffusing substance varies as T_0^a where a lies between 1 and 3. As a consequence of an uptake rate that changes linearly with cell radius and a use rate that changes nonlinearly, the optimum cell size depends on the ambient concentration of nutrients and is larger for cells in high-nutrient conditions than for those in low-nutrient conditions. This requirement explains the empirical observation that bacteria tend to be larger in high-nutrient conditions. In the context of optimal feeding strategies, we may extend the consideration of optimal conditions for a given organism to include risk of predation. As large size increases vulnerability to predators, the optimal size of a bacterium may therefore be smaller than that dictated by nutrient considerations alone.

(b) In the presence of turbulence

In a turbulent water column, the shear created by the turbulence can affect the diffusive flux toward an organism, but (as we saw in section 2.2.14) this effect is negligible for organisms of diameter around 1 μm, which might be taken as typical for bacteria. On the other hand (as will be discussed in section 2.3.5), turbulence increases the encounter rate between a predator and its prey, and in the context of optimal feeding strategy this effect may not be negligible, even for bacteria-size organisms.

Jackson (1989) used a simulation model to investigate whether bacteria could, by active locomotion, improve their chances of encounter with the cloud of leaked organic matter surrounding a phytoplankton cell. It appears that a bacterium cannot move as rapidly as a sinking phytoplankter, unless the latter organism is incorporated in a marine snow aggregate. If a bacterium is found within the cloud of dissolved organic matter secreted by a passing diatom, it could, at most, prolong its stay in this cloud by making active locomotory efforts. Even so, the phenomenon is not believed to make a significant difference to bacterial nutrition under actual conditions encountered in the sea.

Bowen *et al.* (1993) made a similar modelling study, but added the effects of a turbulent shear, which had not been considered by Jackson (1989). In spite of these refinements, the basic conclusion was the same. Clusters of bacteria formed around exuding phytoplankton cells are likely to comprise no more than 15% of the total bacterial population in the area, and are unlikely to significantly alter either the uptake of nutrients by the phytoplankton or the magnitude and fate of bacterial secondary production in the food web.

2.2.17 Bacteria and colloids

Molecules in true solution are usually less than 1 nm in diameter, and are distinguished from particulate matter by their ability to pass through very fine filters. Occupying the size range of 1 nm to a few millimeters are colloids, which have properties that are in some ways intermediate between dissolved and particulate fractions. In their relationship to bacteria, colloids may often be treated as particles.

The colloidal fraction represents a reservoir of organic matter in the ocean that has a far larger mass than either the phytoplankton or the bacteria, though it is not readily accessible to bacterial degradation by virtue of its particle size characteristics (1 nm to a few μm). A number of mechanisms in the ocean tend to cause these colloids to aggregate, however, after which they are more readily attacked by bacteria. The process has been reviewed by Kepkay (1994).

The basic mechanisms causing colloids to aggregate are:

1   Brownian motion, in which particles undergo random walk caused by the impact of solute molecules.
2   Shear, in which particles travelling at different speeds in shear flows collide.
3   Differential settling, in which particles sinking at different rates collide.
4   Diffusive capture, in which a particle is advected until it enters the diffusion-limited boundary layer of another particle. It is effectively captured and will eventually collide with that particle.
5   Surface coagulation, in which colloids are trapped on the surfaces of rising bubbles.

6 Filtration, either by organisms with special mechanisms for filtering very small particles or by sinking macroaggregates known as marine snow (see section 2.2.15), which are often porous, so that water passes through them and colloidal particles are captured.

7 Bacterial motility. A few observations have been made of bacteria moving fast enough and far enough to capture colloidal particles, but it is not known how important this process is relative to those already mentioned.

The first of these mechanisms, Brownian movement, causes small colloidal particles to collide and create small aggregates. These aggregates again collide, until they form particles several μm in diameter, a process referred to as the Brownian pump.

The next three mechanisms lead to still larger aggregations in which the bacteria themselves are included. The resulting aggregate provides an excellent site for bacterial feeding and multiplication. Furthermore, it is large enough to undergo a significant rate of sinking, with consequent renewal of the supply of gases and nutrients at its boundary layer. With the capture of phytoplankton and the arrival of micrograzers, such an aggregation has all the components necessary to fit Goldman's 'aggregate spinning-wheel hypothesis' (see section 2.2.15).

Coagulation on the surface of bubbles has long been suspected as an important mechanism for converting dissolved organic matter to particulate form. Some quantification of this process has been achieved. Kepkay and co-workers have found that coagulation by bubbles leads to a marked increase in food available to bacteria and thereby to increased bacterial respiration. A storm at sea can lead to a 36-fold increase in respiration, lasting for 2 to 4 hours.

Alber and Valiela (1995) used bubbles to coagulate dissolved organic matter released from decaying seaweeds. This material was labelled with ^{15}N (by growing the algae in a solution of $^{15}NH_4Cl$), and it was shown that bay scallops (*Argopecten irradians*) ingested the particles formed in this way and incorporated the labelled N in their tissues. This process demonstrates that dissolved organic matter released by decaying algae can be a significant pathway for the transfer of energy and materials to filter-feeding macro-invertebrates.

2.2.18 Conclusions

We have seen that a phytoplankton cell that remained neutrally buoyant in still water would soon deplete the nutrients in the water surrounding it and create a gradient. Under these conditions, the only mechanism that would tend to replace the nutrients would be molecular diffusion. This mechanism may be adequate for very small cells of the order of 1 μm, but it tends to be extremely limiting of the metabolism of phytoplankton of the scale 10–100 μm. It has been proposed that they overcome this difficulty by either sinking or swimming, rather than floating neutrally buoyant. A detailed study of the dynamics suggests that the observed rates of sinking go some way toward reducing diffusion limitation, perhaps enough to account for one species having a competitive edge over another, but even then the uptake rate of nutrients is likely to be no more than one-third of what the rate would be in the absence of diffusion limitation.

Swimming is found in all classes of organisms from bacteria-size upward. For bacteria and picoplankton it seems that the role of swimming is to move the organisms to regions of higher nutrient concentration. The relative importance of viscosity is so great for these organisms that their locomotion does not materially change their diffusion limitation. For cells of about 100 μm the effect of viscosity is less drastic, and it has been calculated that at observed rates of swimming they have the possibility of reducing diffusion limitation by about one-third.

There has been some study and much speculation about the possibility that the excretion of microbes and animals creates microzones of high nutrient concentration in the sea, and that phytoplankton are capable of taking advantage of them. It has been shown under laboratory conditions that zooplankton as large as *Daphnia* are able to create nutrient-rich zones from which *Chlamydomonas* can benefit, but the ecological applicability of these results is in doubt. Most calculations from field conditions suggest that the nutrient plumes generated by zooplankton are too few, and disperse too rapidly to be of great significance to phytoplankton.

An idea that is still largely untested holds that, especially in oligotrophic seas, phytoplankton cells live in close proximity to bacteria on the surfaces of amorphous organic aggregates formed by the precipitation of dissolved organic matter, or on the marine snow thought to be derived from large gelatinous zooplankton. The bacteria may be consumed by micrograzers such as ciliates and these organisms may be responsible for the excretion of the nutrients needed by the phytoplankton. These ideas are intriguing but await further investigation.

Consideration of the problems facing bacteria in their need to absorb organic molecules in solution shows that uptake from solution in the water column is subject to severe limitations but that the progressive formation of ever-larger aggregations of colloids may lead to the formation of particles to which bacteria become attached and feed by the production of exoenzymes that convert the particulate matter into soluble form for absorption.

2.3 ZOOPLANKTON

2.3.1 Life in a viscous environment

For particles of the size of phytoplankton, inertial forces are very small and by comparison viscous forces are strong and predominate. As a consequence, the relationship between a zooplankter and its food is quite different from anything in our own experience. Viscous flow is reversible and a limb moving backward and forward creates a symmetrical pattern of flow. A copepod may scratch a bacterium from its surface, but if the motion is symmetrical the bacterium will return to the place of origin. A complex asymmetrical movement must be made if the bacterium is to be removed from the cuticle (Strickler 1984). On the other hand, viscous properties can be used to advantage when feeding. Algae captured by copepods and assembled between the setae of the mouthparts will remain there so long as the mouthparts make only symmetrical movements.

When it comes to swimming, zooplankton occupy an interesting transitional range of Reynolds number, between 0.1 and 500 (Fig. 2.03). At the scale of the setae on the feeding appendages, viscous forces are important and inertial forces are trivial. However, some copepods are capable of accelerating their mass to 35 cm s^{-1} and inertial forces can then be exploited (Strickler 1984).

2.3.2 Feeding in a viscous environment

The effect of boundary layers on the action of mouthparts, when viscous forces are predominant, is graphically illustrated by a scaled-up model (Strickler 1984). The medium is honey, the food particle to be captured is a grain of rice, and the mouthparts are a pair of knife blades. Figure 2.10(a) shows that if a food particle is close to the surface of the mouth-part and that mouthpart is moved in a direction parallel to its flat surface, the particle will move with it, because the medium closest to the blade does not slip, and layers of medium progressively further from the blade are drawn along by viscous stress. Figure 2.10(b) shows that if the two mouthparts are brought together in the vicinity of the food particle (a clap) the particle escapes because its inertia is negligible compared with the viscous forces carrying it away in the fluid (in contrast to the situation in which we successfully capture a mosquito in air by clapping). Figure 2.10(c) shows that if the mouthparts move in opposite directions simultaneously they can exert equal and opposite viscous drag forces on the particle so that it remains approximately stationary and is captured. Finally, Fig. 2.10(d) shows a technique that has come to be known as 'fling and clap.' The two mouthparts are flung apart, drawing in a parcel of medium containing the food particle. The edges closest to the inflow then converge behind the particle, effectively trapping it.

The older view of the way in which copepods feed was that the appendages pushed water postero-laterally, creating a large swirl on each side of the animal (Fig. 2.11). The max-illipeds swung outward, then inward, sucking some of the swirling water into the midline of the animal and forward through a filter formed by the second maxillae. The captured food was transferred to the mouth by the endites of the first maxillae. There were several stud-ies of the efficiency of this filtering mechanism for the retention of particles of different sizes. Koehl and Strickler (1981) were at a loss to understand how the water would be forced through the sieve formed by the setae, when it could take the path of least resistance and go around it. They also noted that the feeding mechanism had been studied by holding animals in drops of water under the microscope and suspected that the large lateral swirls were ar-tifacts resulting from the confinement.

Koehl and Strickler (1981) reinvestigated the feeding mechanism by tethering speci-mens of *Eucalanus pileatus* in an optical glass cuvette holding 120 mL of water. They took high-speed cinematographic movies of the feeding process while India ink was released

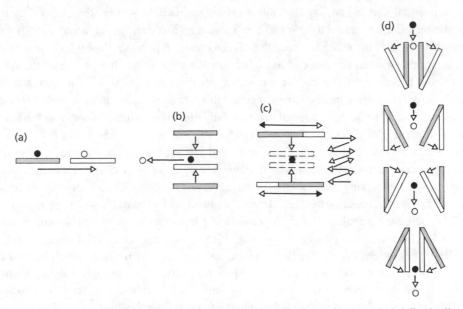

Fig. 2.10 Diagrams of experiments on low Reynolds number flow. Starting positions are shaded. For details see section 2.3.2. Modified from Strickler (1984).

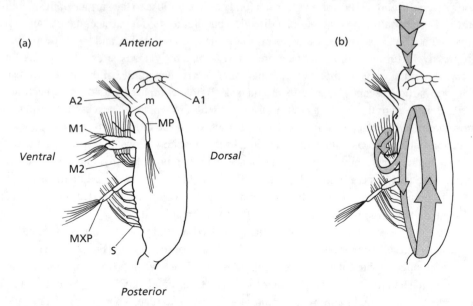

(a)

Anterior

A2 — m — A1

MP

M1

Ventral   Dorsal

M2

MXP

S

(b)

Posterior

Fig. 2.11 *Eucalanus pileatus.* (a) Diagram of an animal in typical feeding position. Only left appendage of each pair is shown. A1, A2—first and second antennae; M1, M2—first and second maxillae; MXP—maxilliped; S—swimming leg. (b) Diagram of feeding currents observed by earlier workers when animal was imprisoned in a drop of water. From Koehl and Strickler (1981).

from a micropipette close to the animal. They found that second antennae, mandibular palps, first maxillae, and maxillipeds combined in action to produce a pulsing stream of water past the ventral side of the animal (Fig. 2.12). When a food particle came within range, the appendages began to beat asymmetrically, simultaneously drawing the particle toward the midventral line and turning the animal toward the particle. At the critical moment, the second maxillae performed a fling and clap movement that drew the alga and a small volume of water into the space between them, then closed in to retain the alga while forcing the surrounding water out through the spaces between the setae. There are several important points to note. First, forcing water continuously through a fine sieve is an energetically costly activity, and the copepods avoid much of this cost by scanning the water moving past them and forcing through the second maxillae only that small volume of water surrounding the food particle. Second, the flow of water around and between the setae of the second maxillae is very different on the small scale and low Re from what we expect from observing events on the larger scale, and it seems that algae are often directed into the mouth without actually making contact with the setae. When they do make contact with the setae, they stick, as mentioned earlier, and they are combed off by the endites of the first maxillae. The large-scale analogue of this operation is using a fork to comb crumbs off another fork when both are immersed in honey (Koehl 1984)—clearly not a trivial task.

The observations that the feeding appendages change the symmetry of their beat when an alga is in the proximity shows that there is a mechanism for sensing the algal presence before it makes any contact with the copepod, and that the animal is performing a 'detect and capture' operation rather than behaving as a passive filter feeder.

Fig. 2.12 Diagrams of the feeding of *Eucalanus pileatus* obtained by placing dye streams from a micropipette close to the animal and using high-speed cine-camera. Heavy arrow indicates movement of second maxillae. Light arrows and circles show movement of a particle during food capture. (a), (c), (e)—view from left side of animal; (b), (d), (f)—view from head end. From Koehl and Strickler (1981).

Using a technique of laser-illuminated video monitoring, Yen *et al.* (1991) investigated the paths taken by algal particles and used them to reconstruct the flow fields around the copepod *Euchaeta rimana* when it is feeding. They noted that the currents generated by the appendages served not only to enable the animal to detect and capture prey, but also to maintain the animal in position in the water column, by overcoming negative buoyancy. They found that the sheared laminar flow was most intense near the paired second antennae, but decreased rapidly as distance from the animal increased. They interpreted this condition as an adaptation to minimizing the information available to potential predators, as the currents and associated patterns of turbulence could be used by the copepod's predators to detect its presence.

The authors also compared energy dissipation generated by the animals with that caused by purely physical processes. Viscous energy dissipation rates within the feeding current were higher than those due to natural turbulence, although the rates lost this advantage when spatially averaged, except possibly in the case of very dense swarms of copepods. The spatially averaged vertical eddy diffusivity caused by feeding currents was several orders of magnitude less than that caused by turbulence.

2.3.3 Detection of food

It seems probable that detection of food particles occurs by chemoreception or mechanoreception. In his review, Strickler (1984) pointed out that structures have been identified that look like chemoreceptors, but there is no proof that they function as such. However, when he offered copepods glass particles that were too large to ingest, they captured them provided that they were simultaneously exposed to an appropriate chemical stimulus delivered by means of a micropipette, but ceased capture and discarded any they had already captured as soon as the chemical stimulus was interrupted. Clearly, chemical stimuli do affect the feeding behavior of copepods. Both taste and smell types of chemosensors have been identified.

Wong (1980) showed that cyclopoid and calanoid copepods can sense moving bodies with low Re flow fields, at 3–4 bodylengths away. This ability suggests the existence of mechanoreceptors. However, algae swim weakly or not at all and are unlikely to produce signals that would be detected by mechanoreceptors more than 1 mm away. Furthermore, Friedman and Strickler (1975) searched for ultrastructural evidence of mechanoreceptors in calanoid copepods and failed to find them, but they speculated that there really had to be mechanoreceptors to facilitate the handling of food after capture.

2.3.4 Possible modes of selective feeding

The methods copepods use to feed selectively on different types of particles are not yet well understood, but various mechanisms can be proposed in the light of our new understanding of the feeding process in a viscous environment (Koehl 1984). It is now no longer possible to regard the second maxillae with their array of setae as simple sieves through which water is forced while particles are retained. The no-slip condition, to which we have repeatedly referred, means that water moving between two closely spaced setae has zero movement at the point of contact with the setae and is subjected to shear deformation in the boundary layer surrounding each seta. Hence, there is great resistance to the movement of water between closely spaced setae. A copepod appendage with an array of setae behaves more like a solid paddle than an open rake. Only when water is trapped and has no other route of escape can any quantity be forced between the setae.

Copepod second maxillae have been found to retain particles that are smaller than the spaces between the setae. Rubenstein and Koehl (1977), using filtration theory, drew attention to five possible mechanisms (Fig. 2.13). The first is conventional sieving, in which a particle is retained because it is larger than the space through which the water is passing. The second is direct interception, in which the water containing a particle passes so close to a seta that the particle collides and, because of the low Re, adheres. The third is inertial impact, in which a stream of water is diverted around a seta, but a particle has sufficient momentum that it carries on in more or less a straight line and collides with the seta. The fourth is diffusion, or motile-particle deposition, in which a particle has a random move-

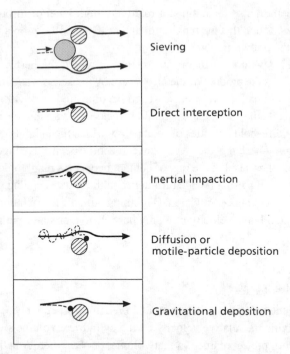

Fig. 2.13 Diagram illustrating five mechanisms of particle retention, as described in section 2.3.4. From Rubenstein and Koehl (1977).

ment in relation to the stream of water carrying it, and is brought into contact with the seta. Finally, there is gravitational deposition, in which the weight of a particle produces a small sinking motion that brings it into contact with the seta. As particle size increases, the filtering element's ability to capture by inertial impaction, gravitational deposition, and direct interception is improved. As particle size is reduced, diffusion is enhanced.

Centropages typicus has second maxillae with widely spaced setae that move almost 20 times faster than those of *Eucalanus pileatus*. Hence, the Reynolds number is very different in the two cases, and the operation of the various mechanisms of particle interception will be very different. Many copepods move their second maxillae in a different way when feeding on small algal cells than they do when feeding on large ones (Price and Paffenhofer 1980). The precise consequences of these various adaptations have yet to be investigated.

2.3.5 Calculation of zooplankton feeding rates

A lack of agreement exists between measurements of zooplankton feeding rates in the laboratory and calculations made on the basis of budgets constructed for field situations. In general, the latter calculations suggest much higher figures. The evidence has been reviewed by Cushing (1975). One of the few laboratory data sets to obtain high rates of feeding approaching those deduced from field observations was obtained by Corner *et al.* (1972). These workers carried out the experiments in glass columns that were rotated in the vertical plane and completed one revolution every 2 minutes. Although the strategy was intended to ensure that the phytoplankton cells remained in suspension throughout the experiment, the method also maintained a relatively high level of turbulence in the containers. Corner *et al.* found that the maximum daily rations of the zooplankton were equivalent to 47.5% of the body nitrogen.

Cushing (1968) developed a theory of grazing in which the herbivore is assumed to search a volume that is the product of the area swept by the sensors and the distance swum. As the animal stops to eat, the volume searched becomes reduced. Others have modified the concept by introducing irregular swimming tracks and varying speeds of swimming, and by introducing threshold densities of food below which the animals did not feed. Many of the assumptions underlying this work have recently been challenged by Rothschild (1988), whose basic thesis is that the rate of contact between a planktonic organism and its food is strongly affected by small-scale turbulence in the water. Feeding studies that do not take this factor into account need revision. Beginning with the formulation of Gerritson and Strickler (1977), the relative velocity of a prey particle to a predator particle is

$$(u^2 + 3v^2)/3v \text{ for } v > u$$

and

$$(v^2 + 3u^2)/3u \text{ for } u > v$$

where u and v are the velocities of the prey and predator, respectively. Contact rate is calculated by multiplying the relative velocity by the cylindrical volume swept by the predator and the number of prey contained in that volume, or $N\pi r$ where N is the density of prey and r is the radius of the cross-sectional area searched by the predator. Rothschild (1988) modified this equation by introducing a term w for the RMS velocity difference of two points in a homogeneous isotropic turbulent medium. With this change, the RMS relative velocity of the predator and prey becomes

$$\frac{u^2 + 3v + 4w^2}{3(v^2 + w^2)^{\frac{1}{2}}}$$

where the predator is faster than the prey. He developed an example using data (Table 2.2) for Georges Bank during the month of May, from Backus and Bourne (1987). Using wind velocities of 10 knots and 20 knots, Rothschild made some assumptions about levels of turbulence in the mixed layer and calculated the RMS turbulent velocities, w, shown in Table 2.3. He then calculated the contact rates shown in Table 2.4, which show that turbulence has an important effect. The impact affects the smallest organisms most dramatically, because their rates of locomotion are small compared with the velocity of the turbulent water. The rates of contact of bacteria with phytoplankton are enhanced 3–4 orders of magnitude by high turbulence, and even the contacts of macrozooplankton with phytoplankton are almost doubled.

Table 2.2 Georges Bank scenario. Densities of organisms approximate those reported for May at the terminus of the spring bloom.

	Number cm^{-3}	Diameter (μm)	Velocity (cm h^{-1})
Bacteria	10^6	1	0.1
Phytoplankton	100	20	0.1
Nannoplankton*	2000	10	36[†]
Pseudocalanus	0.002	1000	720[‡]

*Mostly at surface and heterotrophs. [†]100 μm s^{-1}. [‡]0.2 cm s^{-1}.

Table 2.3 A range of turbulent RMS velocities, w, given wind velocities and particle separation distances. The wind velocities are given only as examples as tides or other sources might also generate turbulent motions. From Rothschild (1988).

Particle separation distance (m)	Wind velocity (cm s^{-1})	
	10 knot wind	20 knot wind
10^{-4}–10^{-3}	0.005	0.01
10^{-2}–10^{-1}	0.1	0.2

Table 2.4 Relative velocity and number of contacts among different kinds of organisms. Zero, low-turbulence (0.005 cm s^{-2}), and high-turbulence (0.2 cm/sec) RMS velocity, w, is incorporated in the computation. From Rothschild (1988)

	Turbulence	Bacteria		Phytoplankton		Nannoplankton	
		Relative velocity	Contacts per hour	Relative velocity	Contacts per hour	Relative velocity	Contacts per hour
Pseudocalanus	zero	720	22 × 10⁶	720	2262	721	4.5 × 10⁴
	low	720	22 × 10⁶	720	2262	721	4.5 × 10⁴
	high	1187	37 × 10⁶	1187	3729	1188	7.5 × 10⁴
Nannoplankton	zero	36	113	36	—	—	—
	low	43	135	43	—	—	—
	high	961	3019	961	0.03	—	—
Phytoplankton	zero	0.1	1	—	—	—	—
	low	24	301	—	—	—	—
	high	960	12,064	—	—	—	—

Several experiments have been designed to test the theoretical developments of Rothschild (1988) and Rothschild and Osborn (1988). Saiz and Alcaraz (1991) followed the development and growth of the copepod *Acartia grani* in replicated 10 liter containers, half of which were turbulent (vertical eddy diffusivity coefficients 1–5 cm^2 s^{-1}) and the other half quiescent, with food present in excess. Growth was accelerated and development time shortened in the turbulent chambers throughout the naupliar stages. The effect was less marked in later development stages, however, and the total development time through to the adult was not significantly different in the two treatments. The authors acknowledged that the effect on naupliar growth could have been caused partly by differences in the composition of the planktonic food community under the different turbulence regimes.

In a study of behavior of another species of *Acartia* at three levels of turbulence, Saiz and Alcaraz (1992) found that the highest level of turbulence increased feeding activity by a factor of 2.7, but also increased a locomotory activity known as 'jumping' by a factor of 1.7–1.9. It was not clear whether the increased metabolic activity of 'jumping' burned off the additional energy acquired through increased feeding.

Food abundance can change the effect of turbulence on zooplankton. Saiz *et al.* (1992) determined feeding rate (food clearance rates and ingestion rates) and gross growth efficiency (from rate of egg production) at two levels of turbulence and in saturating and nonsaturating food conditions. For all three species studied, turbulence significantly enhanced feeding rates in nonsaturating food conditions, but not in saturating conditions. With one species, *Acartia clausi,* feeding declined at higher levels of turbulence, even at low food concentrations. Saiz *et al.* suggested that, while moderate levels of turbulence enhance encounter probability, strong turbulence could have a detrimental effect on the feeding currents used to ingest the prey.

Yamazaki *et al.* (1991) investigated plankton contact rates in a turbulent flow simulated directly from the Navier–Stokes equations, using a Reynolds number of 38. In all cases, a significant increase in contact rates occurred when turbulence was present, compared with the "no turbulence" scenarios.

In general, the theoretically derived ideas of Rothschild and Osborn—that turbulence strongly affects contact rates between zooplankton and their prey—are supported by experimental evidence. However, there are qualifications. The effect is more marked at low food concentrations than at food saturation, and a dome-shaped relationship probably exists, with feeding rate declining when turbulence is too strong.

A further complication is introduced by recent observations on the distribution of marine snow in natural water columns (MacIntyre *et al.* 1995). Marine snow (aggregated particles over 0.5 mm diameter) off central California appears to be more abundant at density discontinuities in the water column. Various mechanisms, including patterns of turbulence, have been suggested to account for this accumulation. The net result is that marine snow aggregations are more vulnerable to grazers than are individual small particles, both because very small particles are aggregated in the marine snow and because the marine snow is aggregated at certain depths in the water column.

2.3.6 Turbulence and food contact rates in larval fish

The influence of wind- and tide-induced turbulence on the encounter rates between larval fish and their zooplankton prey was quantified in simulation models by Mackenzie and Leggett (1991). They found that failure to consider the influence of small-scale turbulence when prey density was less than $35 \, L^{-1}$ could result in as much as an 11-fold underestimate of contact rates, under conditions prevailing at tidal fronts in summer. As a result, they suggested that turbulence may be a factor stimulating food chain productivity at such fronts.

We should not, however, assume that increase in turbulence will enhance larval fish ingestion over the whole range of small-scale turbulence in nature. MacKenzie *et al.* (1994) modelled the probability of ingestion as the product of the probability that the prey is encountered and the probability of successful pursuit. The latter measure was quantified by calculating the probability, for different levels of turbulence, that the prey will remain within the larva's encounter sphere long enough for it to approach and orient toward the prey, and then adopt an effective strike position. This interval is called the pursuit time. The encounter probability was calculated in the usual way from Rothschild and Osborn (1988). They concluded that the relationship of feeding rate to turbulence is dome-shaped. At low levels of turbulence, a rise in turbulence increases the feeding rate; at high levels of turbulence, a rise in turbulence decreases the feeding rate by disrupting the pursuit after initial encounter.

Sundby and Fossum (1990) sought empirical evidence for the effect of turbulence on the feeding rate of cod larvae. They collected data on the number of prey per cod larval gut, the concentration of prey (*Calanus finmarchicus* nauplii), the wind speed, and static stability of the water column, expressed as

$$F = W^3/N$$

where N is the average Brunt–Väisälä frequency between 5 and 25 m depth, and W^3 is the average cube of the wind speed during 8 hours previous to the sampling.

Sundby and Fossum found that as the wind speed increased from 2 m s^{-1} to 6 m s^{-1}, the contact rate increased by a factor of 2.8. This result was in reasonable agreement with the theory of Rothschild and Osborn (1988). Substituting reasonable values for the predator speed, prey speed, and RMS turbulent velocity in their equation gave an increase in contact rate of 2.2 times for the same change in wind speed. They found that when wind speeds were in excess of 4 m s^{-1}, turbulent velocity made a greater contribution to contact rate than larval swimming speed. Sundby and Fossum concluded that wind speed is an important factor in year-class survival and subsequent recruitment.

In further studies (Sundby *et al.* 1994), the feeding rate of cod larvae off northern Norway was found to increase by a factor of 7 as the wind speed increased from 2 to 10 m s^{-1}. Under these conditions, any negative effect of wind speed on feeding rate must occur at levels higher than 10 m s^{-1}. These results do not necessarily contradict the results of modelling studies reported earlier. It is likely that the density gradient factors influencing the Brunt–Väisälä frequency at the study site were not the same as those in the model. Hence, the relationship between wind speed and turbulence would be different.

Landry *et al.* (1995) exposed fathead minnow (*Pimephales promelas*) larvae to four turbulent velocities at a single food concentration (30 brine shrimps per liter) and again to four levels of turbulence at three different prey concentrations. In the first experiment they found a dome-shaped relationship, with enhancement of feeding rate at low levels of turbulence and reduction at high levels. In the second experiment they obtained the same dome-shaped relationship with the highest concentration of prey (500 L^{-1}), but with 50 prey L^{-1} and 15 prey L^{-1} they obtained a depression of feeding rate at all levels of turbulence, compared with the no-turbulence condition.

For fish in coastal upwelling systems (see Chapter 5), Cury and Roy (1989) and Ware and Thomson (1991) have found that survival of newly hatched fish and subsequent recruitment is optimal at average wind speeds of 5–8 m s^{-1} and that a dome-shaped relationship exists between the range of wind speeds encountered in nature and year-class success. The results of Landry *et al.* (1995) provide a possible mechanism for this phenomenon.

In a modelling study, Davis *et al.* (1991) added another component to the relationship of planktonic predators to their prey—namely, microscale (<10 m) patchiness. They found that the growth rate of predators is strongly dependent on the existence of patches in which prey density is well above the large-scale average. Physical turbulence, at levels corresponding to winds of 5 m s^{-1}, tends to homogenize prey patches and reduce the growth rate of predators. Winds of 10 m s^{-1}, on the other hand, lead to more frequent encounters between predators and prey, and have the effect of restoring predator growth rates to low turbulence values.

Herring larvae are particularly abundant in a region of strong tidally generated turbulence off southwest Nova Scotia. Muelbert *et al.* (1994) investigated whether the turbulence assisted prey capture by the larvae. They carried out a dimensional analysis, using the modifications by Rothschild and Osborn (1988) of the Gerritson and Strickler (1977) model of interactions between a moving predator and its prey in a water column. They derived two nondimensional numbers from the existing models and proposed a third that incorporates the minimum encounter rate required for larval maintenance.

Muelbert *et al.* showed that when root-mean-square turbulence velocities are the same order of magnitude as the prey velocities, they have a small influence on encounter rates; when they are an order of magnitude larger than predator velocities, they dominate the encounter process. In tidally well-mixed regions, herring larvae could apparently survive on an order of magnitude lower food concentration than in stratified water where the turbulence is lower. At the food concentrations occurring in the study area in winter, it appears that turbulent, tidally well-mixed areas provide ideal feeding conditions for the herring larvae and that fish in stratified water have a much lower probability of survival. These results appear to support the ideas of Iles and Sinclair (1982) that the spawning of herring occurs predominantly in tidally mixed areas, and that the larvae tend to be retained within these areas during critical stages of development (see section 7.3.2).

Reviewing the effect of small-scale turbulence on processes in the plankton, we see that it influences trophic transfers at all levels from dissolved nutrients to fish. Turbulent shear causes aggregation of colloidal particles into particles that bacteria can digest. It also affects the uptake of nutrients by phytoplankton, enhancing it for some organisms and inhibiting others, depending on the level of turbulence. It seems probable that a dome-shaped relationship exists between turbulence and feeding rates for both zooplankton and fish. At low levels of turbulence feeding is enhanced by increasing rates of contact with food particles; at higher levels the process of food handling may be disrupted, leading to a reduced feeding rate. The optimum turbulence rate differs for various species, and there is a complex interaction of levels of turbulence with food density and with the patchiness of its distribution.

2.4 BENTHIC PLANTS

2.4.1 The problem

The concept applied to phytoplankton applies also to attached benthic plants. If a plant is surrounded by completely still water it will soon produce a zone around itself in which the nutrients are depleted and metabolism is limited by the rate at which nutrients are able to diffuse along the concentration gradient (Box 2.02). The problem is alleviated if there is a movement of water relative to the plant, for the zone of nutrient depletion is partially replaced by less depleted water. We saw that phytoplankton achieve a movement relative to the water by passively sinking or by actively swimming. They may also preferentially occupy zones of locally high-nutrient concentration. For benthic plants, the key to their productivity is the movement of water over them while they are held in one place by their attachment to the substrate.

Benthic plants are limited to areas where the sea floor is within the zone where there is sufficient light to support photosynthesis and it is usual to find a fringe of marine macrophytes associated with the various types of coastline. Seaweeds (i.e., attached algae) predominate in eroding areas where there is a rocky bottom for attachment of the holdfast, while flowering plants such as seagrasses, marsh grasses, and mangroves predominate in sedimenting areas where their roots can take hold and obtain nutrients. In the section that follows we shall explore the question of how dependent benthic plants are on the movement of water around them.

2.4.2 Water movement and productivity

Koehl (1986) has given a good summary of the water flow encountered by benthic marine algae. Giant kelps that colonize water of about 10 m depth and have buoyant fronds near the surface encounter orbital water movement as a result of surface wave motion. Closer to shore, where the water depth is less than half the crest-to-crest distance of the waves, the algae experience a back-and-forth oscillation of the water, parallel with the bottom, as the waves pass overhead. This movement is accentuated in the shallow intertidal zone, where waves break and water rushes up the shore, to recede again after each wave.

Within these broad zones of water movement, algae are able to experience flow micro-habitats, according to their growth habit. Very small forms or encrusting species may be so close to the substrate that they are in the boundary layer of slowly moving water, either the viscous sublayer or the turbulent zone described in section 2.2.8. Because the boundary layer builds up as water moves across a surface, as described by Eq. 2.13, algae further from the leading edge of a structure such as a rock will experience a thicker boundary layer than those living near the edge. Again, algae growing close to one another may be compressed into one large clump at high water velocities, so that water currents are directed around, rather than through the clump, and current speed within the clump is markedly reduced. Full quantification of these processes has not yet been achieved, but there are interesting studies of components of the problem.

It has been shown several times that nutrient uptake and photosynthesis in aquatic macrophytes tend to increase with increasing flow of water over their surfaces (Whitford and Schumacher 1961; Schumacher and Whitford 1965; Westlake 1967; Wheeler 1978, 1980; Gerard 1982). Gerard and Mann (1979) found a trade-off between blade morphology, drag, and productivity in *Laminaria*. In high-energy sites the algae have narrower, thicker blades and solid stipes, which are an adaptation to withstanding drag (see section 2.4.3). At more sheltered sites the algae have hollow stipes that help lift the blades closer to the water surface and thinner, more frilly blades that increase turbulent exchange at their surface. Despite having less water movement over the blades, algae at the sheltered sites showed better growth and productivity. In a subsequent laboratory study, Gerard (1987) found that narrow blades were produced in response to longitudinal stretching, and suggested that increased thickness was a response to frequent bending.

More elaborate, instrumented studies were carried out by Koehl and Alberte (1988), working on the Pacific coast of North America with *Nereocystis*. This kelp also has narrow, flat blades in exposed situations and wider, undulate blades in more sheltered localities. They used electronic current meters to produce continuous records from the vicinity of the plants and from these were able to describe not only the current velocities but also the turbulence intensities. They pointed out that while the mean velocity gives a measure of the advective transport of nutrients, turbulence intensity indicates the importance of small-scale motions that lead to enhanced turbulent diffusion as outlined in section 2.2.7.

Koehl and Alberte (1988) also measured the velocity gradient at the surface of *Nereocystis* blades. They used a measure analogous to the boundary friction velocity u_* discussed in section 2.2. Water velocity profiles along strips of the blades were determined by placing them in a tank through which particles moved while being illuminated by strobe flashes. The distances between the particles as seen on a photographic image gave a measure of water velocity at different positions. Next, u_* was calculated from Eq. 2.13, with $z_0 = 0$ for the case of a smooth strip of algal blade. For blades living in water of different velocities they characterized the average friction velocity by a technique similar to Gerard's (1982) use of plaster buttons. They sewed the candies known as 'Pep-o-Mint Life Savers' to the blades and used weight loss during 10 minutes as an index of friction velocity. They calibrated the technique against the measurement of u_* in a tank. From this result, they were able to clearly demonstrate that the effect of current on photosynthetic rate was through the increase of the velocity gradient. They also found that there were conditions of low flow in nature that led to gradients corresponding to a definite limitation on photosynthetic rate.

From measurements made in a flow tank Koehl and Alberte determined that frilly wide blades 'flapped' with a greater amplitude than narrow blades in response to the current, but caused greater drag on the plant. Narrow flat blades, on the other hand, collapsed into more streamlined bundles in a strong current, but in this position they were subjected to more self-shading. They showed that photosynthetic rate decreased in slow flow, but that blade flapping enhanced it. Hence, the frilly blade morphology, which caused increased flapping, enhanced photosynthesis. There was a trade-off between the photosynthetic advantages of wide, frilly blade morphology and the risk of the plant's breaking, due to increased drag.

Anderson and Charters (1982) placed the bushy alga *Gelidium nudifrons* in a water tunnel in the laboratory and observed turbulence by a dye stream technique. They found that local turbulence was damped at low velocities, but the alga created its own microturbulence at higher velocities. How this difference relates to the realities of life on a rocky shore remains unclear.

2.4.3 Water movement and drag

As we have seen from several examples, seaweeds often have the ability to respond phenotypically to the ambient level of water movement. Those in quiet waters may develop structures variously described as undulant, frilly, or ruffled, while those in exposed situations may be much more streamlined. Since the undulant form has been shown to have a higher rate of photosynthesis in moderate currents, but a greater drag, and since plants that are torn from their holdfasts are often cast up on the beach and die, there is an evolutionary trade-off between photosynthetic capability and streamlining.

An obvious adaptation to minimizing drag in an oscillating current is the ability of a plant to bend in the direction of the flow at a given time, minimizing S, the area presented to the current. However, as Koehl (1986) pointed out, flexibility may have disadvantages as well. A stiff plant may be able to hold its blades above those of its more flexible neighbors, out-competing them for light. On the other hand, a flexible plant, by its lashing about, may shake off epiphytes, or herbivorous animals. Some seaweeds, as they bend in the wave-induced currents, sweep the area of sea floor surrounding each plant, keeping it free from competitors. What degree of flexibility is optimal for this? Koehl's (1986) review gives a great deal of information about the biomechanics of seaweed structures, and interested readers may care to go more deeply into this subject.

An interesting and unusual seaweed is the sea palm, *Postelsia palmaeformis*, which lives intertidally and is sufficiently stiff to stand more or less upright when not bent over by wave action. Holbrook *et al.* (1991) investigated the physical properties that make this possible. The elastic modulus, E, is low enough to permit bending down to the substratum without breaking, and is not very different from that found in other large marine algae. The stipe is wide in relation to its height, however, and this property confers a flexural stiffness. When subject to extreme drag, the stipe is extended 20–25% before breaking. Holbrook *et al.* contrasted this behavior with terrestrial trees. Wood is much stiffer but has a much lower breaking extension. Trees often break in strong winds, while *Postelsia* withstands forces from breaking waves that are about 850 times greater than those on a tree hit by wind moving at the same velocity and acceleration.

2.4.4 Water movement and community structure in kelp beds

Several interesting studies have focused on the way in which kelp beds influence water movement, and through it the structure of the biological community (see Eckman and Duggins [1991] and references therein). Earlier work by G. A. Jackson showed that beds of giant kelp, *Macrocystis*, on the California coast attenuate internal waves passing through them, and reduce longshore currents to about one-third of their value outside the bed. When the water column contains invertebrate larvae that are competent to settle, slowing the current has the effect of concentrating newly settled larvae within the bed.

Eckman, Duggins, and their colleagues made similar studies in beds of smaller kelps, *Agarum* and *Laminaria*, off the coast of Washington State. As expected, they found reduced currents within the beds; they also found a greater rate of deposition of particles, which they attributed to the increase in retention time of the water containing the particles. The light intensity was lower within the beds than outside them, a condition that was associated with reduced cover of microalgae on the substratum. When they investigated the influence of these factors on settlement of planktonic larvae, the results varied widely from one species to another, depending on the duration of planktonic larval life and individual responses to sedimentation and light. Similarly, when the growth rates of invertebrates living outside and inside the kelp beds were compared, barnacles, mussels, and serpulid worms often showed enhanced growth rates within kelp beds, while bryozoans and sponges showed no significant differences.

2.5 BENTHIC ANIMALS

2.5.1 Filter feeding in the benthic boundary layer

As we have seen, benthic plants take up nutrients more rapidly and increase their metabolism as the current flows faster, because the supply of limiting substances increases through increased turbulent diffusion. Benthic animals behave in a similar way. Current speed affects the growth rate of such benthic suspension-feeders as barnacles and bivalves (Crisp 1960; Richardson *et al.* 1980; Wildish and Peer 1983). Wolff (1977) reviewed the evidence for particularly high benthic secondary production in estuaries and suggested that the prevalence of tidal currents was an important factor. He remarked, 'In the Grevelingen estuary, for example, tidal currents in the range 0.1–1.0 m s^{-1} would cause the vertical component of the turbulent diffusion coefficient to be approximately 50–500 cm^2 s^{-1}, implying that surface production becomes available to benthic filter-feeders at a few metres depth within an hour or less.'

Wildish and Kristmanson (1979) extended the idea still further, producing evidence that benthic communities in areas affected by tidal currents formed a spectrum. In waters with very low tidal currents, deposit-ingesting animals such as polychaete worms, nematodes, and certain kinds of bivalves predominated; in areas with more water movement, suspension-feeding bivalves and holothurians were much more abundant. Finally, in areas where the mean tidal currents, measured 1–2 m above the bottom, exceeded 30 cm s⁻¹, they showed that the communities were stressed and the fauna impoverished. Conditions at the two extremes can be understood without much difficulty. In very quiet waters nutritious organic matter accumulates on the bottom, where it is most easily utilized by animals that ingest sediment directly. At the other extreme, high currents erode the sediments, leaving little organic matter and making suspension-feeding difficult because there is so much inorganic matter in suspension. For the intermediate stages, the authors found a gradient in abundance of suspension-feeding organisms that they thought could be explained by interactions between the filter-feeders and the tidal currents. They suggested that the suspension-feeding organisms depleted the concentration of food particles near the sediment–water interface, and moderate tidal currents generated enough turbulence in the bottom boundary layer to increase the supply of food to the suspension-feeders by turbulent diffusion.

In later work the same authors (Wildish and Kristmanson 1984) confirmed in a laboratory flume tank that blue mussels, *Mytilus*, and horse mussels, *Modiolus*, were able to cause a reduction in the living organic matter in suspension at a height of 1 cm above the bottom, compared with 30 cm above the bottom.

A more detailed study of the dynamics of a benthic community in relation to turbulent diffusion in the water column was made by Frechette and Bourget (1985a,b) and Frechette *et al.* (1989). For their study area they chose a dense intertidal bed of the mussel *Mytilus edulis* in the St. Lawrence estuary, Canada. They demonstrated depletion of the particulate organic matter above the mussels and showed that the effect was inversely proportional to current speed. They also showed that animals situated 1.0 m above the bottom had better growth rates than those found on the bottom, presumably because more food was available. The relationship between current speed, vertical profile of density of suspended organic matter, and consumption rate of the mussels was demonstrated in a simulation model.

A study of the feeding behavior of barnacles (Trager *et al.* 1990) has shown how intricate are their adaptations to life in a variable flow. In currents greater than 3.1 cm s⁻¹, they extended their cirri and passively filtered the water. At lower current speeds, the barnacles actively swept the water with their cirri. In an oscillating current, they reversed the direction of their cirri just before the current changed direction. They also swept actively while the current was accelerating and changed to passive filtering when it exceeded 3.1 cm s⁻¹.

Thus, the evidence from community structure, single-species populations, and physiological experiments on individual animals supports the view that food consumption of benthic filter-feeders is a function of current speed in the benthic boundary layer, as well as of the more traditional variables, such as food concentration and temperature, that have been investigated in the past.

In the case of the mussels discussed above, the animals normally live attached to solid substrates in areas where water movement is sufficiently vigorous to prevent the accumulation of sediment. There are also many kinds of benthic animals that live in areas where the intensity of turbulence is sufficiently low to permit the accumulation of sediment. Some of these animals feed exclusively by ingesting the sediment, but many ingest food from the suspension of particles that is found just above the sediment. In the next section we look at the vertical distribution of these particles and its effect on the feeding strategies of animals.

2.5.2 Suspension-feeding benthos

If we envisage a current just above the bottom carrying relatively heavy mineral particles and lighter organic particles, we might guess that the heavy particles would be concentrated close to the bottom and the lighter particles higher up. Muschenheim (1987a) studied this question in theory and in practice. The theoretical argument goes as follows.

The mean flow in the boundary layer above the ocean bottom shown in Figs. 2.04 and 2.05 and described by Eq. 2.13

$$u(z) = (u_*/k) \ln(z/z_o)$$

corresponds to a logarithmic decrease in velocity as one approaches the sediment surface. The equation

$$C(z) = C(a)(a/z)^p \tag{2.18}$$

gives $C(z)$, the concentration of suspended material at height z, relative to the concentration $C(a)$ at height a. Particles of different sizes are differentiated by their settling velocity w_s in the Rouse ratio p given by

$$p = w_s/ku_* \tag{2.19}$$

where u_* is the friction velocity and k is Von Karman's constant. When $p = 1$ the particles are heavy and the settling velocity w_s is high relative to the friction velocity. These particles tend to be concentrated near the bottom. For lower values of p the particles are light and easily stirred up by the turbulence so they are distributed higher in the water column.

The horizontal flux, F, of particles at any height z above the bottom is given by multiplying the horizontal velocity u at the height z by the concentration at z,

$$F(z) = u(z)C(z). \tag{2.20}$$

The horizontal flux of the heavy particles ($p = 1$) is at a maximum quite close to the boundary. For lighter particles ($p < 1$) the maximum horizontal flux is found higher up in the water column. Thus, heavy inorganic particles will have their maximum horizontal flux close to the bottom, while light organic particles will have their maximum flux a few centimeters above the bottom. This result leads to the hypothesis that it is to the biological advantage of benthic animals to place their feeding organs a few centimeters above the bottom.

Muschenheim (1987b) studied a polychaete worm, *Spio setosa*, which builds a sand tube that allows it to feed 4–6 cm above the sediment. Worms taken from the field were found to have gut contents that were enriched in organic-mineral aggregates, relative to the sediment surface. In a flume tank study, half of the worm tubes were cut down close to bed level so that the worms were forced to feed at a height of 0–2 cm, while the remainder were left at 4 cm length so that the worms fed at 4–6 cm. Those feeding at the lower level were found to collect particles with a significantly lower organic content. Measurement of the settling velocity of surficial sediment collected from the field site showed that it had a Rouse number consistent with it being carried in suspension several centimeters above the sediment surface.

These results, therefore, support the view that in a boundary layer above a sediment surface, the differential settling velocities of particles of different densities have a natural sorting effect, such that the maximum flux of organically rich particles occurs at a higher level above the sediment than the maximum flux of inorganic particles. This observation explains why a number of different organisms, especially polychaete worms, build tubes that protrude above the sediment surface.

Abelson *et al.* (1993) generalized the foregoing results. Using a physical model, supported by field experiments, they grouped benthic organisms according to the ratio of height to width in the plane normal to the flow. Organisms with a high slenderness ratio (SR) cause divergent flow around the base, while the upper regions experience a good exposure to fine suspended particles. Organisms with a low slenderness ratio caused the flow lines near the substratum to lift over the top of them and expose them mainly to bed-load particles. Abelson *et al.* offered the generalization that high-SR species are suspension feeders, while low-SR species are coarse-particle or bed-load feeders. For example, a species of crab that inhabits tips of coral branches filter-feeds on fine suspended particles, while another species that lives on low, massive corals feeds on coarse particles.

2.5.3 The boundary layer and larval settlement

In this section we consider the problems encountered by small planktonic larvae as they descend from the water column into the benthic boundary layer prior to settling on the bottom to begin their adult life. Tidal currents above the bottom often greatly exceed the swimming speeds of the larvae, but they decrease in a logarithmic manner as the bottom is approached. A viscous layer exists close to the bottom in which turbulence is minimal. Can the larvae enter this layer and then swim strongly enough to be able to select a site for settlement, or are they forced to locate a settlement site by some more haphazard approach?

Butman (1986) studied the question at a site where the tidal currents 0.5 m from the bottom reached $5-10$ cm s^{-1} on each ebb and flow. A vertical profile of the boundary layer was constructed by deploying current meters at 30, 50, 100, and 200 cm above the sea floor, and using the data to calculate u_* and Re$_*$ (Box 2.05). Calculations were made for smooth-turbulent and rough-turbulent velocity profiles; these measurements were compared with the size and swimming speed of polychaete larvae (Figs. 2.14 and 2.15). In smooth-turbulent flows, larvae of 300 μm would not have room to maneuver by swimming if they found themselves below the horizontal dotted line, and would not be able to swim against the current if they were in the region to the right of the vertical dotted line. When rough-turbulent profiles were calculated, it was concluded that eddies would reach so close to the sea floor that there would be no region in which the larvae could effectively move by swimming.

Hence, it seemed that during about 40% of the tidal cycle it is physically possible for the larvae to swim around near the bed, exploring sites available for settlement. They would be confined to a region within 100 μm of the bed. However, the real world includes the additional complication that wind-driven and density-driven currents are superimposed on the tidal cycle. The author concluded that the larvae probably do not search for preferred habitats by active horizontal swimming near the bed. It seems more likely that they swim vertically in smooth-turbulent flows, going down to test the substrate and up to be advected to another site downstream. This process has been called the 'balloonist technique.'

Fig. 2.14 Smooth-turbulent velocity profiles in the viscous sublayer, calculated for a range of near-bottom flow speeds measured at a site in Buzzards Bay, Massachusetts. Light lines indicate size and swimming speed of settling worm larvae. From Butman (1986).

Fig. 2.15 Two rough-turbulent velocity profiles and a smooth-turbulent profile from a site in Buzzards Bay, Massachusetts. From Butman (1986).

From this study we see how consideration of the physics of the boundary layer creates a totally new perspective on the settlement of planktonic larvae. Results from classical studies conducted in the laboratory are in need of reinterpretation.

Interactions between chemical cues and hydrodynamic effects on larval settlement have been investigated for a reef-building polychaete worm, *Phragmatopoma lapidosa californica* (Pawlik *et al.* 1991; Pawlik and Butman 1993). In still water, larvae of this worm settle preferentially on sand derived from adult worm tubes, or on sand treated with free fatty acids extracted from the cemented sand matrix of the tubes. When the experiments were repeated in flowing water, it was found that larvae settled most readily at 15–25 cm s^{-1}. At slower flows, they left the bottom and swam in the water column. At the highest rate of flow, the larvae were eroded from the bed and carried passively as bed load.

Preference for intermediate flow rates probably reflects the requirement of the settled worms for a current strong enough to bring a supply of sand for their tubes, as well as organic matter for food. Pawlik and Butman (1993) also review other examples of interactions between larval settlement behavior and the near-bed flow regime. Barnacle larvae actively explore a substrate on which they have settled, but exploration occurs only along the line of flow at the surface. The larvae of the worm *Capitella* swim up and down as they are carried by the flow, either accepting or rejecting sediments as they are encountered. The larvae of the bivalve *Cerastoderma edule* become concentrated in the viscous sublayer in relatively fast, turbulent flows, suggesting that fast flow increases their opportunities for habitat selection.

2.5.4 The boundary layer and coral reefs

Primary production on coral reefs results from two main processes: photosynthesis by the zooxanthellae embedded in the tissues of the coral, and photosynthesis of algae attached to the surface. Patterson *et al.* (1991) studied the effect of changing flow rate on heads of the coral *Monastrea annularis*, using a recirculating flow respirometry system deployed from an underwater habitat. Not surprisingly, both primary production and respiration of the coral increased with increasing flow. By plotting the average Sherwood number against the Reynolds number, Patterson *et al.* obtained the relationship Sh = 0.01 Re$^{1.30}$ for dark respiration, and Sh = 0.71 Re$^{0.63}$ for maximal photosynthetic rate. According to White (1988), an exponent less than 0.5 indicates transport through a laminar boundary layer, while values greater than 0.6 indicate increasing importance of turbulent transport. A value of 1.30 indicates a performance better than predicted by available theory, and may relate to the role played by cilia in assisting transport.

Measurements on the productivity of algae attached to the reef surface, under different flow regimes in a chamber, were made by Carpenter *et al.* (1991). They found that both primary production and nitrogenase activity increased over unidirectional flow speeds of 0–10 cm s^{-1}. Introduction of oscillatory flow (as would be experienced as a result of wave action in the water column above) led to even more marked increases in photosynthesis and nitrogenase activity.

Subsequently, Carpenter and Williams (1993) investigated flow profiles on the surface of a reef in the U.S. Virgin Islands. They found that where the algal turf was short (usually as a result of grazing) the boundary layer was dominated by turbulent eddies, but in significant areas (such as depressions) the algal turf was longer and the boundary layer dominated by viscous forces, leading to diffusion limitation on metabolism.

2.5.5 Processes at the sediment–water interface

Benthic sediments are sites of processes important to the functioning of marine ecosystems. Dead organic matter accumulates at those sites, where it becomes subject to microbial decomposition. This degradation requires a supply of oxygen and leads to a release of nutrients to the overlying water. The diffusive sublayer of the benthic boundary layer presents a barrier to the downward flux of oxygen and the upward flux of nutrients.

Jørgensen and Revsbech (1985) used oxygen microelectrodes to investigate the properties of the viscous sublayer. They found that when oxygen was being taken up by the sediments, a layer of water with rapidly decreasing oxygen concentration which they interpreted as being the viscous sublayer, was found close to the sediment surface. It varied in thickness from 0.2 mm to more than 1 mm, depending on the velocity of the flow and the roughness of the sediment surface. This finding is in accordance with the theory presented in section 2.2.8. Using the calculation shown in Box 2.02 and a molecular diffusion coefficient, D, for oxygen of 2.0×10^{-9} m^2 s^{-1}, they found that the mean time for diffusion of oxygen across the boundary layer was 1.2–9 minutes. For sediments with high oxygen demand, this result constituted a serious resistance to oxygen flux.

A near-shore sediment covered with a mat of the filamentous, colorless sulfur bacterium *Beggiatoa* was found by Jørgensen and Revsbech (1985) to have such a steep gradient of oxygen through the viscous sublayer that almost zero oxygen was present at the surface, even though the overlying water in the turbulent layer was near saturation. As water movement over this surface increased, the diffusive sublayer became thinner, the oxygen gradient became steeper, and the rate of diffusion of oxygen increased. The group noted that the small polychaete worm *Pygospio elegans* builds sediment tubes that extend a few millimeters above the sediment surface, penetrating the viscous sublayer and enabling the worm to draw water from the well-aerated turbulent layer above. They also suggested that suspension-feeding ciliates such as *Vorticella* and *Carchesium* (which live on decomposing detritus) have a stalk several millimeters long so as to enable them to reach the oxygen-rich water above the viscous sublayer. Conversely, bacteria requiring anaerobic or near-anaerobic conditions can live on the surface of a piece of detritus or a sediment with a high oxygen demand. Even though the water in the turbulent layer above is rich in oxygen, the bacteria will experience near-zero oxygen at the sediment–water interface.

Further experiments (Gundersen and Jørgensen 1990) have shown that the diffusive boundary layer over a Danish coastal sediment at 15 m depth averaged 0.5–0.7 mm thick, but its thickness was influenced both by turbulence in adjacent water and by harmonic oscillations resulting from surface waves. The layer followed contours on the sediment surface but smoothed out features smaller than 100 µm. The result of these spatial and temporal changes in thickness of the viscous sublayer was an increase in the oxygen flux up to 2.5 times compared with that calculated from a simple one-dimensional diffusion model.

Using a more elaborate and generally applicable model of turbulent transport, Dade (1993) was able to investigate the effect of sediment roughness on transport through a diffusional sublayer. He concluded that, for example, the typical biogenic roughness in a fine-sediment marine environment enhances transport rates threefold over those expected for a smooth bed. This increase occurs despite viscous ponding of 'dead water' among roughness elements.

The general conclusion of these studies is that simple one-dimensional models of flux over smooth surfaces greatly underestimate natural fluxes at sediment–water interfaces.

2.6 SUMMARY: LIFE IN BOUNDARY LAYERS

In this chapter we have explored the influence of the small-scale properties of sea water on the life of plants and animals. Two phenomena crop up most frequently: the existence of a boundary layer around all organisms and the importance of the viscosity of the water at scales of millimeters. We saw that large-scale turbulence generated by ocean currents, tides, and waves cascades down through various scales of turbulence until its energy is dissipated by viscosity in small-scale eddies. This mixing transports essential substances such as carbon dioxide, plant nutrients, and food particles into the general vicinity of the organisms, but the last stage of the journey into the organisms is inhibited by the boundary layer. Water molecules stick to the surface of organisms creating a zone of zero water movement. Beyond this zone, viscosity creates a zone of reduced water movement known as the viscous boundary layer, into which turbulent eddies cannot penetrate so that the passage of essential substances is limited by the very slow rates of molecular diffusion. The thickness of this boundary layer decreases with the speed of the water moving beyond it.

For large plants, such as seaweeds, thinning of the boundary layer is achieved by attaching themselves to a solid surface in a zone where tidal currents and wave action cause vigorous water movement. It has been shown that if the drag on them is excessive the plants are in danger of being broken free and cast up on the shore to die. Many of them are capable of elegant changes in shape during growth, reflecting the trade-off between the need to maximize turbulence close to the plant surface and the need to reduce drag.

For medium- to large-size phytoplankton, thinning of the boundary layer is achieved either by passive sinking or by active swimming. The ability to sink in the water column varies from species to species according to the amount of skeletal material present, and a careful analysis shows that the limitations caused by the boundary layer can be alleviated by up to one-third by either passive sinking or active swimming. Differences in these properties may account for differing growth rates among species, and account in part for the succession of dominant species observed to occur with the progression of the seasons.

Very small phytoplankton and heterotrophic bacteria live in a world where viscous forces predominate and any swimming mechanism that relies on the inertial properties of water is simply unworkable. These organisms must make complex asymmetrical flagellar movements to travel through water. When they do so, they do not reduce their boundary layers significantly. The main advantage of moving about appears to be the increased possibility of encountering a higher concentration of nutrients.

Physical processes ranging from Brownian movement to turbulent strain cause colloidal particles to collide with one another, forming ever-larger aggregates. Bacteria can attach to the surfaces of these aggregates and digest them by secreting exoenzymes. This process is a much more efficient method of feeding than absorption of dissolved organic substances from the water column. It seems likely that phytoplankton that become enmeshed in aggregates are favorably placed to take up nutrients released by bacterial decomposition of organic matter.

On the sea floor, a boundary layer of reduced water movement extends for tens of centimeters above the bottom. Animals living on the sea floor benefit from the turbulence created by tidal currents, which have the effect of increasing downward transport of food material through this boundary layer. In areas of low turbulent energy soft sediments containing both organic and inorganic matter accumulate on the sea floor. When this sediment is transported laterally by tidal or wind-generated currents, a natural sorting mechanism ensures that the denser inorganic materials are in a higher concentration close to the bottom, while the lighter organic materials are concentrated a few centimeters above the bottom. Many animals such as tube worms are adapted to take their food from the upper of these two layers. Microscopic larvae are unable to hold their position above the bottom during the ebb and flow of tidal currents, unless they are in the viscous layer very close to the bottom. This layer is often too thin for them to move about and they are forced to select their sites for settlement by making vertical excursions in and out of the viscous layer. Finally, we see that boundary layers are important in the functioning of coral reefs, and have a key role in sediment–water exchanges.

3

Vertical Structure of the Open Ocean: Biology of the Mixed Layer

3.1 INTRODUCTION

One of the problems confronting phytoplankton in the ocean is that they need light and nutrients for growth and reproduction, but the source of light comes from above, while the source of nutrients is at depth. The sun's energy that reaches the surface waters is absorbed as it passes downward, decreasing exponentially with depth. In a finite layer, the euphotic zone, there is enough light for photosynthesis and growth to take place. In a water column with no turbulence, the euphotic zone would become depleted of nutrients as a result of uptake by the phytoplankton.

The reserve of nutrients in deeper water is constantly replenished by the decomposition of organisms from the euphotic zone that sink and decay. In the situation of zero turbulence that we have envisaged, there would be a very low level of nutrients in surface waters, a high level at depth, and the only mechanism for transfer from one to the other would be molecular diffusion, which, as we have seen, is extremely slow. In the real world, the ocean is filled with turbulent motion, generated by wind stress at the surface, internal waves, and so on. Phytoplankton depend absolutely on this turbulence to bring nutrients into the euphotic zone where they can be utilized in photosynthesis. This chapter is, therefore, concerned with some of the physical processes that affect the vertical distributions of light, heat, and nutrients so as to better understand the dynamics of phytoplankton production.

On average, the situation in the open ocean is less complex than in coastal waters where the influence of fresh water run-off from the land, of tidal currents, and of coastal topography lead to a high degree of complexity. This chapter deals with the open ocean, while coastal waters are deferred until Chapter 4.

3.2 VERTICAL STRUCTURE AND PHYTOPLANKTON PRODUCTION: TROPICAL WATERS

Figure 3.01 shows a vertical profile of temperature, chlorophyll, primary production, and nitrate from the eastern tropical Atlantic. It may be taken as representative of large areas of tropical ocean, and has been referred to as 'typical tropical structure,' or TTS (Herbland and Voituriez 1979). The water column is clearly divided into a warmer, lighter, upper mixed layer, and a cooler, heavier, lower layer, separated from the upper layer by a region of rapid change of temperature and density, known as the thermocline or pycnocline. The mixed layer has a very low level of nitrate, the lower layer has a higher level, and the zone of rapid change is referred to as the nutricline. The difference in nutrient concentration between upper and lower layers was noted in the introduction, but it is not immediately obvious why there is a sharp gradient in properties such as nitrate and a maximum concentration of chlorophyll and primary production at a depth of 20–30 m.

Several explanations have been offered for the subsurface chlorophyll and production maxima. Some, especially the early theories, depend on biological factors such as the sinking of diatoms or the alteration of the sinking rate according to light or nutrient levels. The

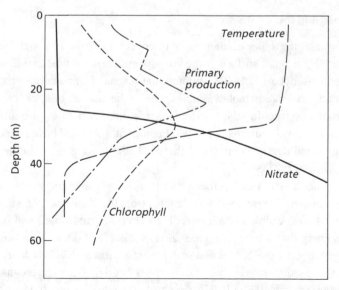

Fig. 3.01 Schematic diagram showing typical vertical structure of the water column in tropical latitudes ('typical tropical structure'—TTS). Note that the thermocline and the nutricline are at the same depth. The peak of primary production is more shallow than the peak of chlorophyll (= phytoplankton biomass).

Fig. 3.02 Spectrum of short-wave radiation reaching the sea surface and four depths. Note the progressive elimination of longer wavelengths as depth increases. From Jerlov (1976).

grazing impact of the zooplankton has also been suggested as a significant process. Recently, the decrease with depth in the vertical mixing rate of the ocean has likewise been demonstrated to cause a subsurface chlorophyll maximum in certain circumstances. Before treating these ideas in detail, we first explore the physics of mixed layers and thermoclines, beginning with the way heat is gained and lost through the surface of the ocean.

3.2.1 Heat gain and loss

The heat that warms the upper ocean comes, of course, from the sun, and the amount that arrives each day at the outer surface of the atmosphere varies in predictable ways with latitude and season (Budyko 1974). The amount that reaches and penetrates the sea surface, however, depends on unpredictable variables such as the amount of cloud cover. Measurements of the radiation reaching the sea can be made directly but they are difficult and expensive. Usually estimates are made by using empirical relationships derived from ships' observations of cloud cover. Examples of these calculations are given in the climate atlas of Isemer and Hasse (1987).

The sun's radiation that reaches the sea surface, often called the short-wave radiation, contains energy over a wide range of wavelengths from the ultraviolet light at 0.3 μm to the far infrared at 2.4 μm. Visible radiation lies between 0.36 μm (violet) and 0.75 μm (red). The amount of energy at each wavelength is as shown in Fig. 3.02, with minimal amounts at the shortest and longest wavelengths and with a maximum at a wavelength of about 0.5 μm.

The short-wave radiation is absorbed as it passes through water and the intensity at each wavelength decreases exponentially with depth, but the rate of decrease is different for each wavelength (Fig. 3.02). The intensity of the light of wavelength λ, at depth z, can be calculated from the intensity of the light at the sea surface, I_0, as

Fig. 3.03 Vertical profiles in sea water of light intensity relative to the surface for three wavelengths and for the total incoming short-wave radiation. For explanation, see text.

$$I = I_o \exp(-\alpha_\lambda z) \tag{3.01}$$

where the absorption coefficient α_λ is a function of wavelength, increasing from 0.004 m^{-1} for blue light to 1000 m^{-1} in the infrared at a wavelength of 1.0 μm. Thus blue light is absorbed much less rapidly than infrared, and at any particular depth the intensity of blue light is a larger fraction of its surface value than is the longer wavelength light.

Figure 3.03 shows three different exponential rates of attenuation with depth, calculated using Eq. 3.01. For purposes of comparison, the intensity at the surface is taken to be 100% for all wavelengths, even though Fig. 3.02 shows this case not to be true. The curves are also drawn on the assumption that the water is clear open ocean water. In other locations, especially coastal areas, the water may be less clear, with the result that α_λ values are much higher and the vertical absorption profiles are very different (see, for example, Jerlov 1976).

Box 3.01 Calculating the temperature increase in the mixed layer

The temperature increase that results from the absorption of short-wave radiation can be estimated with the equation

$$\Delta T = \Delta Q / mc \qquad\qquad (3.02)$$

where ΔT is the temperature increase, ΔQ is the energy absorbed in kJ, m is the mass of the water, and c, the specific heat (i.e., the amount of heat required to raise the temperature of 1 kg of water 1 °C), is $\simeq 4.2$ kJ kg^{-1} °C^{-1}. Over short periods of time, the amount of energy absorbed is estimated from the empirical formulae mentioned above; for periods of a month or more, however, the data may be extracted from climate atlases.

   To give an example, Isemer and Hasse (1987) indicate that over the course of an average July day in the North Atlantic Ocean, at 40° N 40° W, roughly 22,500 kJ m^{-2} of energy are absorbed by the ocean. The heat loss, from evaporation and other processes, over the same period is about 10,400 kJ m^{-2}. If the mixed layer is 5 m deep, Fig. 3.03 indicates 76% or 17,100 kJ m^{-2} of the incoming energy is absorbed in the 5 m mixed layer and the remainder is absorbed below 5 m depth. Since 10,400 kJ m^{-2} is lost from the surface of the water, the net energy gain or ΔQ in the mixed layer over the day is 6700 kJ m^{-2}. To get the temperature increase this ΔQ must be divided by the mass, which for the 5 m mixed layer is 5000 kg, and the specific heat (4.2 kJ kg^{-1} °C^{-1}). This calculation works out to about 0.3 °C.

The variation with depth of the total amount of short-wave radiation is also shown in Fig. 3.03 constructed from data published by Ivanoff (1977). The curve represents the cumulative effect of the different wavelengths being absorbed exponentially with depth at different rates. Although it is made up of many exponential curves, it is not itself an exponential function. In some mathematical thermocline models, such as that of Denman (1973), the rate of attenuation of the total incoming radiation is assumed to be exponential, partly for mathematical convenience and partly because it does not introduce large errors when the mixed layer is tens of meters deep. Woods and Barkmann (1986), on the other hand, use an empirical absorption profile similar to the curve of total absorption in Fig. 3.03 because it is significantly more accurate when dealing with shallow mixed layers.

   Although the absorption of radiation and the resultant gain of heat occurs through the top few meters of the ocean, the loss of heat is almost entirely from the top centimeter. Losses occur mainly through evaporation, infrared (long-wave) radiation, and conduction. Like the estimates of short-wave radiation, estimates of heat losses are calculated from empirical formulae derived from the common meteorological observations obtained from ships at sea. These formulae and their applications are discussed in Smith and Dobson (1984) and Isemer and Hasse (1987) (see Box 3.01).

3.2.2 Temperature increases and mixed layers

Absorption of short-wave radiation causes the temperature of the water to increase by an amount directly proportional to the amount of energy absorbed. Thus, the temperature rise after a given length of time will be very similar to the curve in Fig. 3.03 representing the total absorption of short-wave radiation. The greatest increase will be in the surface layers where the absorption is greatest and will decrease rapidly with depth. Observations of temperature with depth in the ocean, however, do not normally resemble this absorption curve because the upper layer of the ocean is usually stirred up by wind waves or by convection that is generated by the loss of heat at the surface. This stirred layer where the temperature remains constant with depth is, of course, called the mixed layer. Sometimes it is called the surface mixed layer to distinguish it from homogeneous layers in the interior or at the bottom. A method for estimating the increase in temperature in the mixed layer is outlined in Box 3.01.

3.2.3 Deepening the mixed layer

The depth of the wind-mixed layer in the previous section was assumed to stay constant with time. If the wind continues to blow, the turbulence generated by the waves will work to deepen the mixed layer. Imagine a homogeneous 5 m thick mixed layer with a temperature of 2 °C lying on top of a deep homogeneous layer at 0 °C. Now suppose that the stirring of the wind waves deepens the mixed layer from 5 to 10 m. Since the amount of heat does not change, the temperature of the layer is the result of mixing 5 m of 0 °C and 5 m of 2 °C water. Thus, the temperature of the mixed layer will be the average of the two, or 1 °C. Obviously energy must be used to accomplish the mixing since the colder and therefore denser water must be lifted upward against gravity, while the warmer and less dense water is carried downward, against buoyancy. The energy used to mix the water appears as a change in the distribution of mass or, in terms of energy, as a change in the potential energy of the water column. This energy change is discussed more fully in Box 3.02.

3.2.4 The pycnocline barrier

At the base of the mixed layer a density gradient or pycnocline separates the lighter water of the mixed layer from the denser water below (see Fig. 3.04). If a particle of density ρ_2 from below the pycnocline gets displaced above the pycnocline, as shown in the diagram, it will be surrounded by water of density ρ_1. The denser particle will be heavier than the surrounding water and will fall back toward the interface. Similarly, a particle of water from the upper layer displaced below the interface will be lighter than the surrounding water and will move up to the interface. The buoyancy force per unit volume that pushes the displaced particles back to the interface is

$$(\rho_2 - \rho_1)g \tag{3.05}$$

where the constant g is the gravitational acceleration ($\simeq 10$ m s^{-2}) (Gill 1982). The restoring force increases as the density difference ($\rho_2 - \rho_1$) increases and is the same as the force required to displace the particle initially. Since displacing particles from one side of the pycnocline to the other is equivalent to stirring, it follows that it takes more energy to stir water across a pycnocline with a large density difference than one with a small density difference. The fact that energy must be supplied by the turbulent motions for vertical mixing to occur in the pycnocline presents a barrier to vertical transport.

The effectiveness of the barrier increases with increasing density difference ($\rho_2 - \rho_1$). The reduced vertical transport means that the layers on either side of the pycnocline are

---

Box 3.02 Calculating the potential energy increase of a deepened mixed layer

The potential energy of a mass m is defined as the product mgh, where g is the gravitational constant, 10 m s^{-2}, and h is the height of the center of the mass m, above a reference level. For convenience we assume that the potential energy of a mixed layer is zero when the temperature is 0 °C. As heat is absorbed the water expands, which raises its center of mass and consequently its potential energy. The amount that a column of height h increases is

$$\Delta h = \alpha h\, \Delta T \tag{3.03}$$

where α is the coefficient of thermal expansion for water ($\simeq 10^{-4}$ °C^{-1}) and ΔT is the increase in temperature. If h is 5 m before it absorbs the heat and it warms up by 2 °C, Δh is $\simeq 10^{-4} \times 5 \times 2 = 10^{-3}$ m (1 mm). The center of mass rises by half of this amount or 0.5×10^{-3} m. Thus, the increase in potential energy is $mg\Delta h/2$ or $5000 \times 10 \times 0.5 \times 10^{-3} = 25$ J m^{-2}.

After mixing, the layer is 10 m deep but, because heat is conserved, the temperature is 1 °C rather than 2 °C and the potential energy referred to at a constant temperature of 0 °C is 50 J m^{-2}. Thus, the wind mixing has increased the potential energy of the water column by 25 J m^{-2}. This calculation is important when considering changes in mixed layer depths but it is only a tiny amount of energy compared with the 4.2×10^4 kJ m^{-2} of heat associated with the temperature of 1 °C through the top 10 m.

According to Denman and Miyake (1973), the rate at which the energy in the wind becomes available for increasing the mixed layer thickness is

$$E_\mathrm{m} \simeq 0.0015\, \rho_\mathrm{a} C_{10} U_{10}^{3} \tag{3.04}$$

where ρ_a is the air density ($\simeq 1$ kg m^{-3}), U_{10} is the mean wind speed measured 10 m above the sea surface, and C_{10} is the drag coefficient to be used with 10 m winds ($C_{10} \simeq 0.001$; Smith 1980). If $U_{10} \simeq 10$ m s^{-1}, then $E_\mathrm{m} \simeq 1.5 \times 10^{-3}$ W m^{-2}. This measure represents the rate in joules per second at which the potential energy of the water column is changed by the mixing. In the example, the potential energy increased by 25 J m^{-2}, which would be supplied by the 10 m s^{-1} wind in 25(J m^{-2})/1.5 $\times$ 10^{-3}(J m^{-2} s^{-1}) $\simeq 16.7 \times 10^3$ s or 4.6 h.

The turbulent motions arising from convection rather than wind mixing also act to deepen the mixed layer. Deardorff *et al.* (1969), in a laboratory experiment, found the effect to be small and it is ignored in the thermocline models of Denman (1973) and Woods and Barkmann (1986).

---

somewhat isolated from one another. This separation, as we discuss in the next section, has important biological consequences: the phytoplankton grow mostly in the upper layer but the supply of nutrients from the lower layer into the upper layer is, to some degree, blocked by the pycnocline.

A variable defining the strength of the density gradients in the ocean is often required. It could be specified by the density gradient $d\rho/dz$ but it is more usual to use the Brunt–Väisälä or buoyancy frequency

$$N = (g/\rho\ d\rho/dz)^{1/2} \tag{3.06}$$

where g is again the acceleration due to gravity ($\simeq 10$ m s^{-2}), ρ is the density ($\simeq 1000$ kg m^{-3}), and z is the depth. Physically, N is the frequency of the oscillation that results when the pycnocline is displaced, then left to return to its rest position. For example, in Fig. 3.04(a) the pycnocline has been pushed above its position of rest. Buoyancy forces then

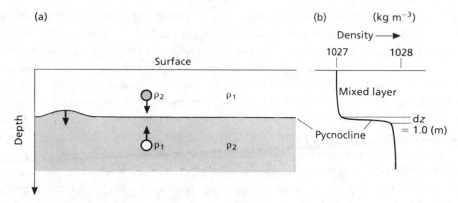

Fig. 3.04 (a) A cross-section through two layers of homogeneous water having unequal density separated by a thin pycnocline. Displacements above or below the pycnocline give rise to the buoyancy forces indicated by the arrows. (b) A vertical profile of the density through the two layers.

act to push it back, but because of inertia the interface overshoots the rest point and moves below the original position. Buoyancy forces then act in the opposite direction and push the interface back up. This process continues and results in an oscillation of the pycnocline, which spreads out as a moving wave. The process is just the same as when waves are made at the surface of the water. The buoyancy frequency is the most popular variable representing the strength of the vertical density gradient because physicists use it as a reference in studies of other types of waves in the ocean. It is calculated by the method shown in Box 3.03.

3.2.5 Phytoplankton production in tropical and subtropical oceans

As we saw in section 3.2, large areas of the world's ocean are permanently stratified, with a high concentration of nutrients below the pycnocline, a low concentration above, and the pycnocline acting as a barrier to turbulent diffusion of nutrients from one area to the other. This situation does not completely inhibit photosynthesis, because the organisms that graze upon the phytoplankton excrete nutrient substances such as ammonia, so that phytoplankton growth and reproduction can be maintained at a certain level with these recycled nutrients. The situation is potentially unstable, however, because any removal of organisms and their contained nutrients will lead to a running down of the production based on recycling.

Box 3.03 Estimating the Brunt–Väisälä frequency

The value of the Brunt–Väisälä frequency is easily found using a profile of the vertical density gradient, such as that shown in Fig. 3.04(b). Here the density difference across the pycnocline is 1 kg m^{-3} and the thickness of the pycnocline is about 1 m. These values give $N = [10(g) \times 1.0(\Delta\rho)/1000.0(\rho) \times 1.0\ (\Delta z)]^{1/2} \simeq 0.1$ rad s^{-1}. The radian frequency is 2π times the cycle frequency f—that is, $N = 2\pi f$ or $2\pi/\tau$ where τ is the period of the waves; thus $\tau = 2\pi/N \simeq 2\pi/0.1 \simeq 1$ min. This is an atypically high gradient and a fast wave for the open ocean. The minimum period in the upper ocean, Phillips (1977), is about 10 minutes, while in the deep ocean or in a well-mixed surface layer the period may be as high as 2–4 hours.

For example, the sinking of dead organisms and excretion by zooplankton that migrate down through the thermocline both deplete the nutrients in the mixed layer. If the food web in the mixed layer of a particular locality consists of phytoplankton, zooplankton, and small forage fish that feed upon the zooplankton, and a shoal of tuna comes through the area, consumes large quantities of forage fish, and moves away, the amount of nutrients available for recycling is diminished. On the other hand, if some physical event occurs that permits the vertical transport of a significant quantity of nutrients from below the thermocline, the nutrient pool in the mixed layer is increased. It is clear that in the long term, the persistence of a phytoplankton community in the mixed layer depends on the losses from the pool of recycling nutrients being balanced by vertical transport from below. There might, of course, be horizontal advection from one part of the mixed layer to another, but it remains true for the mixed layer as a whole that primary production ultimately depends on the vertical transport of nutrients.

Dugdale and Goering (1967), working with nitrogen metabolism, noted that recycled nitrogen is usually in the form of ammonia, while nitrogen transported from below is usually in the form of nitrate. They introduced the term 'regenerated' production for the recycled material based on ammonia and 'new' production for that based on nitrate. More recently, the ratio of new production to total production, the *f* ratio (Eppley and Peterson 1979), has come to be widely discussed in the context of the global carbon budget (see section 10.3.1).

In tropical waters vertical distributions of nitrate, chlorophyll, and primary production are more or less the same in all seasons of the year, as seen in Fig. 3.01. This constancy suggests that the processes maintaining the low nitrate levels in the surface mixed layer and the maxima of chlorophyll and primary production near the pycnocline must be in balance. This steady-state condition is quite unlike the situation found in temperate waters, where the vertical structure changes from season to season. The tropical structure, however, is very similar in appearance to the late summer structure observed in the temperate zones, and the following tentative explanation of the tropical profiles is extrapolated from the more complete body of knowledge that exists for the temperate areas. This work is described more fully in section 3.3.4.

To begin with the simplest case, consider the idealized vertical profile of density in Fig. 3.05(a). Here a homogeneous mixed layer lies over a thin pycnocline and a homogeneous deep layer. The level of vertical mixing, as symbolized by the column of circles, is vigorous enough to keep the upper layer homogeneous and the pycnocline thin. The light intensity, Fig. 3.05(b), allows the phytoplankton growth to a depth somewhat below the pycnocline. Any nitrate diffusing up through the pycnocline is quickly spread by the turbulence throughout the mixed layer and is consumed by the phytoplankton, which is also distributed evenly through the layer. If the nitrate level is low in the mixed layer, the phytoplankton production will be low and limited by the vertical flux of nitrate. Below the pycnocline, the phytoplankton are mixed into the lightless deep layer by the turbulence, and lost. This arrangement, then, is not capable of exhibiting chlorophyll maxima.

In the more realistic situation illustrated in Fig. 3.05 (c), it is possible to create the maxima because the intensity of vertical mixing decreases with depth gradually, rather than suddenly as in the simpler model. Nitrate diffusing up is not quickly spread through the mixed layer but moves slowly up through the pycnocline. At the depth where the light intensity is high enough, the photosynthesis begins. The nitrate flux is sufficient to support photosynthesis through a few meters of the water column but there is not enough nitrate to keep phytoplankton growing

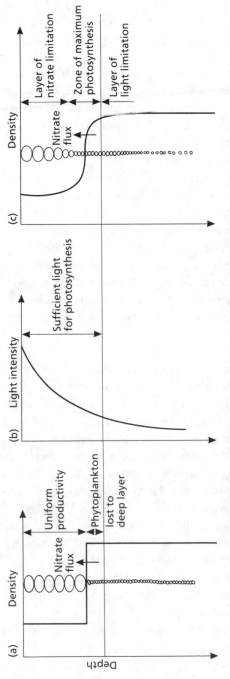

Fig. 3.05 (a) A vertical density profile through a thin pycnocline separating two homogeneous layers. The rate of vertical mixing, symbolized by the circles, changes abruptly through the pycnocline, and is high above and low below the pycnocline. (b) A vertical profile of the average light intensity with an indication of the depth where phytoplankton growth is light-limited. (c) A vertical density profile similar to (a) except that the density increases and the vertical mixing decreases gradually through the pycnocline.

through the whole upper layer. The vertically diffusing nitrate is consumed in a layer of production bounded on the bottom by low light and on the top by low nitrate.

This explanation of the subsurface chlorophyll and productivity maxima in the tropics must be viewed skeptically. It is not, as we show in section 3.3.4, the only possible mechanism, and it is neither quantitative nor rigorous. Although it has been extrapolated from work carried out in temperate oceans and may not apply in the tropics, it is a promising starting point from which improvements can be devised. The most important factor in the argument, the vertical variation in the turbulent diffusion coefficient, K_v, is very difficult to measure directly and has only recently been done in a few special locations. The main thrust of investigations of vertical profiles in the tropics has been to determine: (i) how much of the primary production is new production and how much is based on recycled nutrients; (ii) the rate of nitrate consumption in the euphotic zone; and (iii) the vertical flux of nitrate through the pycnocline. We now describe these efforts.

One group of studies established that in the oligotrophic ocean gyres and some coastal waters the vertical eddy diffusivity is normally low, on the order of $0.3 \times 10^{-4}\,\mathrm{m^2\,s^{-1}}$. From this finding we may infer that the normal rate of delivery of nitrate to the mixed layer from below is extremely low, and the tropical oceans probably do have a low rate of primary production, despite claims to the contrary (see sections 8.5.1 and 10.3.1).

King and Devol (1979), working in the eastern tropical Pacific, used nitrate gradients at the top of the nutricline (which coincided with the thermocline) along with phytoplankton nitrate uptake rates in the mixed layer to calculate vertical eddy diffusion coefficients. Following Okubo (1971) they used Eq. 2.11, the eddy diffusion equation

$$F = K_v dN/dz \tag{3.07}$$

where F is the flux of the property N, and K_v is the vertical eddy diffusivity. Putting values of the integrated nitrate-uptake rate by photoplankton in the mixed layer as the vertical flux, F, and the vertical gradient of nitrate dN/dz in the nutricline into the equation enabled them to calculate eddy diffusion coefficients ranging from 0.05 to $1.10 \times 10^{-4}\,\mathrm{m^2\,s^{-1}}$. They found that these values were in good agreement with coefficients calculated by others on the basis of heat flux through the thermocline, and with those found using radiochemical tracers. They are also similar to those given in section 2.2.7.

Eppley *et al.* (1979) made similar calculations for coastal waters of southern California and came up with eddy diffusion coefficients of $0.01–0.3 \times 10^{-4}\,\mathrm{m^2\,s^{-1}}$ in the thermocline. The agreement between the vertical eddy diffusion coefficients calculated on the basis of the nitrate uptake of the phytoplankton and those calculated from nonbiological considerations lends support to the view that in these conditions vertical eddy diffusion is the responsible agent in the transport of nitrate to the mixed layer and is a controlling factor in primary production. The agreement also supports the expectation that the turbulent or eddy diffusion constant, in Eq. 3.07, is the same for all properties such as heat, salt, and nitrate. This result is unlike molecular diffusion, in which different properties have different molecular diffusivities, as we discussed in section 2.2.6.

Eppley *et al.* (1979) also found that at the various stations studied, the rate of regenerated production was proportional to new production. This finding implies that increased new production leads to increased production of consumers, and hence to increased recycled nitrogen. In other words, vertical transport of nitrate is a controlling factor for total primary production.

A detailed study of vertical nitrate fluxes at a site in the oligotrophic eastern Atlantic approximately 1000 km southeast of the Azores was made by Lewis *et al.* (1986). They determined the turbulent kinetic energy dissipation rate (ε) from repeated measurements of microscale velocity shear and found low values with a mean of 1.7×10^{-9} W kg^{-1}. When combined with the buoyancy frequency, as in Eq. 3.12, they were able to calculate profiles of eddy diffusivity, K_v. The mean value was 0.37×10^{-4} m^2 s^{-1}. Using Eq. 3.07 they obtained a vertical flux value of 0.14 mmol N m^{-2} d^{-1} (95% confidence interval 0.002 to 0.89), which is stoichiometrically equivalent to a carbon fixation rate of 0.9 mmol C m^{-2} d^{-1}. This extremely low figure was used to contribute to the continuing debate about the level of new production in the oligotrophic oceans and its importance in the global atmospheric carbon budget (see section 10.3.1).

During the same study Lewis *et al.* (1986) made direct measurements of nitrate uptake and obtained an integrated value for the water column of 0.807 mmol N m^{-2} d^{-1}. This value, though higher, is not significantly different at the 95% confidence interval from the nitrogen flux calculated from physical parameters.

The conclusion to be drawn from this work and that of others in the Pacific is that in the absence of special upwelling situations the level of new production in oligotrophic ocean waters is extremely low, and is limited by the stable vertical structure with its pycnocline barrier and the low level of vertical transport of nitrate and other nutrients. The results also do not contradict the simple explanation given above for the chlorophyll and productivity maxima being located in the pycnocline rather than at the surface.

On the other hand, Dandonneau (1988) showed that seasonal and longer-term changes in the climate of various tropical regions, and the advection of water masses from coastal upwelling areas, cause relatively frequent deviations from the general picture of an oligotrophic mixed layer in tropical waters. This matter will be discussed more fully in connection with large-scale ocean circulation.

3.2.6 Equatorial upwelling and domes

In many parts of the tropical oceans, the concepts of a stable mixed layer and a low coefficient of vertical eddy diffusivity in the thermocline do not apply. The first of these areas is at the equator, where extensive upwelling occurs. The detailed physics of the process is best presented in connection with major ocean currents, but a general description is needed here.

At the equator, westward-blowing trade winds give rise to a westward flow along the equator in the mixed layer (Gill 1982; Leetmaa *et al.* 1981). In the Pacific (Fig. 3.06), this westward current is the south equatorial current found roughly between 5° S and 5° N. A short distance away from the equator the Coriolis force (section 4.2) causes the current north of the equator to be deflected north (i.e., to the right in the northern hemisphere), while the current south of the equator is deflected south (i.e., to the left in the southern hemisphere). A divergence or spreading of the water is thus set up in the surface layers at the equator. The lost water is replaced by water that upwells from beneath the equator.

Because of the continuous flow of surface water from east to west along the equator, the mixed layer is deeper in the west than in the east. At 155° W it is 80 m or more while at 100° W it is only 10–20 m thick. The water upwelled along the equator comes from the nutrient-rich waters below the shallow pycnocline in the east, but further west, where the

Fig. 3.06 Schematic diagram of equatorial upwelling system. Based on information in Vinogradov (1981). Water upwelling at the equator diverges north and south. Water travelling north eventually converges with water from the north equatorial counter-current and sinks. For explanation of the biological zones, see text.

mixed layer is deep, the upwelled water comes partly from the nutrient-depleted water in the mixed layer. As Sorokin *et al.* (1975) have pointed out, this difference leads to greater biological productivity in the east (Table 3.1). The trade winds blow from the northeast and generate the westward-flowing north equatorial current between 10 and 20° N. In the band between this current and the south equatorial current, from 5 to 10° N, is the north equatorial counter-current, so called because it is counter to the prevailing winds. Right at the equator at the top of the thermocline is another counter- or eastward-flowing current. This current is the Cromwell in the Pacific and the Lomonosov in the Atlantic. They are about 100 m thick and 200 km wide, and have top speeds of $\simeq 1 \text{ m s}^{-1}$.

Equatorial upwelling leads to a biological community with a striking pattern that runs parallel with the lines of latitude (Vinogradov 1981). Consider the fate of a parcel of water upwelling at the equator. It is rich in nutrients but not in phytoplankton. As time passes it moves away from the equator and the phytoplankton population grows rapidly. A zone of maximum biomass is found between latitudes 0.5 and 1.0° from the equator. After a further period of time the water has moved further from the equator and a population of herbivorous zooplankton has developed. Still later, and further from the equator, a population of carnivorous zooplankton reaches its maximum. Finally, by the time the parcel of water is

Table 3.1 Biomass of organisms in equatorial upwelling systems at various longitudes. From Vinogradov (1981).

Group of organisms	155° W	140° W	122° W	97° W
Phytoplankton	4.8	4.5	5.1	46.5
Bacteria	2.9	5.2	2.8	16.5
Flagellates	0.51	0.91	0.83	1.7
Ciliates	2.1	0.54	1.7	0.24
Other protozoans (radiolarians, foraminifera)	0.05	0.12	0.25	0.04
Total protozoans	2.6	1.6	2.8	2.0
Fine filterers, metazoans (Appendicularia, Doliolidae, small calanoids)	1.7	2.0	1.1	2.7
Copepod nauplii	0.5	0.3	0.4	1.6
Coarse filterers (calanoids, juv. euphausids)	1.1	1.1	0.87	3.3
Total noncarnivorous metazoans	3.3	3.4	2.37	7.6
Cyclopoids	0.39	0.91	1.4	2.3
Predatory calanoids	1.4	0.49	0.38	0.83
Other predators (Chaetognatha, Polychaeta, Amphipoda, etc.)	0.69	0.52	0.38	1.5
Total mainly carnivorous metazoans	2.5	1.9	2.1	4.6
Total zooplankton (including protozoans)	8.4	6.9	7.27	14.2
Total plankton	16.1	16.6	15.17	77.2

*Biomass (wet weight g) of the components of plankton communities along the equator in the Pacific Ocean in the 0–120 m layer in January 1974.

close to the boundary with the north and south equatorial counter-currents, a mature community including macroplankton and young fishes has developed. While these biological events are in progress, the water mass also travels westward with the surface drift. Vinogradov (1981) estimated that water containing a fully developed community has been displaced 250–450 km away from the equator and 1800–2500 km west since being upwelled. Predatory fish such as tuna congregate near the convergence of the two currents to feed on the products of the equatorial upwelling system. A line of high phytoplankton biomass along the equator can easily be seen in satellite images (Plate 1).

A seasonal study of the equatorial upwelling system has been carried out at 5° W (Vinogradov 1981). Between February and May, a warm season occurs that is characterized by a slackening of the southeast trades. The surface layers reach their highest seasonal temperature and upwelling is at a minimum. This event coincides with the period of minimum phytoplankton productivity, the equatorial area showing little difference from the surrounding oligotrophic ocean. In June–July the southeast trades strengthen to force 4–5 and upwelling becomes very vigorous. After a bloom of phytoplankton the process of development of a large biomass of zooplankton follows, reaching its peak between September and November. As a result, the production in these systems may show as much seasonal variability as in temperate latitudes.

As the equatorial counter-currents approach the coast on the eastern boundaries of the major oceans, part of the water turns to the left and part to the right, eventually joining the north and south equatorial currents, respectively. In doing so, they create cyclonic gyres that have ascending water masses in their centers. A well-studied example is the Costa Rica dome, so named for the tendency of the mixed layer to be thin and the thermal structure below it to be dome-like (Wyrtki 1964). The ascending velocity within the dome is estimated at 10^{-6} m s^{-1}, with the upwelling water originating between 75 and 200 m. Physically, chemically, and biologically the waters in the domes are similar to the water in the equatorial divergence (Black-burn 1981), with a rich development of phytoplankton downstream of the upwelled water.

At the eastern end of the Atlantic equatorial counter-current is found the Guinea dome. Unlike the Costa Rica dome, it is probably seasonal. A similar feature, again seasonal, exists in the Indian Ocean south of Java (Wyrtki 1962).

3.2.7 Magnitude of equatorial phytoplankton production

The eastern half of the equatorial region of the Pacific is particularly productive. As the satellite picture (Plate 1) shows, there is a triangular region that clearly has higher chlorophyll biomass than the central gyres. Chavez and Barber (1987) made an estimate of its contribution to global phytoplankton productivity. They recognized a 'cold tongue' area of equatorial upwelling extending from 90 to 180° W, extending 10° in meridional width, with a total area of 1.3×10^{10} km^3. They found that the mean primary productivity was 0.54 g C m^{-2} d^{-1}, or 197 g C m^{-2} y^{-1}. For the total area Chavez and Barber estimated annual production at 1.9×10^{15} g C y^{-1}. Using the model of Eppley and Peterson (1979), they estimated that 0.85×10^{15} g C y^{-1} is new production. As will be seen in Chapter 5, the Peru upwelling system, which has been widely regarded as a major contributor to global fish production, has a total primary production that is an order of magnitude less than the eastern Pacific equatorial upwelling system. The rate of production per unit area in the coastal upwelling system is much higher, but the area is only a minute fraction of that of the equatorial upwelling system. Taken together, the two contribute about 10% of the global primary production as estimated by Shushkina (1985) or 18% of that estimated by Platt and Subba Rao (1975). The old idea that Pacific waters are mostly nutrient-deficient and unproductive is no longer tenable.

3.2.8 The paradox of high nutrients and low productivity: the iron limitation hypothesis

Throughout this chapter it has been taken as axiomatic that if nutrients are present in the mixed layer and the phytoplankton is not light-limited, nutrients will be taken up until the concentrations reach a very low level. This assumption does not hold for some areas of the ocean where significant amounts of nutrients persist for long periods, while phytoplankton biomass and productivity remain low. These conditions exist in parts of the eastern equatorial Pacific, in the Southern Ocean, and in parts of the sub-Arctic Pacific.

Two explanations have been suggested for this paradox (Cullen 1991). The grazing hypothesis says that environmental stability permits the evolution of a balanced food web in which phytoplankton biomass is held at a low level by the grazers. The phytoplankton is never allowed to become sufficiently abundant to use all the available nutrients. The second hypothesis suggests that the standing crop of phytoplankton is constrained by the availability of iron. If more iron became available, the standing crop of phytoplankton would increase and deplete the major nutrients, despite any grazing pressure.

The iron hypothesis has been tested in each of the three major oceanic areas by adding iron to shipboard incubations of natural phytoplankton. In each case iron enrichment enhanced the final yield of phytoplankton, but the results have been challenged on the grounds that events *in vitro* may not adequately reflect events in nature.

A large-scale experiment was mounted in 1993 to observe the effect of iron enrichment of an area of the Pacific just south of the Galápagos Islands (Martin *et al.* 1994). Iron and a highly sensitive tracer were pumped into the propeller wash of a research vessel as it steamed back and forth across an 8×8 km area over 24 hours, raising the iron concentration from about 0.05 nM to about 4 nM. In the first three days plant biomass doubled, chlorophyll increased threefold, and plant production increased fourfold. This response was of short duration, however, and was small compared with changes observed in incubation experiments. Macronutrients were not significantly depleted, and a relatively small decrease was observed in the concentration of carbon dioxide.

The experiment was repeated in May-June 1995 (Falkowski 1995) and a very different result was obtained. A similar total amount of iron was added, but in several batches over a period of 17 days, giving final concentrations between 1 and 2 nM. The chlorophyll concentration rose 30- to 40-fold, nitrate was drawn down from 15 to 4 μM, and carbon dioxide in the water decreased about 50 ppm. The bloom consisted mainly of diatoms, and zooplankton abundance increased markedly. It is believed that there was a very significant export of organic carbon to the depths of the ocean.

These results reinforce the view that variations in the supply of iron dust from the atmosphere could have been an important factor in past changes in atmospheric carbon dioxide, for example, during the most recent ice age (Kumar *et al.* 1995). They also lend support to those who propose that large-scale seeding of the ocean with iron could substantially reduce atmospheric carbon dioxide and influence the course of global warming. This subject will be discussed further in sections 3.3.9 and 10.3 in connection with the Southern Ocean.

3.3 VERTICAL STRUCTURE AND PHYTOPLANKTON PRODUCTION: TEMPERATE AND POLAR WATERS

We discussed the mixed layer of the open tropical ocean as if it were relatively constant and unchanging, except in special upwelling areas. This assumption is a reasonable first approximation. In temperate and polar waters seasonal change is very evident, however, and it is now being realized that diurnal changes are also important.

As we move poleward from the stable tropical and subtropical waters, we find that in the winter season downward mixing by convection of cooled surface water and wind-driven turbulence combine to cause a progressive deepening of the mixed layer. Turbulence penetrates deeper and deeper into the zone of high-nutrient concentration and brings nutrients up into the euphotic zone. On the other hand, phytoplankton cells are carried deeper and deeper by the turbulent mixing, and spend a large proportion of their time below the euphotic zone, where photosynthesis is no longer possible. At the end of winter, when surface warming begins, the reverse process occurs. The mixed layer becomes shallower and phytoplankton cells trapped above the pycnocline spend more and more of their time in the euphotic zone. The net result is a great burst of phytoplankton activity that we call the spring bloom (see section 3.3.2). Reference to Fig. 3.07 shows that at latitude 41° N deepening of the mixed layer continued until late March, but there was a rapid shallowing during the early part of

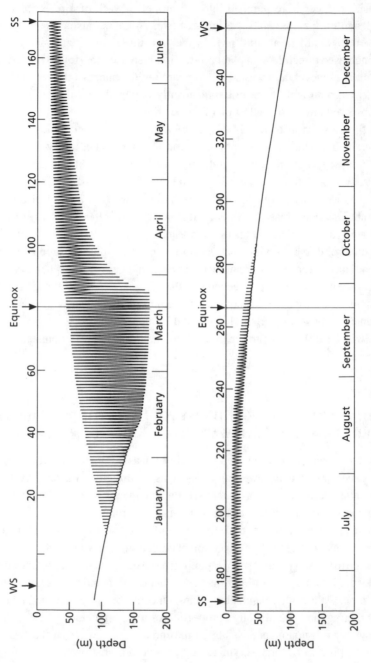

Fig. 3.07 Model simulation of the diurnal variation of the mixed layer depth over the year. Note the diurnal movement of the thermocline from January to October. WS, winter solstice; SS, summer solstice. From Woods and Barkmann (1986).

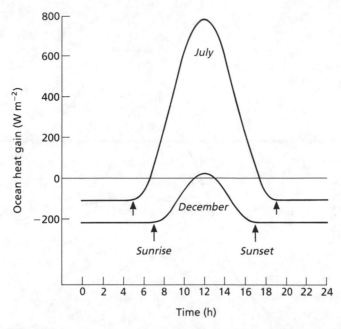

Fig. 3.08 Diurnal changes in the net heat gain during an average day in July and December at 40° N 40° W.

April. The figure also shows that from January onward, a diurnal oscillation occurred in the depth of the mixed layer. In the next section we consider the physics of these processes.

3.3.1 Diurnal and seasonal changes in mixed layer depth

The discussions in sections 3.2.2 and 3.2.3, concerning changes of temperature and depth of mixed layers, assumed the inputs of heat and mixing energy to be constant in time. These factors, however, vary over the day and the year, giving rise to diurnal and seasonal changes in the temperature and depth of mixed layers.

To illustrate diurnal changes, the variation of net heat gain through the ocean surface at 40° N 40° W, mid-Atlantic site, through the course of an average day in both July and December are shown in Fig. 3.08. In July the sea loses heat at the rate of $\simeq 120\,\mathrm{W\,m^{-2}}$, mostly by evaporation and long-wave radiation. The losses continue at much the same level throughout the day because the controlling factors, such as wind speed, air–sea temperature difference, and cloud amount, tend to remain fairly constant.

Heat gain, on the other hand, depends mostly on the altitude of the sun, which varies greatly throughout the day. Heat input varies from zero during the night to almost 900 W $\mathrm{m^{-2}}$ at noon. Under these circumstances a mixed layer of constant depth becomes warmer during the day and cooler at night but, because there is a net gain of energy over the day, there will be a net increase in temperature over the day. If the wind decreases, the mixed layer becomes thinner and the temperature increase in the layer will be greater. This change occurs because most of the radiant energy is absorbed in the top few meters as indicated in Figs. 3.02 and 3.03, and because weaker winds allow the heat to be concentrated in a thinner layer. If the wind increases, the mixed layer becomes deeper and the heat input is spread throughout a greater depth, leading to a smaller temperature rise over the day.

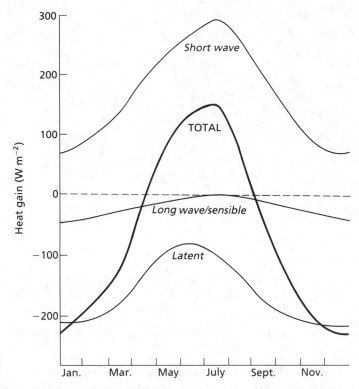

Fig. 3.09 Seasonal changes in net short-wave radiation, net long-wave radiation, sensible heat exchange, and latent heat exchange, and the total at 35° N 48° W.

In December the sun is above the horizon each day for only 10 hours at 40° N 40° W and delivers a total of only 4300 kJ m^{-2} over the day, while the heat losses total 19,000 kJ m^{-2}. Under these circumstances not enough heat is gained during the day to form a warm surface layer for even part of the day.

The annual variations of the heat budget components at the site of ocean weather ship *Echo,* 35° N 48° W, are illustrated in Fig. 3.09. Included are the heat input from short-wave radiation, and the losses from radiation, evaporation, and conduction. These data, extracted from the atlas of Isemer and Hasse (1987), show that short-wave radiation is the only source of heat and that the latent heat of evaporation is responsible for most of the heat loss in all months. The sum of all the factors indicates that the sea is gaining heat through the sea surface at this location only from April to August, and over the year more heat is lost than is gained, which indicates that heat must be advected into this region to keep the mean temperature the same from year to year.

The curve of total gains and losses suggests that all the heat gained through the surface in the summer would likely be lost by the following January or February. By that time no seasonal thermocline would remain in the upper layers of the ocean.

Woods and Barkmann (1986) have created a numerical model to predict the depth of the mixed layer that incorporates both the diurnal and seasonal variations in the heat gains and losses. They used input parameters from 41° N 27° W (north of the Azores) and the result of their calculations over a year is shown in Fig. 3.07. In mid-December at the winter solstice the model indicates there is no diurnal thermocline, but that the depth of the seasonal thermocline is approaching 100 m. In early January a diurnal thermocline begins to be formed. Its minimum depth decreases from about 100 m in January to 10 m in July. The maximum depth of the diurnal thermocline, which is the depth of the seasonal thermocline, continues to increase from December through late March. It then quite suddenly decreases when the heat gained during the day is neither lost at night nor mixed down to great depths by the wind.

Through the spring and early summer the upper mixed layer becomes more stable. The depth of the seasonal thermocline decreases, as does the difference between the minimum and maximum depths of the diurnal thermocline. From late summer through early winter the amount of heat lost to the atmosphere from the sea steadily increases, as does the mixing energy from the wind. The diurnal thermocline tends to disappear as the seasonal one moves to greater depths.

3.3.2 The mechanism of the spring bloom

Early biological oceanographers on both sides of the North Atlantic became aware that, in temperate coastal waters and in the open ocean, there was a marked increase in abundance of phytoplankton in spring time. This growth was christened the spring bloom. In this section we follow the development of our understanding of this process. We now know that no distinct bloom occurs in large areas of the temperate Pacific Ocean, but rather a gradual increase in phytoplankton abundance from mid-winter to mid-summer (see section 3.3.8).

Working in the waters off England and Scotland, Atkins (1928) suggested that the spring bloom was triggered by the spring increase in solar radiation. Marshall and Orr (1928), on the other hand, found that enough light was present for vigorous phytoplankton growth at the latitude of the British Isles even in winter. In the course of an investigation into the productivity of the Bay of Fundy, Gran and Braarud (1935) suggested that stabilization of the water column by thermal stratification could have the effect of greatly stimulating primary production. The idea was that in winter time in temperate latitudes the phytoplankton cells are being circulated to the full depth of the mixed layer; hence, they spend a large proportion of their time in regions lacking enough light for growth. With the onset of surface warming and the formation of a much shallower mixed layer, phytoplankton cells are held for longer periods in the euphotic zone (i.e., the zone in which there is sufficient light for growth and cell division). Mills (1989) traced the evolution of this line of thinking through earlier publications by Gran (e.g., Gran 1931).

We shall see in later chapters that this mechanism—by which turbulent mixing of the entire water column brings nutrients to the surface waters, after which stratification sets in and the phytoplankton are held in the euphotic zone so that they can multiply rapidly—is the key to understanding phytoplankton productivity in a variety of situations including estuaries, upwelling areas, and tidal fronts. We shall henceforth refer to it as the Gran effect.

The idea was further developed by G. A. Riley and used to explain the spring bloom on Georges Bank (Riley 1942). It was also incorporated into a major model of the phytoplankton dynamics of the North Atlantic (Riley *et al.* 1949). However, the classic quantitative exposition of the theory that we present here in detail is by Sverdrup (1953) and is entitled 'On Conditions for the Vernal Blooming of Phytoplankton.'

Consider the situation (Fig. 3.10) of a water column in which the mixed layer extends to depth *d,* the turbulence is sufficiently strong to distribute the phytoplankton cells

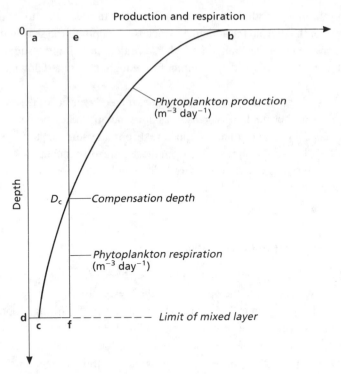

Production and respiration

Fig. 3.10 Diagram illustrating theoretical distribution of phytoplankton production and phytoplankton respiration. After Sverdrup (1953).

uniformly through it, nutrients are not limiting, and the extinction coefficient of light (α in Eq. 3.01) is a constant. Because photosynthesis is proportional to light intensity, and assuming that the light intensity at the surface is not high enough to inhibit photosynthesis, the expected distribution of the daily rate of phytoplankton production will be as shown, with a maximum at the surface, decreasing logarithmically to a low level at depth. The rate of respiration, on the other hand, being more or less independent of light, will be constant throughout the mixed layer. At some depth D_c the daily rate of photosynthesis is just balanced by the daily rate of respiration. This level is the compensation depth. Experiments have shown that for commonly occurring diatoms the light energy flux I_c at the compensation depth is $\simeq 1.5$–1.7 W m^{-2}.

This information, together with the daily irradiance at the surface and the extinction coefficient, permits calculation of the compensation depth, D_c. Taking the daily mean surface irradiance of I_0 as 200 W m^{-2} and the extinction coefficient α as 0.4 m^{-1} we have, from Eq. 3.01, $I_c/I_0 = 1.5/200 = \exp(-0.4D_c)$. Taking the natural logarithm of each side gives $-4.9 = -0.4D_c$, from which $D_c \simeq 12$ m.

Now consider a phytoplankton cell circulating in a random manner throughout the mixed layer. When it is near the surface, the rate of photosynthetic production will greatly exceed respiration. When it is near the bottom of the mixed layer, the reverse will be true. The condition for net positive population growth is that the integrated production, represented by the area *abcd,* be greater than the integrated respiration, represented by the area *aefd.* As the depth of the mixed layer increases, respiration increases proportionately but production increases by a small amount, or not at all. There is, therefore, a critical depth of the mixed layer at which integrated production just equals respiration. At shallower mixed layer depths the phytoplankton population has an excess of production, and therefore grows. At greater depths of the mixed layer it fails to grow. Note that the effect of grazers has not been considered up to this point.

Sverdrup (1953) showed that the critical depth of the mixed layer D_{cr} can be calculated by algebraically equating the areas representing production and respiration. Assuming that daily photosynthetic production at any depth is proportional to the mean daily light energy at that depth, we may regard the line *bc* in Fig. 3.10 as the distribution with depth of the mean daily light energy or the mean daily photosynthetic production. The length of the line *df* or *ae* is equal to the light intensity at the compensation depth, D_c, and is proportional to the rate of respiration at all depths in the mixed layer. From the discussion above, the intensity of radiation I_c, or the length of *df* or *ae*, at the compensation depth D_c can be expressed following Eq. 3.01 as

$$I_c = I_o \exp(-\alpha D_c) \tag{3.08}$$

When the depth of the mixed layer is equal to the critical depth D_{cr}, the area *aefd* representing the total respiration is given by $I_c D_{cr}$. The area *abcd* representing the total photosynthesis is given by

$$I_0 \int_{D_{cr}}^0 \exp(\alpha z) dz = I_0 [1 - \exp(\alpha D_{cr})]/\alpha \tag{3.09}$$

Equating the two areas gives

$$I_c D_{cr} = I_0 [1 - \exp(\alpha D_{cr})]/\alpha \tag{3.10}$$

which can be solved graphically for D_{cr}.

Sverdrup then proceeded to calculate the values of D_{cr} for ocean weather ship *M* in the Norwegian Sea in the spring of 1949. When plotted with data for the actual depth of the mixed layer and the numbers of phytoplankton and zooplankton (Fig. 3.11), it became clear that the critical depth increased from about 30 m at the beginning of March to nearly 300 m at the end of May, and that the major increase of populations of both phytoplankton and zooplankton occurred when the depth of the mixed layer was much less than the critical depth.

Detailed examination of Fig. 3.11 reveals the following: (i) through March the amounts of phytoplankton and zooplankton were very small, and during this period the depth of the mixed layer was much greater than the critical depth; (ii) the mixed layer thickness changed from more than 300 m on 2 April to 50 m on 4 April, which probably indicated advection of a different water mass containing abundant phytoplankton, rather than stabilization by warming; (iii) the development of zooplankton populations in April undoubtedly reduced the size of the phytoplankton populations; and (iv) during May, when the depth of the mixed layer was much less than the critical depth, phytoplankton populations reached high densities in spite of the presence of sizable zooplankton populations. The overall picture was in accordance with the theory that the spring bloom is strongly dependent on the stability of the water column. We may note that this model takes no account of diurnal variation in the thickness of the mixed layer.

Sverdrup's (1953) paper was, then, a quantitative exposition and a field check of the phenomenon first described by Gran in 1931. In later chapters we shall find that

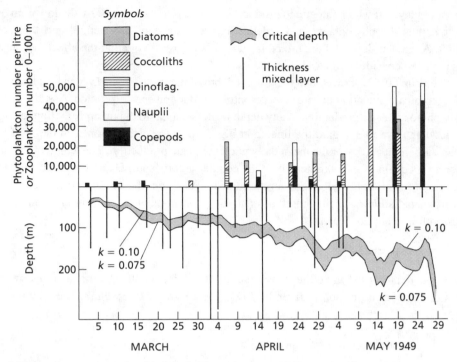

Fig. 3.11 Data from ocean weather ship *M* in the Norwegian Sea on plankton, mixed layer thickness, and critical depth. Note that the mixed layer is always deeper than the critical depth until early April, when a large increase in phytoplankton biomass occurs. From Sverdrup (1953).

enhancement of primary production by stabilization of the water column after a period of strong mixing is a phenomenon that occurs in many situations, in estuaries, and in coastal upwelling areas, for example. Many oceanographers refer to it as the Sverdrup mechanism, but we prefer to call it the Gran effect.

It is important to remember that Sverdrup's model applies to temperate latitudes where, after the winter deepening of the mixed layer, surface waters contain a good supply of nutrients, and where spring warming leads to a shallowing of the mixed layer. We may now proceed to follow the fate of the phytoplankton as the year progresses. During the period of stratification the pycnocline presents a barrier to the vertical diffusion of nutrients and the phytoplankton progressively depletes those nutrients present in the mixed layer. At the same time, zooplankton are consuming phytoplankton and excreting nutrients. The budget for total nitrogen contained in the mixed layer has the following terms:

Depletion by:

1   uptake of nitrogen by phytoplankton;
2   sinking of phytoplankton out of the mixed layer; and
3   consumption of phytoplankton by zooplankton. These animals produce fecal pellets that tend to sink rapidly through the pycnocline. Zooplankton may also descend below the mixed layer at night and release nitrogen by excretion and defecation.

Addition by:

4   excretion by zooplankton and other heterotrophs; and
5   vertical diffusion.

A common result of these processes is a rapid depletion of the dissolved nitrogen content of the mixed layer and a fall in phytoplankton biomass, marking the end of the spring bloom.

During the summer period conditions may be similar to those described in section 3.2.5 for tropical oceans—i.e., there may be stable stratification and very limited diffusion of new nutrients into the mixed layer. Phytoplankton production is then confined to that which can be sustained by recycled nitrogen and the small amount of diffusion, and it is common to find that the phytoplankton community changes from one dominated by diatoms to one dominated by flagellates. (This change occurs because the flagellates have a better surface:volume ratio and are able to perform locomotion out of micro-environments with low nutrient concentrations. See section 2.3.4 and Margalef 1978a). As the days shorten and surface waters begin to cool, however, convective and wind-driven processes once more begin to cause a deepening of the mixed layer, so that new nutrients are entrained from below the thermocline, and an autumn bloom may occur before the depth of the mixed layer once again exceeds the critical depth. Deep winter mixing leads to increasing concentrations of nitrate in surface waters, and these levels reach a maximum in March. Spring warming and a shallowing of the thermocline lead to a spring bloom in late April and early May, as shown by the peak in chlorophyll concentration. Nitrate concentration declines rapidly during the spring bloom. During summer, primary production is moderately high and the chlorophyll low, because zooplankton remove the phytoplankton biomass almost as fast as it is produced. Most of the primary production is made possible by the regeneration of ammonia through excretion of the consumers.

The timing of spring and autumn blooms are dependent in part on latitude. At high latitudes, where the spring warming comes late and autumn cooling begins early, it is common to find just one bloom in mid-summer. This bloom is often more pronounced than in temperate waters and two factors undoubtedly contribute (Cushing 1975). The first is the longer daylight characteristic of high latitudes in summer, which permits the phytoplankton populations to grow for up to 24 hours per day. The second is the low temperature, which slows the metabolism of the grazers and allows the phytoplankton populations to grow in spite of them. Photosynthesis is much less sensitive to temperature than are respiration and growth in zooplankton.

During the early 1990s there was a reevaluation of the usefulness of the Sverdrup (1953) formulation (summarized in Platt *et al.* 1994). Several reports had been made of the initiation of blooms in the absence of stratification (e.g., Eilertsen 1993). A typical scenario involves clear water with good light penetration and a low level of turbulence. A rapid increase in phytoplankton biomass precedes the shallowing of the mixed layer. Platt *et al.* (1994) showed that these examples would be consistent with the Sverdrup formulation if one of the following were true: (i) the loss terms generated by grazing or sinking were unusually low; (ii) the light flux increased rapidly due to increasing day length; or (iii) the actual depth of the vertically mixed layer was less than that indicated by the profile of conductivity and temperature. Except for day length, none of these parameters had been measured when the observations were made on blooms preceding the shallowing of the mixed layer, so it was concluded that the Sverdrup model had not been violated.

The feedback that can occur between the biological events and the physics of the mixed layer has also spurred discussion in relation to the Sverdrup model. Stramska and Dickey (1993) summarized earlier examples of a heating effect on the mixed layer due to absorption of light in the phytoplankton cells. They also contributed new data from the North Atlantic and a model, showing that the increase in phytoplankton abundance induced an increase in sea surface temperature of about 0.2 °C. This increase, in turn, led to stronger near-surface thermal stratification and shallower mixed layers. It was a factor in the formation of a bloom that preceded the formation of the seasonal thermocline. This topic is discussed further in section 11.3.

3.3.3 Large-scale turbulence and phytoplankton performance

In Chapter 2 we discussed the relationship between phytoplankton and small-scale turbulence. In this section we turn our attention to turbulence on the scale of meters or tens of meters that is characteristic of the mixed layer. The exponential decay of light intensity with depth means that as a phytoplankton cell is carried passively from near-surface to depth and back again it is exposed to a changing light environment. There has been a controversy over the question of whether phytoplankton cells in a fluctuating light regime are more productive than they would be if exposed to light of a constant average value. Mara (1978) produced experimental evidence that seemed to show a strong effect, but Falkowski and Wyrick (1981) and Gallegos and Platt (1982) concluded that the effect was negligible.

Resolution of the controversy was provided by Lewis *et al.* (1984a), who pointed out that the metabolic machinery of a phytoplankton cell is complex and there is no reason to suppose that all of its component parts would adapt to changes in light level at the same rate. They proposed a model in which properties that adapted rapidly to the changing light levels would show marked differences between cells taken from the top and from the bottom of the mixed layer, while properties that adapt slowly would show minimal differences between surface and depth. They suggested that this variation would explain the difference in the results of earlier investigations.

To test this model Lewis *et al.* (1984b) made a series of measurements of the turbulent energy dissipation rate (as a measure of vertical motion through the light gradient) in the mixed layer of a coastal inlet on different days, and simultaneously measured two photoadaptation properties of the phytoplankton on those same days. As an example of a relatively slowly adapting parameter they chose P_M^B, the maximum potential photosynthetic rate normalized to chlorophyll-a. The difference between surface samples and those from the base of the mixed layer was labelled ΔP_M^B. There was a marked decrease in ΔP_M^B with increase in turbulent energy dissipation rate. For their example of a property with a rapid photoadaptive response they chose *in vivo* fluorescence per unit chlorophyll. Its value was less at the surface than at depth on all days except one, which was the day of highest turbulence.

The results were in accordance with the model. At slow rates of turbulent mixing, and hence of light fluctuation, there was a marked difference in values between samples from the top and the bottom of the mixed layer for the slowly adapting parameter, as well as for the fast-adapting parameter. As the rate of light fluctuation increased, the differences between top and bottom in the slowly adapting parameter decreased while those in the fast adapting parameter persisted.

Lewis *et al.* (1984b) then calculated the time scale of the turbulent mixing. We saw in Box 2.04 that the approximate time for vertical diffusion to occur over the distance L can be estimated using the vertical eddy diffusivity K_v from $\tau = L^2 K_v^{-1}$. Following Osborn (1980), the authors substituted $\varepsilon/4N^2$ for K_v (where ε is the energy dissipation and N is the Brunt–Väisälä frequency), so that they were able to calculate the time scale of turbulent mixing from the energy dissipation:

$$\tau = 4N^2 L^2 \varepsilon^{-1} \tag{3.11}$$

They found that on the various days of observation the mixing time scale ranged from about 1 to more than 10 hours. Since it had previously been shown that the time scale for adaptation of the *in vivo* chlorophyll fluorescence is about 1 hour, while the time scale for photoadaptation of P_M^B is 6–10 hours, the results were in good agreement. Hence they concluded that the time scale of turbulence has strong effects on some of the parameters of photosynthetic performance, but is less strong for others. This important result will undoubtedly be incorporated in future models of phytoplankton productivity.

3.3.4 The oligotrophic phase in temperate waters

Much of the current controversy about the magnitude of primary production in the world's oceans revolves around the mechanisms controlling production in the subtropical oligotrophic basins that occupy a large proportion of the total area (see sections 8.3 and 10.5). Work on the transition from bloom conditions to oligotrophic conditions in temperate waters greatly assists in understanding the basins that are permanently oligotrophic.

We have seen that after the termination of the bloom in temperate oceans the water is stratified and the vertical diffusion of new nitrogen through the pycnocline is restricted. Production in the mixed layer is then largely confined to that which can be supported by nitrogen regenerated by the grazers. During this phase the profiles of nutrients, chlorophyll, and production resemble those of an oligotrophic tropical system, and the similarity extends to the presence of a subsurface chlorophyll maximum. In the course of the debate about what mechanism causes the formation of a subsurface chlorophyll maximum in temperate waters, much light has been thrown on the mechanisms that sustain production at the permanent chlorophyll maximum in tropical waters.

There are two processes that may produce the chlorophyll maximum. If the plants forming the spring bloom sink, they carry the chlorophyll maximum with them to greater depths. Variations on this basic model constituted the sinking rate hypotheses that were the first explanations of the existence of a chlorophyll maximum at some depth below the surface layer. A second group of theories argued that the chlorophyll maximum occurs because the cycle of phytoplankton growth and decay is faster in the upper mixed layer than in and below the pycnocline. At the top of the water column, the intensity of light is high and growth will be rapid, but so will nutrient depletion and decay of the population. At deeper levels, the light level is lower than at the surface, and growth and decay of the phytoplankton population is slower, reaching a maximum after the growth at the surface has died out. Thus, the chlorophyll maximum appears at ever increasing depths as the maximum in growth passes down the water column. This process also depends on the fact that turbulent mixing decreases in and below the pycnocline. This decline slows down exchange between the well-mixed surface layer and the layers below, and allows the growth and decay of the phytoplankton in the various layers to proceed relatively independently of one another. In this and the next section we examine some of these theories in more detail, starting with the sinking theories.

Riley *et al.* (1949) were among the first to present a theory based on the sinking of the phytoplankton. They assumed that the dense concentrations of phytoplankton formed in the spring bloom sank to greater depths, and in their model of production in the western North Atlantic they assumed a constant sinking rate for the phytoplankton. Steele and Yentsch (1960) showed that an improved model of vertical distribution of chlorophyll and production resulted if the sinking rates were assumed to decrease with depth. They showed experimentally that *Skeletonema* decreased its sinking rate when exposed to dark, nutrient-rich waters, and proposed this factor as the explanation of the formation of the deep chlorophyll maximum. They suggested that cells reduced their sinking rate when they came close to the base of the euphotic zone and encountered the nutricline. More recently Bienfang *et al.* (1983) found that reduction in light alone produced a decrease in sinking rate. For *Thalassiosira,* reduction of light levels to those characteristic of the chlorophyll maximum was enough to cause the diatom to decrease its sinking rate. They inserted their data into a modified form of the model of Riley *et al.* (1949). Setting phytoplankton biomass constant with depth in the mixed layer, ignoring grazing, and allowing sinking rate to vary with light according to the experimental results, the authors found that phytoplankton biomass increased by 31–100% between 40 and 86 m, the magnitude of the increase being negatively correlated with the value of K_v. This result was in good agreement with field data and supports the view that in temperate waters having a shallow mixed layer, a well-developed pycnocline, and phytoplankton populations characterized by the abundance of large-celled organisms, reduction in sinking rates is a prime factor in the formation of deep chlorophyll maxima. They emphasized that both modelling and experimental data indicate little need to invoke this mechanism to explain chlorophyll maxima in oligotrophic subtropical waters where the phytoplankton cells are smaller, and where the response to decreasing light is sufficient to account for the observed increase in chlorophyll with depth.

Jamart *et al.* (1977) produced a thorough model that includes all conceivably important factors. They have terms representing the seasonal change in light intensity, the phytoplankton sinking rate, the rate of grazing by zooplankton, the phytoplankton respiration, and the decrease with depth in the turbulent diffusion coefficient (K_v). They succeeded in reproducing the main features of the developing chlorophyll maximum through the spring and summer of the temperate zone. In a subsequent paper (Jamart *et al.* 1979) they showed by means of sensitivity analysis that changes in sinking rate were not important in their model. The decrease in turbulent diffusion with depth was the main factor inducing a progressive deepening of the chlorophyll maximum.

The simple explanation for the subsurface chlorophyll and productivity maxima in tropical waters, presented in section 3.2.5, is based primarily on this model by Jamart *et al.* (1977). We assumed that the decrease in the turbulent diffusion with depth combined with the decreasing light intensity are the two most important factors that the authors incorporated. The profile in Fig. 3.05(c) that is intended to represent the tropical situation is similar to that found in late summer in the model.

The effects of short-term light level changes in the production of chlorophyll were examined by Taylor *et al.* (1986). They modelled the vertical distribution of phytoplankton in temperate waters under stratification and showed that a marked reduction of incident light, as in very cloudy weather, may reduce the photosynthetic activity in the deep chlorophyll maximum to the point where upwelling nutrients are not utilized immediately but are allowed to diffuse upward into the mixed layer. Banse (1987) considered whether the same mechanism would operate in a 'typical tropical structure' (TTS). Under these conditions the deep chlorophyll maximum is normally located deeper in the pycnocline (see Fig. 3.01) than it is in the temperate waters considered by Taylor *et al.* (1986). The conclusion was that severe cloud cover would be expected to lead to enhanced nutrient flux into the upper parts of the pycnocline, but not into the mixed layer.

Summarizing to this point, two processes have been proposed to explain the existence of the subsurface chlorophyll maximum in temperate waters. The first is mainly biological and depends on the sinking rates of phytoplankton. Increase of biomass is assumed to begin near the surface and to sink subsequently where it is observed as a subsurface maximum. The second process is physical in nature and depends on the gradual decrease in turbulent mixing through the pycnocline coupled with the decrease in light intensity with depth. As the light decreases, the time for phytoplankton to grow, deplete nutrients, and decay increases. This increase, together with the smaller vertical mixing, allows the maximum in production to occur at different times at different depths. It seems likely that both the biological and physical processes are important, but which dominates will depend on the circumstances.

3.3.5 The results of Lagrangian modelling

A new type of model of phytoplankton growth was introduced by Woods and Onken (1982). They pointed out that all previous models had studied a cloud of individual plankters relative to a continuous property of sea water—for example, chlorophyll concentration. They called this the Eulerian-continuum method. This procedure suffers from the disadvantage that processes such as cell growth are assumed to proceed at the average rate. The authors suggested that it was important to understand the interactions of the physiology of each phytoplankton cell with the diurnal changes in mixed layer depth, light intensity, and nutrient concentration. For this purpose they proposed to follow an ensemble of individual phytoplankters. At each time step of the model, the growth or decay of each cell was determined from the light and nutrient environments it was in. This incremental change was then added to or subtracted from the existing cell. They called this process of following the cells through their individual lives the Lagrangian-ensemble method.

The authors drew particular attention to the diurnal changes in the depth of the mixed layer. As we saw in section 3.3.1, increased cooling at night causes a progressive deepening of the mixed layer. This process continues until about one hour after sunrise, when solar heating begins to exceed the rate of surface cooling and the mixed layer gets progressively shallower, continuing until noon. This diurnal cycle changes with the seasons (Fig. 3.07). The net result is that the water column can be divided into four zones that vary in thickness with the seasons. They are: (i) the mixed layer; (ii) the diurnal thermocline; (iii) the seasonal thermocline; and (iv) the interior of the ocean.

Fig. 3.12 Diagrams illustrating terms used in the model of Woods and Onken (1982): (a) the temperature profile; (b) the turbulent kinetic energy profile; and (c) the mixing regime. L, largest scale of mixing; H, thickness of mixed layer; T, time scale of turbulence; z, depth.

In a version of the Lagrangian-ensemble model (Wolf and Woods 1988), which simulates a water column in the open ocean at 41° N 27° W, the seasonal variation in the mixed layer depth was as shown in Fig. 3.07. In the model it was assumed that the time scale of turbulence in the mixed layer was 30 minutes (Fig. 3.12). Since the model ran with time steps of 1 hour it was assumed that at the end of each period a particle in the mixed layer had equal probability of lying at any depth between the surface and the base of the mixed layer. Particles lying near the bottom were left isolated below the diurnal thermocline and were subject to only their normal sinking of 2 m d^{-1}. Most of them were re-entrained into the mixed layer when the thermocline descended the next night, but at times when the diurnal mixed layer was becoming shallower, a proportion were not re-entrained and entered the seasonal thermocline, where they remained until the following spring. The model simulated well the course of events in nature.

Wolf and Woods (1988) pointed out that phytoplankton cells in the model arrived at a particular depth and time with differing histories of exposure to light and to nutrients and hence differing degrees of adaptation to the prevailing conditions. This 'adaptation diversity' makes the model performance significantly different from that of Eulerian models. The broad conclusions were that the chlorophyll maximum formed as a consequence of the surface exhaustion of nutrients, with growth limited to cells in and below the nutricline. Cell division was nutrient-limited in the upper reaches of the nutricline and light-limited below. In these respects the model is similar to the one discussed in the previous section by Jamart *et al.* (1977). The chlorophyll maximum moves down through the water column because of the change in the turbulent mixing rates between the mixed layer and the pycnocline. Also as in the Jamart *et al.* (1979) model, the chlorophyll maximum in the model descends through the seasons without the invocation of a change in phytoplankton sinking rate.

In the standard model with no upwelling the nutricline moved deeper and deeper below the bottom of the mixed layer throughout the summer period. With moderate upwelling, at a speed less than the sinking rate of the phytoplankton, production increased but the nutricline and chlorophyll maximum stayed well below the mixed layer. Upwelling rates greater than the sinking speed caused the nutricline and chlorophyll maximum to move upward until they entered the mixed layer and produced an explosive burst of phytoplankton growth. The authors proposed that episodic upwelling events are responsible for the transient patches of high surface chlorophyll ('hot spots') seen in satellite images of ocean color (see section 3.2.6).

Considering this and the previous sections, it appears that the nutricline is formed as a result of progressive depletion of nutrients in the mixed layer during the oligotrophic phase in temperate waters, and that the subsurface chlorophyll maximum forms in and around the nutricline, with a magnitude of production that depends on the intensity of upwelling. While experimental evidence shows that reduction in light intensity and/or increase in nutrient concentrations brings about a reduction in phytoplankton sinking rate, this mechanism appears not to be essential to the formation of a subsurface chlorophyll maximum. It is enough that the growth and decay of phytoplankton occur more slowly in the pycnocline than in the mixed layer.

McGillicuddy (1995) compared the Lagrangian model of Wolf and Woods (1988) with a Eulerian model of Lande and Lewis (1989) at two levels of mixing. For relatively weak mixing, the population growth rates predicted by the two models differed by less than 5%, but in vigorously mixed conditions Lagrangian effects caused a 20% reduction in population growth rate. The key to the variation lies in the differing parameters used for photoadaptation in the two types of models. If similar formulations were used, the two models would be reconciled. The literature appears to show that the values used by Lande and Lewis are more appropriate.

3.3.6 The poleward migration of the spring bloom

The use of undulating towed sensors, often referred to as 'batfish,' has made possible higher horizontal resolution of vertical profiles along transects than was possible with traditional 'bottle stations.' Strass and Woods (1988) reported on a series of such transects running 2000 km north–south in the mid-Atlantic from the Azores (38° N) past ocean weather ship *C* (52.5° N, 35.5° W) into the cyclonic sub-Arctic gyre at 54° N. The batfish cycled between 10 and 200 m every 1.4 km. Representative transects were obtained in late April 1985, June–July 1986, and August–September 1984. The records of near-surface chlorophyll fluorescence (Fig. 3.13) showed patchy bloom conditions to about 49° N in late April, a bloom centered on 49° N in June–July, with oligotrophic conditions well established south of 42° N, and oligotrophic conditions over the whole transected except for bloom conditions north of the polar front at 52° N in August–September.

Examination of the vertical structure showed that the horizontal migration of the near-surface chlorophyll maximum did not keep pace with the northward movement of the region of mixed layer shallowing, but followed the slow propagation of the 12 °C isotherm outcrop. This trend suggests that, in addition to the need for stable stratification before a bloom can occur, there may be a temperature limitation on bloom processes.

Strass and Woods (1988) also found that during summer the deep chlorophyll maximum descended through the seasonal thermocline at approximately 10 m month^{-1}. They interpreted these data as indicating that phytoplankton were using new nutrients from greater and greater depths. Since events occurred at progressively later dates as one moved north,

Fig. 3.13 Near-surface chlorophyll fluorescence at three times of the year on a transect lying between the Azores and Greenland, showing poleward migration of chlorophyll maximum. From Wolf and Woods (1988).

the August–September transect showed the chlorophyll maximum sloping upward from south to north (Fig. 3.14).

The sites where the highest concentrations of chlorophyll were found—between 1150 and 1300 km, for example—were sites where the isopycnals were displaced upward and closely spaced, indicating frontal upwelling probably associated with a cyclonic gyre or eddy. In the next section we shall review the evidence for the occurrence of events inducing localized upwelling.

3.3.7 Transient upwelling events

In addition to the biological enrichment that occurs in open waters as a result of upwelling at divergences at relatively fixed locations, satellite observations show that there are transient pulses of increased surface chlorophyll that indicate meso-scale (10^3–10^5 m) up-

Fig. 3.14 Vertical distribution of chlorophyll on a transect between the Azores and Greenland in late summer. Note that the zone of maximum chlorophyll gets progressively more shallow as one moves north. From Wolf and Woods (1988).

welling events widely distributed in otherwise oligotrophic waters (Gower *et al.* 1980). Two main mechanisms have been proposed to explain them: internal waves, and fronts. Fasham and Pugh (1976) proposed that patches of high chlorophyll they encountered were best explained by the presence of internal waves, and Holligan *et al.* (1985) showed that at a station with a water depth of 4000 m there was upwelling of nutrient-rich water into the surface layer caused by large soliton-like internal waves that had been generated more than a day earlier at the shelf edge (see section 6.4).

Woods (1988) has pointed to meso-scale patches of high chlorophyll concentrations with horizontal dimensions of about 10 km, which are a frequent component of satellite images, and has proposed that they are caused primarily by the jets associated with meso-scale fronts. Shelf-edge processes are considered in more detail in Chapter 6, and open-ocean fronts in Chapter 8.

3.3.8 Seasonal events in the plankton of the North Pacific

Most of the early work on seasonal events in the plankton was performed in the North Atlantic, so that we now have a fairly detailed understanding of the processes leading to a peak of phytoplankton production in the spring of the year and almost no production in the winter. The North Pacific produces a different seasonal pattern of events because phytoplankton production, accompanied by zooplankton, continues throughout the winter.

This continuous production is possible because the surface mixed layer in the North Pacific does not deepen in winter as much as in the Atlantic (120 m versus 400–600 m). The reasons for this variation have been examined by Warren (1983), who points out the greater density difference between the upper mixed layer and the deep water in the Pacific, which

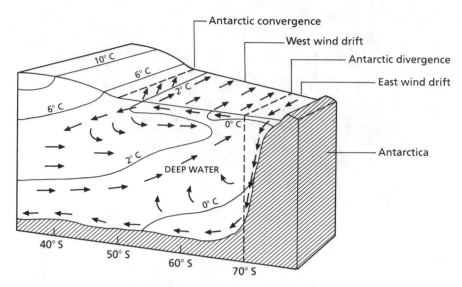

Fig. 3.15 Diagram of the Southern Ocean to show the Antarctic divergence and the Antarctic convergence. Antarctica is on the right, and north is to the left.

limits the depth of wintertime deepening. The cause of the greater density contrast lies in the salinity of the surface layer, which is low relative to the deep water. For example, in the Atlantic the salinities of the surface and deep layers are about 34.9 and 34.9–35, respectively; in the Pacific they are 32.8 and 34.6, respectively, so that even if the surface waters in the Pacific were cooled to freezing they would not be dense enough to sink as deeply as those of the Atlantic. Warren (1983) suggests two fundamental reasons for the lower salinity. First, the evaporation rate in the Pacific is lower. Second, the surface circulation in the Pacific is, compared with the Atlantic, relatively closed to the input of high-salinity water from the subtropics (see also sections 8.6 and 10.2.3).

Because of the continued biological production throughout the winter, the zooplankton population is ready and able to prevent a buildup of phytoplankton biomass when the phytoplankton production increases in spring with the shallowing of the mixed layer. This idea was first proposed by Evans and Parslow (1985) and has been confirmed by recent detailed studies (Boyd *et al.* 1995 a, b). Boyd and co-workers carried out the first biological rate process measurements in the winter season. At Ocean Station P, they found a phytoplankton standing stock of about 20 mg C m^{-3}, dominated by pico- and nano-phytoplankton, which was growing at about 50% of its maximum capacity. Mesozooplankton biomass in the mixed layer was only 1–4% of its summer value, but a strong component of microzooplankton was judged able to prevent the formation of a spring bloom. Because many mesozooplankton prey on microzooplankton, the low winter biomass of the former was considered to be an important factor in maintaining a high microzooplankton biomass and eliminating the spring bloom.

3.3.9 The Antarctic divergence

The waters of the Southern Ocean exhibit a phenomenon not found in temperate or polar waters elsewhere—a region of strong upwelling known as the 'Antarctic divergence,' caused by two wind systems circling Antarctica.

Close to the Antarctic continent the prevailing winds are from the east (Fig. 3.15), giving rise to a westward flowing coastal current known as the 'east wind drift.' Since the Coriolis force is to the left in the southern hemisphere, this water tends to flow toward the coast, where it piles up and causes downwelling. The cause of the easterly winds close to Antarctica is strong katabatic winds (McNider *et al.* submitted). The long winter nights and the summer combination of low sun angles and high albedo produce a negative heat flux at the surface of the Antarctic continent through most of the year. This cooling generates strong and persistent gravity-driven katabatic winds that descend the coastal slopes to the sea, creating some of the windiest places on earth. The Coriolis force deflects these offshore winds to the left, resulting in strong easterly winds over the coastal waters. McNider *et al.* (submitted) modeled this process and showed how the easterly winds help drive the upwelling at the Antarctic divergence, and carry nutrient-rich water into the region of ice-melt, close to shore. The stability conferred by the low-salinity water, coupled with the poleward advection of nutrients (including trace metals such as iron), leads to high biological productivity close to shore.

Five to ten degrees of latitude north of Antarctica one enters the 'west wind drift,' or 'Antarctic circumpolar current,' flowing in the opposite direction under the influence of strong westerly winds. Again the Coriolis force causes flows to the left of the wind, toward the equator. Hence, the surface waters diverge between the east and west wind drifts, and the Antarctic divergence constitutes a site of upwelling that circles the globe at a latitude of approximately 70° S. As Fig. 3.15 shows, a major flow of Antarctic surface water extends northward as far as the 'Antarctic convergence' or 'polar front,' which lies at a latitude of approximately 50° S and constitutes the northern boundary of the Southern Ocean.

The upwelled water at the Antarctic divergence is rich in nutrients, with nitrates around 25×10^{-6} M. As Holm-Hansen (1985) has pointed out, one would have expected that in the course of the northward drift to the Antarctic convergence, which takes about 200 days, the phytoplankton would strip most of these nutrients from the water. In practice the nitrate concentrations tend to remain relatively high all the way from the Antarctic divergence to the Antarctic convergence. Silicates, on the other hand, drop sharply over the same distance. If surface water is isolated in a container and exposed to ambient light and temperature, the phytoplankton grow and strip the nitrates in 15–20 days. Holm-Hansen (1985) concluded that there must be a considerable amount of upward mixing of nitrate into the photic zone throughout the Southern Ocean. He also observed that about 75% of the production in the Southern Ocean is based on recycled ammonia, and the work of Biggs (1982) indicates that most of the ammonia for this process is produced by microbial organisms , 200 mm in diameter. Holm-Hansen suggested that the drop in silicate concentrations is caused by uptake by diatoms and silicoflagellates, which sink out of the euphotic zone, leaving behind phytoplankton with a low silicate requirement.

An earlier view of the Southern Ocean was that the relatively high levels of dissolved nitrate in surface waters, and the known abundance of consumers such as krill, seals, and whales, all indicated a high level of productivity during the short Antarctic summer. Holm-Hansen *et al.* (1977) and El-Sayed (1978) have pointed out that many of the primary productivity measurements tending to support this view came from in-shore areas. On the basis of the most recent measurements in the open ocean they concluded that daily primary production averages about 0.134 g C m^{-2} d^{-1} during the growing season of about 120 days and that average annual production is only about 16 g C m^{-2} y^{-1}.

As mentioned in section 3.2.8, the Southern Ocean is one of the areas where addition of iron to shipboard incubations of the natural plankton has been found to lead to a marked increase in phytoplankton biomass and productivity. This observation has led to the suggestion that productivity of the Southern Ocean is iron-limited, a view supported by the observations of high nutrients and low chlorophyll in surface waters. De Baar *et al.* (1995) have provided additional support for this hypothesis. They studied natural levels of productivity in regions of the Southern Ocean with differing iron abundance. The waters of the southern part of the Antarctic circumpolar current (= west wind drift; see Fig. 3.15) have sufficient iron to sustain moderate primary production, but not enough to permit blooms to develop. At the polar front (= Antarctic convergence; see Fig. 3.15), however, a fast-flowing, iron-rich jet operates at a depth of 60–100 m, within which spring blooms produce phytoplankton biomass an order of magnitude higher than in the Antarctic circumpolar current.

A different point of view was put forward by Sakshaug *et al.* (1991). As a result of modelling studies they suggested that deep mixing caused by the passage of strong wind storms may occur with sufficient frequency to severely retard the development of diatom blooms and prevent the depletion of nutrients in the mixed layer. The diatoms would, in effect, be light-limited, even at the time of highest insolation. They suggested that this mechanism might cause any enhancement of productivity by iron enrichment to be short-lived.

3.3.10 The Arctic polar front

Sakshaug and Slagstad (1992) summarized a large body of information about primary production in the Barents Sea, which is divided into a northern and a southern part by the polar front. North of the polar front, the melting of ice in spring induces stability in the upper 20 m of the water column, initiating a spring bloom. Nutrients rapidly become depleted and the strong pycnocline prevents the upwelling of nutrients by wind mixing. The short-lived bloom moves steadily north with the melting ice.

South of the polar front, the events associated with a classic spring bloom occur. After the bloom, periodic vertical mixing of nutrient-rich water driven by wind events stimulates further bursts of diatom production. Total annual productivity is, therefore, about 115 g C m^{-2} north of the polar front and about 165 g C m^{-2} south of it.

3.4 NEITHER TROPICAL NOR TEMPERATE: THE DISTINCTIVE SUBTROPICAL PATTERN

Longhurst (1995) has drawn attention to the fact that extensive subtropical areas, comprising almost half the area of the world ocean, have a seasonal pattern of pelagic production that conforms with neither the classic pattern of temperate/polar waters (section 3.3) nor the typical tropical pattern of stability with little seasonal variability (section 3.2). The dynamic principles invoked by Gran and Braarud (1935) and Sverdrup (1953) are applicable, but the relatively light winter winds ensure that the mixed layer is deepened for only a short period (Fig. 3.16). Since the sun is high in the sky, and phytoplankton biomass is low, phytoplankton is never light-limited. The effect of the brief period of mixed-layer deepening is to recharge the upper layers with nutrients, so that the rate of primary production increases through the winter and peaks in late spring before nutrients again become limiting. Phytoplankton biomass does not reflect the peak in primary production, probably because it is removed by zooplankton grazing. Instead, a broad peak occurs in mid-winter.

Fig. 3.16 Multiyear averages of the seasonal cycles in photic depth, mixed layer depth, total primary production, primary production at subsurface chlorophyll maximum (SCM), and chlorophyll concentration in the Bermuda region. From Longhurst (1995).

A well-documented example of this concept is the Sargasso Sea of Bermuda, as observed by Menzel and Ryther (1960) and modelled by Fasham *et al.* (1993). The areas showing this pattern comprise the north and south subtropical gyres of both the Atlantic and Pacific oceans, with differences between the two oceans reflecting the permanent, relatively shallow halocline of the North Pacific (see section 3.3). The sub-Antarctic province, girdling the globe in the area between the subtropical convergence and the polar front, also exhibits this seasonal pattern of production (see section 3.3.9). The strong westerly winds of this region lead to a deeper mixed layer than in the subtropical gyres, however. The seasonal fluctuations in both primary production and biomass accumulation are relatively small.

3.5 AN INTEGRATED WORLD VIEW OF PRIMARY PRODUCTION

The technique of measuring surface ocean chlorophyll from satellite radiometer data has stimulated new attempts to estimate global primary production (Platt and Sathyendranath 1988; Longhurst *et al.* 1995) and new syntheses of planktonic food web processes (Banse and English 1994; Longhurst 1995).

(a) Definition of characteristic subdivisions of the world ocean

As Longhurst has said, the starting point for a rational inquiry about functions within pelagic ecosystems must be the identification of the marine equivalents of forests, tundra, and grasslands. He recognized four primary ecological domains. The first, characterized by the processes described in section 3.3 for a spring bloom in open temperate waters, he called the *westerly winds domain*. The second, in which the same basic mechanism of bloom formation occurs but is constrained by the presence of ice and by the low-salinity water that results from ice melting, is called the *polar domain*. The third major grouping, corresponding approximately with section 3.2.5 on tropical and subtropical oceans, Longhurst called the *trade wind domain*. Stratification is a permanent feature, with the depth of the mixed layer being influenced by changing trade winds, and algal blooms being produced by Ekman transport, eddy upwelling, and other processes. Finally, in the *coastal boundary domain* algal blooms are produced by classical seasonal changes in insolation and wind, as well as coastal upwelling, shelf break and tidal fronts, river discharges, and other forces. These mechanisms are discussed later, in Chapters 5, 6, and 7.

For purposes of the global descriptions and calculations, each of the four domains was further categorized into biogeochemical provinces based on the seasonal patterns of primary production and chlorophyll accumulation. A total of 56 provinces were identified and, with the use of existing data sets, the mean seasonal cycle of events in the plankton was described for each of them. On the basis of the pattern of primary production, eight distinct groups of seasonal cycles emerged (Fig. 3.17). A clear distinction must be made between primary production, which is most clearly related to physical events, and the pattern of chlorophyll accumulation, which is also influenced by factors such as grazing and sinking rates. The satellites see only the chlorophyll accumulation at the surface. The list of domains and their seasonal cycles is as follows:

Polar domain
Model 1—Primary production has a midsummer peak mediated by polar irradiance. Phytoplankton biomass peaks in the summer and again in the fall. Two explanations are possible for the fall peak: reduced grazing as the zooplankton sink to overwintering depths, or reduced phytoplankton sinking rates.

Westerlies domain
Model 2—A spring production peak is limited by nutrients, which is characteristic of four poleward provinces within the domain. In the North Atlantic drift region, the pattern of chlorophyll accumulation follows the classical pattern, with a spring peak and a subsidiary peak in the fall, except that the North Pacific sub-Arctic gyre experiences no peak in spring and chlorophyll reaches its maximum abundance there in the fall. (For discussion of this pattern, see section 3.3.8.)

Model 3—Winter–spring production occurs with nutrient limitation. This pattern is characteristic of a large part of the subtropical open ocean at mid-latitudes. Productivity is not light-limited and increases during winter with the deepening of the mixed layer. Nutrient limitation occurs in early summer and a subsurface chlorophyll maximum forms. Total chlorophyll peaks in winter (see section 3.4).

Trade wind domain
Model 4—A small-amplitude response occurs with trade wind seasonality. This temporal pattern is associated with the structure described in section 3.2 as typical tropical structure (TTS).

Fig. 3.17 Diagrams of the seasonal cycles in rate of primary production (Pt), pigment concentration (Chl), and mixed layer depth using temperature (TZm) or density (σZm) criteria. For explanation of models 1–8, see text. From Longhurst (1995).

Coastal domain

Model 5—A large-amplitude response occurs with monsoon reversal. Strong winds cause coastal upwelling, which controls the pattern of primary production. Chlorophyll accumulation follows primary production. This model has been most intensively studied in the Arabian Sea and western Indian Ocean.

Model 6—This pattern includes the classical nutrient-limited spring and fall blooms of the mid-latitude continental shelves. Problems of cloud cover and suspended sediment make it impossible to observe these blooms by satellite, so shipboard data were used.

Model 7—Similar to model 6, production in model 7 is continued into the summer because locally strong winds interact with the topography to force the upwelling of nutrients. This pattern is characteristic of three provinces (Alaskan shelf, New Zealand, and Falkland), all of which are special cases.

Model 8—Intermittent production occurs in coastal upwelling systems (see Chapter 5).

In addition, six provinces are recognized in which the annual cycle of production is forced by the seasonal pattern of river discharge. The pattern is highly variable and cannot be generalized. The principles are discussed in Chapter 4.

(b) Estimating global primary production

Longhurst *et al.* (1995) used the measurements of near-surface chlorophyll from the radiometer carried aboard the Nimbus 7 satellite from 1979 to 1986 as a basis for calculating global primary production in the sea. To convert these chlorophyll measurements to estimates of primary production, they used an approach developed by Platt and Sathyendranath (1988). Surface chlorophyll values were combined with local data previously obtained from ships to evaluate: (i) the depth of the chlorophyll maximum; (ii) the standard deviation around the peak value; and (iii) the ratio of the chlorophyll peak height at its maximum to total peak biomass. These figures were used to construct a chlorophyll profile for each grid point at each time step. Surface radiation was computed from the relevant sun angle and information on cloud cover climatology. Finally, appropriate values of the initial slope of the photosynthesis–light curve and the assimilation number, which represented polar, mid-latitude, and low-latitude conditions, were taken from earlier experimental work. All of these factors were combined in a local algorithm to generate an estimate of the rate of primary production for each chlorophyll profile.

The local parameters of the chlorophyll profiles were obtained by analyzing more than 26,000 profiles from about 60 different sources. These profiles were partitioned among the 57 biogeochemical provinces, and, by inserting estimated values where none were available, the authors succeeded in making estimates for four quarters of a year, centered on the 15th days of January, April, July, and October. This calculation led to weighted monthly mean production values for each province, which were then integrated over the year for the whole ocean to give the global water column primary production.

The grand total was a global net primary production of 50.17 g t C y^{-1}. A major cause of uncertainty is the error caused by suspended sediment in coastal waters. If it is assumed that turbid waters contain only 25% of the chlorophyll indicated by the color scanner, the global estimate drops to 44.70 g t C y^{-1}.

Previously published estimates based on extrapolation from shipboard estimates of production using the radiocarbon technique are in the range 20–31 g t C y^{-1}, but Martin *et al.* (1987), by extrapolation from a small number of ^{14}C incubations made under ultra-clean conditions, had estimated 51 g t C y^{-1}. It is interesting to note that the estimate by Longhurst *et al.* (1995) discussed here is independent of all previous ^{14}C estimates, yet is of the same order of magnitude.

3.6 SECONDARY PRODUCTION AND THE MIXED LAYER

3.6.1 Oligotrophic waters

So far in this chapter we have paid scant attention to the role of the zooplankton. The truth is that, lacking readily available automated equipment to obtain continuous records of the abundance of zooplankton, we have far less information about interactions between zooplankton and physical processes than we have about phytoplankton. The relationship between vertical physical structure and zooplankton production is complicated by the ability of zooplankters to make extensive vertical migrations. At one time of year most of the population may be at great depth, surviving a period of food shortage by living a lethargic life in cold water. At another time they may be concentrated in the mixed layer, making use of abundant phytoplankton food and higher temperatures to grow rapidly and reproduce.

There have been many laboratory studies of the feeding rate of zooplankton, but turbulence of the water has seldom been a factor that was controlled. Recently, evidence has been brought forward suggesting that small-scale turbulence has a powerful effect on the rate of contact between a zooplankter and its food. If this is found to be true, classical studies of feeding rates will have to be reevaluated.

Longhurst (1981) has provided a review of the vertical distribution of zooplankton. As before, it is most convenient to start with the relatively stable oligotrophic tropical systems. A very distinct pattern of vertical distribution is found in stable water columns with very shallow mixed layers, such as those found in low latitudes in the eastern parts of oceans, or in mid-latitudes toward the end of summer. The abundance of zooplankton is much greater in the mixed layer than below it, and within the mixed layer the greatest abundance is found just above the thermocline. Near the bottom of the thermocline there is typically a sharp decrease in the abundance of zooplankton, in a region that the specialists refer to as the planktocline.

Debate has arisen about whether the maximum of the zooplankton biomass occurs at the maximum of phytoplankton *biomass* or at the maximum of phytoplankton *productivity*. Ortner *et al.* (1980, 1981) favored the former, while Longhurst (1976; Longhurst and Herman 1981; Longhurst and Harrison 1989) favored the latter. It now seems possible that macrozooplankton feed at, or close to, the phytoplankton production maximum, while at other times and places microzooplankton predominate and feed at the biomass maximum (King *et al.* 1987; Le Fèvre and Frontier 1988).

3.6.2 Zooplankton and seasonal changes in vertical structure

Longhurst and Williams (1979) showed that at the ocean weather ship *India* located south of Iceland (59° N 19° W) the situation at the end of winter (late March) was that most of the zooplankton biomass lay at depths greater than 350 m. By day, 85% of the biomass was found at the location, and at night, when many species made upward migrations, 56% of the biomass was still below 350 m. During April, when the spring bloom was initiated, there was a major upward shift of the zooplankton biomass, brought about chiefly by the upward shift of the *Calanus finmarchicus* population, so that by May *Calanus finmarchicus* had 94.5% of its population in the upper 100 m. At the same time, *Calanus* began rapid reproduction so that between late March and early May the proportion of copepodites I–IV changed from 1 to 65% of the individuals present. Longhurst and Williams (1979) pointed out that any realistic model of the development of a phytoplankton bloom in temperate waters must allow for changes in grazing pressure due to two factors, the upward translocation of biomass and the growth of the zooplankton population *in situ*. Before the extent of the grazing pressure can be calculated it is necessary to know the feeding rate of the individual zooplankters. This problem was discussed in section 2.3.

In the life history of the American lobster, *Homarus americanus*, early larval stages are planktonic and at stage IV the larvae settle to the bottom. Boudreau *et al.* (1992) showed experimentally that the presence of a thermocline tends to inhibit larval settlement. They proposed this as a mechanism for ensuring that larvae produced offshore of Nova Scotia in summer settle close to shore and provide recruitment to inshore stocks. The larvae remain in the mixed layer until they drift close to shore where the water column is mixed to the bottom by tidal or wind action. They then settle out and continue development as juveniles.

If, during the shoreward migration, the larvae are driven into deep water by the turbulent mixing associated with strong winds, this would be expected to adversely affect recruitment to the lobster stocks. Boudreau *et al.* (1991) found such a negative correlation between the incidence of strong winds and recruitment to the lobster stocks of the Gulf of St. Lawrence.

3.7 CONCLUSIONS

In this chapter we have focused attention on vertical processes on a scale of meters and tens of meters, generated by wind stress at the sea surface, convective circulation resulting from surface cooling, and turbulence at the pycnocline generated by internal waves and other forces.

We have seen that phytoplankton cells are at times carried from the surface to great depth on a time scale of 1–10 hours, and are thus exposed to a fluctuating light regime. It appears that some biochemical processes in the cells adapt rapidly to the changing light, while others change too slowly for there to be any noticeable response. In any case, the metabolic response of cells exposed to fluctuating light is not very different from the response of those exposed to the same amount of light delivered at a constant rate.

More than half of the world's ocean is in tropical or subtropical latitudes, and a typical tropical profile of the water column shows a warm, stable, mixed layer separated from the cooler underlying water by a sharp pycnocline. Nutrient concentrations are very low above the pycnocline but relatively high below it. The peak of phytoplankton biomass, the chlorophyll maximum, is found at the nutricline. It appears that nutrients are slowly transported through the pycnocline and are rapidly taken up by the phytoplankton cells in the chlorophyll maximum. Light intensity is relatively low at the chlorophyll maximum and increases exponentially as one moves toward the surface. As a result, the greatest rate of primary production is usually found a few meters above the chlorophyll maximum.

A useful distinction can be made between the primary production that results from the utilization of upwelled nitrate, called new production, and that resulting from the utilization of ammonia excreted by heterotrophs in the mixed layer, called regenerated production. The ratio of new production to total production is called the *f* ratio. In steady-state conditions the nitrogen removed from the mixed layer by sinking, or by predators that leave the area, must be balanced by the vertical transport of new nitrogen.

The chief exception to the 'typical tropical structure' (TTS) of the open tropical ocean is found in the equatorial upwelling zone, where major ocean currents cause upwelling and divergence of large volumes of nutrient-rich water from below the thermocline. Associated with this circulation pattern are several major cyclonic gyres that cause a dome-like structure in the isopycnals and an enhanced rate of vertical transport of nutrients. Other deviations from the TTS occur irregularly but not uncommonly and are associated with seasonal and longer-term climatic changes in tropical latitudes.

In temperate and sub-Arctic latitudes of the North Atlantic, there is typically a burst of primary production known as the spring bloom. The mechanism underlying this condition is shallowing of the mixed layer after a period of vertical transport of nutrients. It was first recognized and described by Gran (1931), and was put on a quantitative physical basis by Sverdrup (1953). It has been shown that a wave of high chlorophyll biomass migrates poleward in the Atlantic in spring, approximately coincident with the 12 °C surface isotherm. Investigation of the vertical structure of the ocean while this migration is in progress shows that shallowing of the mixed layer to less than a critical depth is a precondition for the spring bloom. We shall find that the alternation of a period of strong vertical transport of nutrients with a period of stratification is a recipe for high primary production in a wide variety of marine habitats.

In the temperate North Pacific, a spring bloom does not occur (see section 3.3.8). The presence of a strong pycnocline at about 100 m depth keeps the phytoplankton in a well-lighted zone through the winter. Primary production, therefore, continues and a well-developed zooplankton population persists through the winter and prevents the development of a springtime accumulation of phytoplankton biomass.

4

Vertical Structure in Coastal Waters: Freshwater Run-off and Tidal Mixing

4.1 INTRODUCTION

Coastal waters (which we take as extending from the edge of the continental shelf to the high water mark) are subjected to the same seasonal cycles of warming and cooling as the open ocean, and in temperate climates the mixed layer may alternate between being shallow and deep in the same way as in open water. However, the process is greatly complicated by factors peculiar to the coastal zone. The first of these is the shallowness, which leads to a situation in which a relatively shallow mixed layer may extend to the bottom. Since dead biological material, detritus, tends to accumulate and decompose on the bottom, the nutrients released by it may be carried to the surface waters and rapidly used in photosynthesis. The second important factor is the presence of tidal currents that create turbulence in the water. If the depth is not too great in relation to the strength of a tidal current, tidally induced mixing may extend all the way to the surface. The third peculiar feature is the barrier to advection posed by the coastline itself. For example, if surface water is driven by wind action away from the coast, the only way for it to be replaced is by upwelling from below. Since there is a good chance that the upwelled water has been enriched in nutrients, an upwelling area is likely to be a site of enhanced biological production. When stratification first sets in, some phytoplankton are trapped in a well-lighted mixed layer and production is enhanced. In an area where stratification has been present for a relatively long time, the nutrients in the mixed layer may become depleted and their renewal from below is inhibited by the pycnocline, so that primary production tends to be depressed. Hence, stratification may act in a positive or a negative way on primary production, but its effects are always important.

In the open ocean, stratification is almost always induced by temperature differences between the layers. In coastal waters a very important additional factor must be considered—namely, the flow of fresh water from the land. Having a salinity close to zero, it is much lighter than sea water and by lying on top of the sea water creates a stratification that can be independent of temperature differences between the layers. Furthermore this surface layer, being less dense, rides higher at the sea surface, creating a slope along which water flows. Since these flows depend on the buoyancy, they are known as buoyancy-driven currents. One way of trying to understand the complex relationships existing between physical and biological processes in coastal waters is to view freshwater run-off as a mechanism tending toward greater stratification while wind-driven and tidal currents are mechanisms tending to cause turbulence in the water column and to break down stratification. In the open ocean, horizontal gradients are small and vertical processes control the distributions of heat, salt, nutrients, and other factors. In estuaries and on continental shelves, however, horizontal movement of water and large horizontal gradients tend to determine the property distributions. The Coriolis force and geostrophic balance are important features of horizontal flows, so we begin by reviewing them.

4.2 THE CORIOLIS EFFECT

The Coriolis effect, as mentioned in the last chapter, is an apparent deflection to the right in the northern hemisphere and to the left in the southern hemisphere of objects travelling over the surface of the earth. The deflection is caused by the rotation of the earth as illustrated in Fig. 4.01. Consider first the person at O in Fig. 4.01(a) who throws a ball at the sun as it rises in the east (E). (Let us ignore the friction of the air and the gravity that would bring the ball down to earth.) After an hour (Fig. 4.01(b)), the sun has risen above the horizon along a path slanting up to the right in the northern hemisphere and up to the left in the southern hemisphere. Is the ball that was thrown toward E still heading toward E or is it heading toward the sun?

Because there is no friction between the earth and the ball, the ball moves in a straight line relative to the sun and stars while the earth rotates underneath. The ball continues to head straight toward the sun while the horizon moves with the earth as shown in Fig. 4.01(b). Each point on the eastern horizon moves down and to the left of the ball, and the point E on earth is in a different direction relative to the fixed stars than when the ball was thrown.

The rate of the deflection may be calculated as follows. The point E, like all other points on the earth, moves around the axis of the earth once a day in a circle that lies perpendicular to the axis. The motion along this circle in degrees or radians per second is indicated in Fig. 4.01(b) by the arrow labelled Ω. This motion may be split into two components, one parallel and one perpendicular to the horizon. The parallel component is $\Omega \sin \phi$, where ϕ is the latitude of the observer's position, while the component perpendicular to the horizon

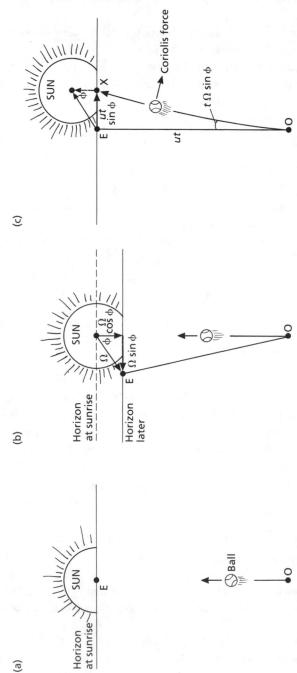

Fig. 4.01 (a) The person at O has just thrown a ball toward the sun rising in the east (E). (b) An hour later the earth has rotated on its axis and moved the horizon to the left and down relative to the sun and stars. The ball continues toward the sun because it is not connected to the earth. (c) If the horizon is assumed to remain motionless, the rising of the sun appears to be due to the sun's own motion rather than the rotation of the earth, and the fact that the ball turns toward the sun is due to the 'fictitious' Coriolis force that pulls the ball to the right in the northern hemisphere but to the left in the southern hemisphere.

Box 4.01 The Coriolis parameter

The angular rate of the deflection cannot be used directly in the equations of motion that require that the effect be described in terms of a force or acceleration per unit mass. The acceleration that must exist to account for the observed deflection can be derived following Pond and Pickard (1983) with help from Fig. 4.01(c). Assume the ball is originally thrown from O toward E but is deflected to the right to land at X. The distance from O to E is the speed of the ball, u, times the time of flight, t. The angle EOX is the angular rate of the deflection, $\Omega \sin \phi$, times the time of flight, t. The deflection from E to X is then the product of the angle, $t\Omega \sin \phi$, and the distance, ut:

$$EX = d = ut^2 \, \Omega \sin \phi \tag{4.01}$$

The acceleration that exists to produce this displacement starting from rest can now be calculated from the formula relating displacement, d, and acceleration, a, which is usually discussed in elementary physics courses, namely, $d = \frac{1}{2} at^2$. Substituting $d = ut^2 \, \Omega \sin \phi$ in this equation gives the acceleration or force per unit mass as $2u\Omega \sin \phi$. This equation is usually written fu where $f = 2\Omega \sin \phi$ and is called the Coriolis parameter.

The expression $2\Omega \sin \phi$ shows that the Coriolis effect is a maximum at the north pole where $\phi = 90°$, and zero at the equator where $\phi = 0$. In the southern hemisphere ϕ, the latitude, is negative and the Coriolis force is in the opposite direction; objects moving over the earth's surface are deflected to the left. The direction of the earth's rotation is not different in the two hemispheres but the direction of gravity does change and what looks like rotation to the right in the north is rotation to the left in the south.

---

is $\Omega \cos \phi$. The component along the horizon is the one of interest to us because it is the cause of the apparent deflection of the ball. The ball moves in a straight line, while the horizon rotates toward the left at an angular rate of $\Omega \sin \phi$.

Earthlings, however, find it much more convenient to ignore the fact that the earth rotates and assume that it remains motionless while the sun and stars move across the sky. The fact that objects, like the ball, are deflected from their original course is accounted for by a force that was invented by mathematicians to represent the effects that are due to the rotation of the earth. This imaginary force, the Coriolis force, results in the situation illustrated in Fig. 4.01(c). Here the earth remains motionless while the Coriolis force pushes the ball to the right, following the rising sun, at the angular speed of $\Omega \sin \phi$. The derivation of the Coriolis force from the angular rate of the deflection is given in Box 4.01.

The 'thought' experiment suggested in Fig. 4.01 is, of course, impossible to perform because gravity brings the ball back to earth before the sun has moved an appreciable amount, but the deflection is a very important factor in the flight of artillery shells that are separated from the earth for significant periods of time. Another common way of demonstrating the earth's rotation is with Foucault's pendulum (Fig. 4.02). When carefully set up, as is often done in science centers, there is no friction between the long swinging pendulum and the rotating earth. Since there is no force to twist the plane of the pendulum's swing to follow the rotation of the earth, it stays fixed relative to the sun and stars as the earth rotates underneath. In the north the plane of the pendulum rotates clockwise relative to the earth at the rate $\Omega \sin \phi$, which is the same as the component of the sun's motion parallel to the horizon noted above.

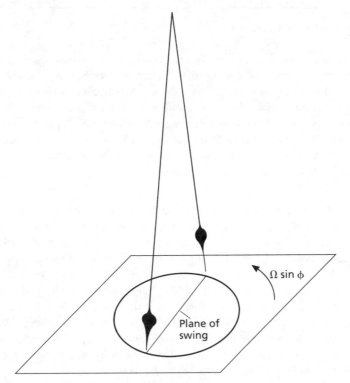

Fig. 4.02 The plane of the pendulum swing stays fixed relative to the stars, but appears to rotate to the right as the room rotates to the left with the earth at $\Omega \sin \phi$ rad h^{-1}.

The time for a complete revolution of the pendulum relative to the earth is called the pendulum day and is $24/\sin \phi$ hours long. At the pole the pendulum day is 24 hours, at 45° it is 34 hours, and at the equator it is infinitely long, when $\sin \phi = 0$. As with the thrown balls, it is hard to see, at first glance, how the pendulum day must be longer than the day. One must, of course, be careful to separate the two components of the rotation that are each less than the total rate of rotation and remember that the period of each component is the inverse. The periods of the components are not components themselves and do not add up to the period of the whole. This situation is a good illustration of the fundamental difference between vectors (rotation) and scalars (period).

4.3 THE GEOSTROPHIC BALANCE

The Coriolis force is an important factor in ocean currents because friction in the water is so small that water can move, like the ball in Fig. 4.01(a), over the earth's surface and remain isolated from the spinning earth. The Coriolis force is not the force that makes the water move but is the apparent force that deflects the flow when the water is moving relative to the earth. Suppose, for instance, that all the surface water in a region in the northern hemisphere starts to move south because there is a pressure gradient from north to south. The Coriolis force deflects the flow around to the right of its southward path. Over the course of about one pendulum day the direction of flow comes around to the right by 90° and is toward the west. At this point the water is moving west while the north to south pressure gradient that started the water on its flow is still in the same direction and the Coriolis force

Box 4.02 Equations of the geostrophic balance

Mathematically, geostrophic motion is expressed by two equations that define the balance of forces for the east–west (x axis) and north–south (y axis) directions. In the x direction the equation is

$$fv = \frac{1}{\rho}\frac{dP}{dx} \tag{4.02}$$

where the Coriolis force (fv) along the x axis is balanced by the x component of the pressure gradient on the right side of the equation. The equivalent equation for the y direction is

$$fu = -\frac{1}{\rho}\frac{dP}{dy} \tag{4.03}$$

Here the y component of the pressure gradient is balanced by the Coriolis force arising from the flow in the x direction.

These equations express only a balance of forces. There are no sources or sinks of energy included in them. Because the flow is perpendicular to the forces, the forces do no work and the motion goes steadily on forever. This scenario may seem a little unrealistic as friction must eventually convert the motion to heat, but that is a very slow process leaving the geostrophic approximation as very accurate, especially in the deep sea where friction is small.

---

is toward the north or 90° to the right of the flow direction. The Coriolis force and the force of the pressure gradient are now in opposite directions and balance one another. This condition is the geostrophic balance.

The geostrophic balance is an everyday occurrence in the atmosphere and is evident in the daily weather charts that display the distribution of isobars. The flow of air around the low- and high-pressure systems is parallel to the isobars—not across them—because the air is close to being in geostrophic balance. The mathematical expressions defining the geostrophic balance are discussed in Box 4.02.

4.4 ESTUARIES

'An estuary,' according to the definition of Pritchard (1967), 'is a semi-enclosed coastal body of water which has a free connection with the open sea and within which sea water is measurably diluted with fresh water derived from land drainage.' Estuaries are economically important features of the ocean because of their high biological productivity, their proximity to large cities with their wastes, and their increasing use as sites for aquaculture. In the following three sections we examine the processes that contribute to the typical circulation patterns and describe the different types of estuaries.

4.4.1 Estuarine circulation

From the definition it is evident that an estuary is just a bay, part of a bay, or a narrow inlet in which freshwater flow from the land has reduced the salinity of the ocean water. The flow

Fig. 4.03 The partially mixed estuary. Fresh water entering at the head of the estuary on the left creates a near-surface flow of light water out of the estuary and a compensating deep flow of saltier water up toward the head of the estuary. The upper layer gets thicker as it moves away from the source of fresh water because salt water is entrained from below.

of fresh water causes a characteristic circulation pattern to be set up in the estuary in which fresher, and therefore lighter, water flows out of the estuary in the surface layer and a deeper flow brings water from the open sea into the estuary (Fig. 4.03). This 'estuarine circulation' is observed to be the dominant circulation pattern in estuaries where run-off is moderate and where mixing by tidal currents is weak.

The main force that drives the estuarine circulation is a horizontal pressure gradient created by the density difference between the newly added fresh water and the resident salt water. The force lies in the surface layer and is directed down the estuary, away from the source of fresh water. The resulting flow, being dependent on the buoyancy of the fresh water, is usually referred to as a buoyancy-driven flow.

The origin of the pressure gradient is illustrated in Fig. 4.04, in which a layer of fresh water of density (ρ_1) overlies water of higher density (ρ_2). We assume there is no horizontal pressure gradient at the bottom of the low density layer or at any deeper level. Thus, the pressure at 'A' at the bottom of the upper layer is the same as the pressure at 'B,' which is on the same level as A but beyond the freshwater layer. If the pressures are to be the same, the weight of the water above the two points must be the same and the column of the lighter water above A must, therefore, be higher than the column of denser water above point B. The difference in the height of the water surface in the two regions (Δh) can be calculated by equating the pressure at A with the pressure at B. Pressure is calculated by multiplying the density of the water times the height times the gravitational acceleration—i.e., $P = \rho g h$. Above A the density is ρ_1 and the height is $h + \Delta h$, so the pressure is $\rho_1 g (h + \Delta h)$, and above B the density is ρ_2 but the height is only h, giving a pressure of $\rho_2 g h$. After equating these expressions, a little manipulation leads to the formula

$$\Delta h = h(\rho_2 - \rho_1)/\rho_1 \tag{4.04}$$

Fig. 4.04 A pool of light water, ρ_1, lies on top and beside water of greater density ρ_2. If the pressures at A and B are the same, the height of the sea surface above A must be higher than above B.

As an example let the freshwater layer with a density $\rho_1 = 1000$ kg m^{-3} have a thickness $h = 2$ m over the deeper layer of density $\rho_2 = 1025$ kg m^{-3}. Equation 4.04 states that the water above point A, in Fig. 4.04, is about 5 cm higher than above point B.

Since the pressures at A and B are the same there is no force to move the water horizontally between the two points. At all points above A, however, the pressure is greater than at the corresponding level above B. This higher pressure pushes the water in the upper layer to the right in the figure over the top of the denser water.

As the lighter water flows over the heavier layer toward the open sea, turbulence in the water causes the two layers to mix together, which reduces the differences existing between the two layers. The upper layer becomes thicker, increases in density, and slows down. The importance of this process to the circulation in the estuary is explained in Box 4.03.

4.4.2 Sources of turbulence: the Richardson number

The turbulent motions that mix the fresh upper layer of an estuary with the saltier layer below may be generated by a number of different processes. Breaking waves at the surface and rough flow over the bottom and sides of the estuary could generate significant amounts of turbulence, but the largest contributions are thought to be provided by the turbulence generated by the breaking of waves on the interface between the two layers (Farmer and Freeland 1983). These internal waves are generated on the density interface by wind waves at the surface, pressure fluctuations in the atmosphere, flow over irregularities in the bottom bathymetry, and other factors.

The likelihood that these internal waves will become unstable and break up into turbulence is usually estimated with the Richardson number. This number is a ratio of two opposing forces—one that tends to keep the density interface level and one that tends to increase the size of the waves. The first is generated by the density change across the interface, which creates a buoyancy restoring force (section 3.2.4) that tends to limit the amplitude of the waves. The energy to increase the size of the waves is generated from the difference in velocity between the two layers. The strength of the buoyancy force is usually represented by the square of the Brunt–Väisälä frequency (Eq. 3.06) while the energy available in the shear flow is represented by the square of the velocity gradient. The Richardson number is the ratio of these two values:

$$\mathrm{Ri} = \frac{g}{\rho}\frac{\mathrm{d}\rho}{\mathrm{d}z} \bigg/ \left(\frac{\mathrm{d}u}{\mathrm{d}z}\right)^2 \tag{4.09}$$

If this number is high the buoyancy forces associated with the density stratification are strong, and waves on the interface do not grow and break but tend to be damped out. In fact, if the ratio is greater than 0.25, waves of all wavelengths are stable (Turner 1973). If the density difference decreases or the shear increases such that $\mathrm{Ri} < 0.25$, waves will grow in amplitude by taking energy out of the velocity shear and eventually become unstable. This Kelvin–Helmholtz instability causes the waves to roll up and disintegrate into a patch of turbulence. Theoretical and experimental results, also reviewed by Turner (1973), show that as the Richardson number decreases, the most unstable waves have a wavelength that is about eight times the thickness of the interfacial layer between the two layers of unequal density. Some interesting *in situ* photographs of this process were obtained in the Mediterranean by Woods (1968).

Box 4.03 The balance of forces in an estuary

The driving force of the circulation is the difference in the pressure that is created down the estuary by the fresh water in the surface layer. This force (per unit mass) is expressed mathematically as

$$\frac{1}{\rho}\frac{dP}{dx} \qquad (4.05)$$

The balancing force is generated by the mixing between the upper and lower layers. When fast-moving upper water is exchanged with slow-moving lower water, a force is generated that slows down the upper layer and speeds up the lower layer. A mathematical expression for the force starts with the equation

$$K_e \frac{d\bar{u}}{dz} \qquad (4.06)$$

which is the diffusion equation for the vertical flux of momentum and is analogous to Eq. 2.11. K_e is the eddy diffusivity and $d\bar{u}/dz$ is the vertical gradient in the mean velocity. The formula calculates the vertical flux of momentum through a horizontal layer of thickness dz. As long as the same amount of momentum is leaving the layer as is entering it, no force is being exerted on the layer. A force arises when the layer is gaining or losing momentum, which happens when the vertical flux is not constant with depth. This situation occurs when either the eddy diffusivity or the velocity gradient varies with depth. Usually K_e is considered constant and the convergence or divergence of momentum is attributed to variations in the vertical current shear. Since the shear is just the velocity gradient, the gradient of the shear is the gradient of the gradient of the velocity, which is commonly called the second derivative of the velocity. Thus if K_e is constant the force is

$$K_e \frac{d^2\bar{u}}{dz^2} \qquad (4.07)$$

and the balance between the pressure gradient down the estuary, and the drag of the underlying water is expressed by

$$\frac{1}{\rho}\frac{dP}{dx} = K_e \frac{d^2\bar{u}}{dz^2} \qquad (4.08)$$

The equation does not include the Coriolis force because we assume that the estuary is too narrow for this force to be important. The flow is considered to be two-dimensional or the same along any section cut lengthwise along the estuary. One problem with the equation is that it is not possible to go to an estuary and directly measure the magnitude of either the pressure gradient or the drag force. It is also well known that the eddy diffusivity is not a constant but a function of time and of position. The equation is, however, very useful because it can be used, in association with the other required equations, to construct a mathematical model of the estuarine flow (Rattray and Hansen 1962; Dyer 1973; Nunes Vas 1994). Then the flow and the property distributions predicted by the model can be compared with the observed distributions. By adjusting the model to produce distributions that are similar to those observed it is possible to determine which are the most important factors controlling the circulation.

One feature of estuarine circulation that is observed in the models is the deep flow of saline water into the estuary. This flow is a direct consequence of the turbulent exchange between the upper fresh layer and the layer immediately below. As the water in the upper layer proceeds out of the estuary it becomes slower and thicker because of the vertical diffusion of momentum. At the same time diffusion causes the fresher upper water to be exchanged with the deeper saltier water and the surface layer becomes saltier as it progresses out of the estuary. Salt then is transported out of the estuary by the outward flow in the upper layer. In time this movement would cause the water in the estuary to become completely fresh, which does not happen because the deep flow develops and brings salt into the estuary.

4.4.3 Different types of estuaries

The well-behaved estuarine circulation outlined above is found in many estuaries but there are also many processes that can interfere to alter the basic pattern. In some estuaries the depth slopes gently from where the river reaches sea level and the river flow is strong enough to drag the sea water layer further out than it would be if there was no freshwater flow. In these cases the upper and lower layers appear as in Fig. 4.05. Because of the wedge-like shape of the lower layer, these layers are called salt wedge estuaries. The position where the lower layer meets the bottom changes with the strength of the flow in the upper layer, and in the case of the Mississippi this point can vary by more than 200 km throughout the year. Calculations to determine the shape of the salt wedge are reviewed by Dyer (1973).

Changing salinity at the boundary of the salt wedge leads to flocculation and sinking of fine particles carried in suspension. These particles get carried landward in the deeper waters and concentrate at the tip of the salt wedge, thus producing a zone of high turbidity, the 'turbidity maximum,' which also tends to move up and down the estuary according to the strength of the freshwater flow. In this way, sinking particles are distributed over a wide area and lead to the formation of intertidal mud flats so characteristic of estuaries.

In some estuaries friction between the bottom and the tidal currents produces enough turbulence to completely eradicate the estuarine circulation. These tidally mixed estuaries (Fig. 4.06) tend to be shallow and in areas of high tidal currents. The rate at which turbulent energy is converted from the tidal flow into turbulent flow can be estimated with the equation given by Pingree *et al.* (1978a) (for further details see Box 4.04):

$$\varepsilon = 2.5 \times 10^{-7} \times \frac{u^3}{h} \tag{4.10}$$

Here u is the absolute speed of the tidal flow in m s^{-1} averaged over one tidal cycle and h is the depth of the water in meters. The value 2.5×10^{-7} is an average drag coefficient, and

Fig. 4.05 The salt wedge estuary. At the mouths of shallow rivers, high run-off may cause the salty water to be pushed outward in a wedge shape.

Fig. 4.06 The tidally mixed estuary. Highly turbulent estuaries tend to be well mixed vertically. Isohalines, or lines of equal salinity, are nearly vertical, with highest values near the sea and lowest values near the river.

can be expected to differ according to the nature of the sea floor. The production of turbulent energy ε is in units of W kg^{-1}. Substituting 2 m s^{-1} for u and 10 m for h the energy conversion works out to 2.2×10^{-7} W kg^{-1}, which, as was pointed out in Chapter 2, is a relatively high level of turbulence in the ocean. Direct measurements of turbulence in the Strait of Gibraltar (Kinder and Bryden 1988) indicate dissipation rates as high as 10^{-4} W kg^{-1}. Thus, suitable conditions exist to create very high levels of turbulent energy that will tend to destroy the vertical stratification created by the fresh water, so that buoyancy-driven flow is eliminated. In this situation the turbulent flow that replaces the estuarine circulation is then the main agency for the transport of properties up and down the estuary. The movement of salt up the estuary, for example, is not by advection, as in the estuarine flow, but by turbulent diffusion.

A fourth type of estuary has a deep inner basin separated from the continental shelf by a shallow sill near the mouth. The waters in the deep basin are isolated from the waters of the continental shelf and are flushed only at irregular intervals when there is an influx over the sill of water dense enough to replace them (Gade and Edwards 1980). In the upper part of the water column, above the sill depth, fjords often exhibit the classic estuarine circulation. Areas noted for an abundance of fjords include the coasts of western Canada, Norway, and Chile.

The longitudinal section of the estuary in Fig. 4.07(a) shows the bathymetry of a fjord with a sill at 'B' separating the waters of the continental shelf from the deep basin at 'A.' The density of the water at the points A and B are plotted against time in Fig. 4.07(b). In the deep basin the density decreases slowly due to vertical turbulent diffusion, which is the only mechanism tending to exchange water between the light upper layer and the denser deep water. The density of the water coming into the fjord over the sill at B, on the other hand, fluctuates in time following changes in tidal currents, run-off, and wind stress. When these factors cause the density at B to be greater than the density at A, the inflowing water sinks down and replaces the water in the deep basin.

The influence of the tidal currents on the density of the water coming over the sill at B depends its ability to create turbulence, which tends to mix the light outflowing water with the denser inflowing water. As the amplitude of the tide decreases between the spring and neap tides (see Chapter 7) the generation of turbulence and the degree of mixing decrease. The stratification at the sill, at this time, is more likely to be maintained and the density of the inflowing water is less likely to be diminished by turbulent mixing. Renewal of the deep water is therefore most likely during neap tides. This process is discussed by de Young and Pond (1988) in relation to a fjord on the west coast of Canada.

The density outside the sill can also be affected by wind stress at the surface, which can produce an increase or decrease in the depth of the isopycnals. According to the studies of Muench and Heggie (1978), this stress is the main mechanism causing deep water renewal

Fig. 4.07 (a) Longitudinal section of a fjord showing the sill at B and the deep interior basin. The estuarine circulation is confined to the waters above the sill. (b) The density in the deep basin at A decreases slowly with time because of vertical diffusion. The deep water is replaced and increases in density when the water coming over the sill at B is denser than the water in the deep basin.

in late summer in the deep silled and low run-off fjords of Alaska. Other aspects of water exchange in estuaries induced by passing storms are addressed by Lewis and Platt (1982) and Goodrich (1988).

The rate of run-off is also an important factor limiting the inflow of dense water into fjords. A high run-off can create a thick, rapidly flowing upper layer that completely dominates the processes at the sill and that blocks any inward flow of dense water from outside. In fjords in cold climates with high run-off, deep water renewal tends to occur in the winter, when the run-off is a minimum.

Let us now examine some real-life examples of the various mechanisms described above. We shall begin with the effect of the freshwater run-off, then consider the effect of tidal mixing.

4.5 THE EFFECT OF FRESHWATER RUN-OFF ON BIOLOGICAL PRODUCTION IN ESTUARIES

4.5.1 Primary production in Chesapeake Bay

In a recent review of the influence of physical processes on the biology of the Chesapeake Bay estuary, Brandt *et al.* (1986) demonstrated very clearly how the spring run-off at the head of the bay leads to a very strong pycnocline with seaward-flowing surface waters and landward-flowing deeper waters. The combined effect of salinity and temperature is integrated into the plots of sigma-*t* in Fig. 4.08. (Sigma-*t*, or σ_t, is the density of the water minus 1000 kg m^{-3}.) They show clearly the movement of the pycnocline from a region near the head of the bay in February to a layer at 8–12 m depth extending about 160 km along

Fig. 4.08 Evolution of the pycnocline stratification in Chesapeake Bay in the spring of 1984. In February it is mainly between 200 and 300 km from the mouth. In May it extends almost the whole length of the bay. After Brandt *et al.* (1986).

the bay in May. In an open-ocean situation this flow would pose a formidable barrier to the vertical movement of nutrients, but the strong tidal currents in combination with net mean flows in opposite directions in the two layers result in strong internal wave activity, as revealed by acoustic records of the plankton (Fig. 4.09). The internal waves cause nutrients to be mixed into the surface waters (see section 4.4.2 on Kelvin–Helmholtz instability). The landward-flowing counter-current is instrumental in causing both phytoplankton and zooplankton to move in a cyclical manner within the estuary. For example, Brandt *et al.* (1986) have documented the movement of high concentrations of the dinoflagellate *Prorocentrum* from near the mouth of the Chesapeake Bay to the shallow waters 250 km landward, travelling with the bottom waters. At the head of the bay they were mixed up into surface waters and caused a 'red tide.'

Webb and D'Elia (1980), working in the York River Estuary, a tributary of the Chesapeake Bay system, found that on neap tides the estuary was stratified and at this time the

Fig. 4.09 Time series of acoustic backscatter intensity versus depth near Chesapeake Bay Bridge, 31 May 1984. Note evidence of vertical movement by internal waves. From Brandt *et al.* (1986).

water below the pycnocline became oxygen depleted while the nitrogen and phosphorus content rose. A few days after the peak of the spring tides the estuary became fully mixed by tidal currents, replenishing the oxygen at depth and upwelling large quantities of nitrogen and phosphorus. There was an inverse linear relationship between nitrogen and oxygen in the water, showing that benthic organisms were adding nitrogen in proportion to the rate that they were consuming oxygen. The relationship of oxygen to phosphorus was not linear, reflecting the fact that phosphorus binds to sediment particles but is released at low oxygen concentrations. The study showed very clearly how the balance between the stabilizing effect of freshwater run-off and the mixing effect of tidal currents can be so delicate that it shifts between neap and spring tides. This situation, alternating between upwelling of nutrients and stratification, provides the conditions for very high primary production. Once again, the parallel with the Gran effect is very striking.

Further work on the plankton of the York River estuary was carried out by Eldridge and Sieracki (1993). They concentrated on cyanobacteria and the protozoans that graze on them. The abundance of cyanobacteria in the estuary peaked during neap tides, when stratification set in, and decreased during spring tides when tidal currents were strong enough to cause vertical mixing. Using appropriate growth rates and feeding rates, they simulated the interaction between the cyanobacteria and their grazers, showing that during the neap tide cycle the growth of the cyanobacterial populations saturated the grazers, but that during the spring tide cycles the protozoan grazers contributed significantly to reduction in numbers of cyanobacteria.

4.5.2 Primary production in South San Francisco Bay

A bloom of phytoplankton occurs every spring in South San Francisco Bay. Cloern (1991) showed that maximum freshwater run-off occurs in the spring and that the bloom is initiated when the buoyancy of the surface layer inhibits tidal mixing and permits stratification. The greater the volume of freshwater, the greater the intensity of the bloom. When day-to-day events in the bloom cycle were analyzed, it was found that the increase in phytoplankton biomass always occurred in the neap tide cycle and the decrease during a spring tide cycle. A clear correlation existed between strength of tidal currents and rate of growth of phytoplankton biomass (Fig. 4.10).

Fig. 4.10 The daily rate of increase or decrease of phytoplankton biomass (vertical axis) in San Francisco Bay plotted against daily maximum current speed (horizontal axis) averaged over the week preceding each sampling date r = −0.87. From Cloern (1991).

4.5.3 Freshwater run-off and secondary production in estuaries

A variety of animals have adapted to the characteristic circulation of stratified and partially mixed estuaries, with a seaward flow of low salinity water and a compensatory landward flow of bottom water. By making vertical migrations at the appropriate times they can travel upstream for part of the time and downstream for another part, maintaining themselves in the estuary year-round or entering and leaving the estuary on a seasonal basis. Since the net seaward and landward flows have the ebb and flow of tidal currents superimposed on them, organisms also have the possibility of making saltatory movements—e.g., moving landward on the rising tide and sitting still in the bottom boundary layer when the tide is ebbing. In this way they will progress faster upstream than if they were in the water column continuously. Bousefield (1955) invoked these mechanisms to account for the distribution of barnacles in the estuary of the Miramichi, in eastern Canada, and Grindley (1964) proposed that copepods in Southampton Water, United Kingdom, remained in the landward-flowing water by ceasing vertical migration when they encountered the lowered salinities of the seaward-flowing surface waters. Wood and Hargis (1971) demonstrated for oyster larvae in the James River estuary (part of the Chesapeake Bay system) that the increasing salinity associated with increasing upstream flow near the bottom on each tidal cycle was used as the cue by the larvae to rise into the water column and get carried upstream. An example of seaward transport of larvae is provided by Provenzano *et al.* (1983), who found that at the mouth of Chesapeake Bay the maximum abundance of larvae of the blue crab, *Callinectes sapidus,* occurred in surface waters at night, while the salinity was falling on an ebbing tide. From this and other evidence they concluded that the crabs undergo their development offshore and reinvade the estuary as megalopa larvae or at juvenile stages.

Turning to fish, Graham (1972) presented evidence to suggest that larval herring were retained within the Sheepscot estuary of Maine by travelling upstream with the landward net tidal flow near the bottom. On reaching the limit of landward penetration, they rose into surface waters and travelled with the net flow to near the mouth of the estuary before descending again into the landward net flow. A comparative study of the young stages of three fish species living in Cape Fear River estuary, North Carolina, was made by Weinstein *et al.* (1980). They found that Atlantic croaker, *Micropogonias,* was concentrated near the bottom at all times and became concentrated near the head of the estuary, while spot, *Leiostomum,* and flounders, *Paralichthys,* made nocturnal excursions into surface waters during flood tides, which served to carry them into the salt marsh creeks. Flounders were thought to also use the inflowing bottom currents to make saltatory movements, as described earlier.

Lawler *et al.* (1988) modeled the physical oceanography of the Cape Fear estuary using a salt-budget technique, then combined it with life-cycle parameters and behavior of spot and Atlantic croaker. They were able to confirm that the observed distribution of the fish could be accounted for by the behavioral mechanisms postulated earlier by Weinstein *et al.* (1980).

A particularly detailed study of the transport of fish larvae in a two-layered estuary was made by Fortier and Leggett (1982, 1983), working in the upper estuary of the St. Lawrence. The dominant pelagic species are capelin (*Mallotus villosus*) and Atlantic herring (*Clupea herengus herengus*). The capelin spawn on tidal flats and beaches of the upper estuary and after emergence are flushed out toward the most productive zones of the lower estuary. The herring spawn demersally at sites downstream of the sampling location and are carried landward in the bottom water, then retained in the upper reaches of the estuary. Fortier and Leggett (1982) showed that the initial stage of dispersal is a passive one, with the capelin larvae concentrated in the upper 20 m of the water column, while the herring larvae less than 10 mm long were concentrated in the deep layer (40–60 m). Herring larvae greater than 10 mm made diurnal vertical migrations through the pycnocline and ceased to move systematically upstream, since their average vertical position was close to the depth of zero longitudinal velocity.

In an intensive set of hourly samples for 129 hours it was found that with increasing length capelin larvae congregated closer to the surface, so that there was a length-dependent acceleration of the seaward drift. Post-larval herring made vertical migrations that brought them into the surface layer during the flood tide, thus minimizing seaward drift during their time in the upper layer. There was evidence that the lower layer was undergoing a cyclonic circulation in the upper estuary, and that the herring post-larvae were retained within it. Fortier and Leggett (1984) found that after yolk-sac absorption a coherence developed between the patterns of variation in abundance of herring and abundance of microplankton. The herring followed the semi-diurnal vertical displacements of the early stages of copepods over distances of a few tens of meters, but it was not possible to tell whether the total pattern of coherence was a result of herring actively searching for higher food concentrations or from differential mortality by starvation of those fish that found themselves in regions of low food concentrations.

Rowe and Epifanio (1994) quantified the transport of weakfish *Cynoscion regalis* in Delaware Bay, United States. In a part of the estuary that was 10 m deep and fully tidally mixed, they sampled current velocity at two depths (2 and 7 m above bottom) and fish larvae at three levels (surface, mid-depth, and near-bottom) over a 6-day period in 1988 and again in 1989. Larvae with the yolk sac still present did not show any differential pattern of vertical migration or abundance with tidal stage, but post-yolk-sac larvae were significantly more abundant in the upper water column during flood tides. The authors developed a model that calculated physical flow and larval abundance at each 1 m depth interval by linear extrapolation from the two current meters, and by linear regression on the larval abundance data from three depths. From these data, they calculated for each 1 m depth interval the instantaneous flux of larvae (F_{zt}). For each depth and each flood or ebb tide they obtained the time-averaged water velocity U_z, time-averaged larval abundance D_z, and time-averaged larval flux F_z. They then divided time-averaged larval flux by time-averaged larval abundance to give a term (U_z^*) that they called larval velocity. If larvae are well mixed in the water column, U_z^* is identical to the time-averaged water velocity; if the larvae migrate between levels during the time series, however, U_z^* differs from U_z.

Flow patterns in the water column differed in the two years of observation. In 1988, time-averaged flow was positive (landward) at all depths; in 1989, a seaward flow was observed at the surface and landward flow near the bottom. U_z^* for post-yolk-sac larvae was always more positive than time-averaged water velocity. For early-stage post-yolk-sac larvae it was about 5 cm s^{-1} greater, but for late-stage larvae it was about 5 cm s^{-1} greater near the bottom, increasing to 25 cm s^{-1} near the surface. Landward transport of larvae was significantly greater on flood tides than on ebb tides, indicating that the larvae were taking advantage of flow patterns to maximize landward transport. In 1989, when the surface and deep waters were moving in opposite directions, the net landward flux of larvae was still positive, but 1–2 orders of magnitude lower than in the previous year.

Summarizing this section, we see that the secondary producers in an estuary are often adapted to the two-layered estuarine circulation that is found wherever the influence of freshwater run-off predominates over the influence of tidal mixing. A common form of adaptation is for the behavior of young stages to be related to the circulation pattern in a way that optimizes their opportunities for feeding and survival. The details are complex and not fully understood. One lesson is clear: a major change in the circulation pattern of an estuary brought about by damming the freshwater flow, a tidal dam, or other engineering projects may well have far-reaching effects on the primary and secondary productivity of the system. Examples will be given in section 4.8. We may note in passing that organisms are capable of adapting their life histories to circulation patterns on many scales, even up to the scale of the Gulf Stream, where eels and squid, for example, ride with the current and return by another route.

4.6 THE BIOLOGICAL EFFECTS OF TIDAL MIXING

4.6.1 Physics of tidal fronts

The idea that the turbulence generated by high tidal currents keeps the water over some shallow regions mixed all year while the quieter regions in the deeper waters become stratified in summer was first advanced about 70 years ago. Bigelow (1927) suggested the process to explain the fact that the water over Georges Bank in the Gulf of Maine remained mixed all year while the water over the deeper regions became stratified in summer. Subsequently, Dietrich (1950) (quoted by Le Fèvre 1986) suggested that the summertime front across the western end of the English Channel exists because of an increase in the turbulent mixing in the shallower regions of the Channel. (Fronts are regions of strong gradients of temperature and other properties in the sea. The various types of fronts are discussed in Chapter 6.) Later Simpson and Hunter (1974) constructed an energy argument that has proved very success-ful in predicting the locations of these tidally mixed fronts. Before we try to understand the processes that may increase the biological activity at these fronts, however, we first take a look in some detail at the theory that predicts the locations of such fronts.

Simpson and Hunter (1974) postulated that a front would be found where the intensity of turbulent mixing was just enough to continuously overcome the barrier to mixing pre-sented by the stratification. They estimated the amount of energy required to mix a strati-fied column by calculating the difference in the potential energy of the water column before and after mixing. The details of this calculation are presented in Box 4.04. The main result of the analysis is summarized by Eq. 4.21, which can be used to locate positions of tidal fronts. The result of one such calculation by Loder and Greenberg (1986) for the whole of the Gulf of Maine is shown in Fig. 4.12. These authors first determined the tidal velocities u and v with a numerical model, then estimated values of h/D_t. The contours of the diagram are $\log_{10}(h/D_t)$ and the transition between mixed and stratified water occurs at the contour 1.9. Values less than this number indicate well-mixed conditions and values greater than 1.9 will be stratified.

Pingree *et al.* (1978a) approached tidal fronts from the point of view of the turbulent-energy dissipation rate due to tidal flow (Eq. 4.10). They calculated an index $E = \log_{10}\varepsilon$ and showed (Fig. 4.13) that in areas where $E > -1.0$ the waters were tidally mixed throughout the year, and where $E < -2.0$ the waters were stratified in summer. This tech-nique is the inverse of the Loder and Greenberg (1986) method, which explains why lower values define stratified conditions rather than higher values as in Fig. 4.12.

Between the stratified and the well-mixed regions Pingree *et al.* (1978a) define a transition zone in which the degree of tidal mixing varies according to the stage of the lunar tidal cycle (Chapter 7). They showed lines at $E = -1.5$ that they predicted would be the average positions of the fronts between stratified and tidally mixed waters. These predictions were, in general, confirmed by field observations. Extensive areas of tidal mixing occur in the southern North Sea, the English Channel, and the southern Irish Sea. The Celtic Sea is a stratified region. Pingree *et al.* (1976) studied the development of the spring bloom in this region and found that the thermocline established first in the region of weakest tidal currents, then gradually spread to areas of stronger tidal currents. The spring bloom of phytoplankton coincided with thermocline formation. The ten-dency of tidally generated turbulence to oppose stratification caused by surface heating is very clear.

Because the mixing energy of the tidal flow D_t is proportional to the cube of the tidal velocity, the criterion expressed by Eq. 4.21 is often expressed by the ratio h/U^3, where U is understood to be $(u^2 + v^2)^{1/2}$, and the fronts are referred to as '*h* over U cubed' fronts.

The theory of Simpson and Hunter (1974) has been refined by Simpson and Bowers (1981) and Loder and Greenberg (1986) to include the effects of wind mixing and of vari-able levels of mixing efficiency. Bowers and Simpson (1987) compared the various theo-retical approaches with the observed positions of tidal fronts on the European-shelf seas and concluded that in shallow water the models incorporating wind mixing effects are most

Box 4.04 The Simpson and Hunter calculation

Simpson and Hunter (1974) first calculated the potential energy on the stratified and mixed sides of the front. The potential energy (PE) of an object (Box 3.02) is its mass (m) times the gravitational acceleration (g) times the height (h) of the mass's center above some reference level, and usually written $PE = mgh$. Profile (a) in Fig. 4.11 represents the density of the stratified column before mixing, while profile (b) is the profile after mixing. Before mixing, the potential energy of the lower layer, whose thickness is h_2, is $mgh_2/2$, since $h_2/2$ is the height of the layer's center of mass above the bottom reference level. The mass of the column of water is given by the density of the water times the volume, which, for a column 1 meter square, is given by the height $h_2 \times 1 \times 1$. Thus the potential energy of the lower layer per square meter becomes

$$PE = \frac{1}{2} \rho_2 g h_2^2 \tag{4.11}$$

The potential energy of the upper layer is the density times the volume ($\rho_1 h_1$), times g, times the height of the center of mass ($h_2 + h_1/2$), and therefore the total potential energy of the stratified 1 meter square column is,

$$PE_s = \frac{1}{2} \rho_2 g h_2^2 + \rho_1 g h_1 (h_2 + h_1/2) \tag{4.12}$$

After the water column is mixed (Fig. 4.11) the potential energy of the whole column is

$$PE_m = \frac{1}{2} \rho' g (h_2 + h_1)^2 \tag{4.13}$$

where ρ' is the density of the mixed water. If we assume that the equation of state for water is linear and that salt plays no role in the stratification, the height of the water column will not change because of the mixing, and the rectangles labelled (c) and (d) in Fig. 4.11 will be equal in area. This conservation of mass is expressed by

$$\rho'(h_1 + h_2) = \rho_1 h_1 + \rho_2 h_2 \tag{4.14}$$

From this expression, ρ' can be expressed in terms of ρ_1 and ρ_2, which can be substituted in Eq. 4.13 to obtain

$$PE_m = \frac{1}{2} g (\rho_1 h_1 + \rho_2 h_2)(h_2 + h_1) \tag{4.15}$$

Therefore, we can write the difference in the potential energy between the mixed and stratified state as

$$PE_m - PE_s = \frac{1}{2} g[(\rho_1 h_1 + \rho_2 h_2)(h_2 + h_1) - \rho_2 h_2^2 - 2\rho_1 h_1(h_2 + h_1/2)] \tag{4.16}$$

which becomes after simplifying

$$PE_m - PE_s = \frac{1}{2} g h_1 h_2 (\rho_2 - \rho_1) \tag{4.17}$$

If density (ρ) is a linear function of temperature (T)—that is, if $\rho = \alpha T$, where α is the thermal expansion coefficient of water ($\approx 10^{-4}$ °C^{-1}), which we saw in Eq. 3.03—then the change in density ($\rho_2 - \rho_1$) = $\Delta \rho = \alpha \Delta T$. Also by using Eq. 3.02 ($\Delta T = \Delta Q/mc$) relating the temperature change to the amount of heat (ΔQ) added to the water, the mass of the water (m), and the specific heat (c), we can write $\Delta \rho$ as $\alpha \Delta Q/mc$ and then Eq. 4.17 becomes

$$\Delta PE = \frac{1}{2} g h_1 h_2 \alpha \Delta Q / mc \tag{4.18}$$

and replacing the mass m by $\rho_1 h_1$ leads to the expression derived by Simpson and Hunter (1974):

$$\Delta PE = gh\alpha\Delta Q/2c \tag{4.19}$$

where h, the total depth $(h_1 + h_2)$, has replaced h_2, which it approximately equals if we assume that h_1 is small relative to h_2.

Equation 4.19 then represents the difference in the potential energy between the stratified and mixed water columns and is equal to the amount of mixing energy that must be supplied to overcome the stratification created by the addition of the heat ΔQ into the layer of thickness h_1.

The turbulent energy that is required to keep the water mixed is generated by the friction between the tidal currents and the bottom. Following Loder and Greenberg (1986) we write the rate at which energy is extracted from the tidal currents as

$$D_t = \rho C_D <(u^2 + v^2)^{3/2}> \tag{4.20}$$

where u and v are the horizontal components of the tidal velocity, ρ is the density of the water, and C_D is a drag coefficient ≈ 0.0027. The angled brackets indicate that the velocity values are averaged over a tidal cycle. Not all of this energy, however, will be used to mix the water. By calibrating the theory in the field, Loder and Greenberg (1986) find that only a fraction, $\varepsilon_t \simeq 2.6 \times 10^{-3}$, is used to change the stratification. Thus the energy required to keep the water mixed will be provided when the energy defined by Eq. 4.19 is equal to $\varepsilon_t D_t$. That is, when

$$\varepsilon_t D_t = gh\alpha\Delta Q/2c$$

then, by rearranging terms,

$$h/D_t = 2c\varepsilon_t/\alpha g\Delta Q \tag{4.21}$$

By substituting $g = 10$ m s^{-2}, $c = 4.2$ kJ kg^{-1} °C^{-1}, $\alpha = 1.6 \times 10^{-4}$ °C^{-1}, $\varepsilon_t = 2.6 \times 10^{-3}$, and $\Delta Q = 170$ W m^{-2}, $h/D_t \simeq 80$ for the region where the stratified water meets the mixed water.

accurate. In deeper water the wind effects decline in importance and no theory is significantly superior to the others at predicting the front positions.

4.6.2 Tidal mixing and phytoplankton production

In light of the postulated dependence of the spring bloom on thermocline formation, what is the pattern of phytoplankton production in tidally mixed areas where no summer stratification occurs? The lack of stratification would be expected to decrease phytoplankton productivity. On the other hand, dead organic matter is continually decomposing on and in the sediments, providing a source of nutrients that can be mixed back into the water column to stimulate phytoplankton production. Georges Bank, in the Gulf of Maine, is one such tidally mixed area that has been intensively studied (Bumpus 1976; Backus and Bourne 1987). Chlorophyll-a concentrations are homogeneously distributed in the water column (O'Reilly *et al.* 1981) and the depth of the euphotic layer in summer averages about 50% of the depth

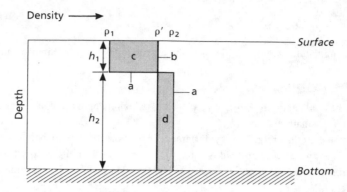

Fig. 4.11 Vertical profiles of density before and after mixing. Before mixing (a) the upper layer of thickness h_1 and density ρ_1 overlies a layer of thickness h_2 and density ρ_2. After mixing (b) the water is homogeneous at a density ρ'. The regions labelled (c) and (d) are assumed to have the same area.

of the water column. Under these conditions primary production continues throughout the year, with no clearly marked seasonal peak. O'Reilly and Busch (1984) reported fluctuating *production* levels with some indication of a peak in October. On the other hand, Riley (1941) and others had found a phytoplankton *biomass* peak in April. This finding is explained by the low level of zooplankton grazing at this time of year. As the season progresses, phytoplankton production remains at a high level, but most of the biomass is removed on a daily basis by the grazers.

Current estimates of total primary production (^{14}C uptake) on Georges Bank are 450 g C $m^{-2} y^{-1}$ on the shallow part (<60 m) and 320 g C $m^{-2} y^{-1}$ on the deeper part (Sissenwine *et al.* 1984). This level is higher than many other tidally mixed coastal areas in temperate waters, and the explanation offered is that a clockwise gyre (see Chapter 6) retains water on the bank, nutrient-rich water from the slopes is advected onto the bank, and the continuous mixing of the waters makes nutrients regenerated from the sediments available to the phytoplankton.

Fransz and Gieskes (1984) reviewed data for both biomass and productivity in various parts of the North Sea. Their biomass data confirm the earlier results, that offshore sites within the tidally mixed area have well-marked spring and autumn peaks while coastal sites have relatively high biomass throughout the spring and summer. Their field data (Fig. 4.14) show that coastal waters have a summer productivity almost twice as high as the offshore areas but a rapid falling off in autumn. This difference appears to be attributable to a larger flux of nutrients from the sediments and rivers into the tidally mixed water column (Mommaerts *et al.* 1984) but a poorer penetration of light in the autumn on account of the sediment load. The estimated annual productivity (^{14}C uptake) for the North Sea is considerably lower than for Georges Bank, 250 g C $m^{-2} y^{-1}$ offshore and 200 g C $m^{-2} y^{-1}$ in coastal waters.

In Naragansett Bay (Rhode Island, United States) the waters are tidally mixed at all times of the year (Kremer and Nixon 1978), and this results in a marked winter–early-spring bloom of phytoplankton followed by a series of summer blooms. Regeneration of nutrients from the benthos accounts for about half the needs of the phytoplankton, the remainder coming from allochthonous inputs and from regeneration within the water column.

The conclusion from this section is that tidally induced mixing in relatively shallow coastal waters prevents stratification of the water column, but the potentially adverse effects on phytoplankton productivity are more than compensated for by the increased

Fig. 4.12 Top: the bathymetry and coastline of the Gulf of Maine used in the numerical model of Loder and Greenberg (1986). Reproduced with permission from Pergamon Press. Bottom: the contours of $\log_{10}(h/D_t)$. Tidal fronts are expected at contours of 1.9. For explanation see text.

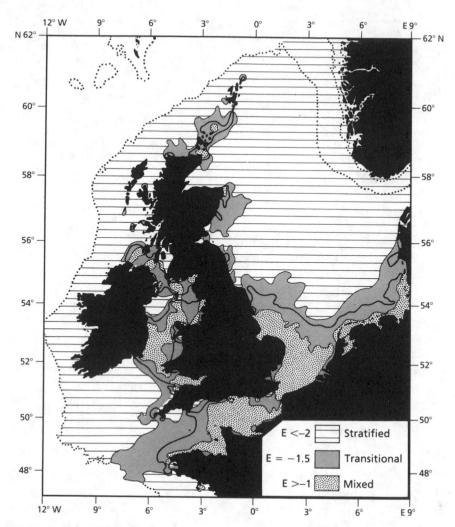

Fig. 4.13 Distribution of $E = \log_{10}(\varepsilon)$, where ε is the tidal energy dissipation per unit mass (erg g^{-1} s^{-1}). The continuous line E $= -1.5$ represents the predicted positions of frontal boundaries. Shading on either side indicates the transition zone between well-mixed and well-stratified waters. Waters that are stratified in summer are represented by light horizontal lines. Reproduced with permission from Pingree *et al.* (1978b), Pergamon Press.

nutrient flux to the water column from the sediments, so that annual primary productivity in tidally mixed areas tends to be above the average for coastal waters. If the phytoplankton is being transported by turbulence throughout the water column and is exposed to abundant nutrients, the onset of the bloom in spring depends mainly on the seasonal increase in light, and not at all on the onset of stratification. The net result is that the annual production cycle appears to start earlier in tidally mixed areas.

4.6.3 Consequences of tidal mixing for zooplankton

We saw in section 3.4.2 that in the open ocean initiation of the spring growth in the zooplankton population is often triggered by the upward migration of a large biomass of adult and late-stage copepods that have spent the winter at depths greater than 350 m. Reproduc-

Fig. 4.14 Coastal waters of the southern North Sea. (a) Biomass of phytoplankton (line) and zooplankton (points). (b) Estimated production of phytoplankton (heavy lines) and zooplankton (fine lines). The estimates for zooplankton are of two types: 'potential production' and 'lowest estimate.' For details see Fransz and Gieskes (1984).

tion begins as soon as this population reaches surface waters and grazing pressure on the phytoplankton develops relatively early in the season. In part this process is made possible by the rapid warming of the surface waters after stratification. In tidally mixed waters, warming is delayed by the lack of stratification, and there is no population from deep water that can ascend to the euphotic zone and begin to feed on the phytoplankton. The net result is that tidally mixed waters tend to have a relatively slow growth of the zooplankton population, which often does not peak until early or mid-summer. It appears that much of the biomass of the spring bloom escapes grazing by the copepods and sinks to the bottom. For example, Fransz and Gieskes (1984) determined biomass and productivity of both phytoplankton and copepods in the coastal region of the North Sea adjacent to Belgium (Fig. 4.14) and concluded that only in June and July was there a match between the productivity of the phytoplankton and the food consumption of the copepods. It should be noted that Gieskes was concerned only with zooplankton larger than 50 μm and that others have noted high abundances of microzooplankton such as ciliates, especially in the late summer and autumn in many inshore locations.

4.7 RIVER AND ESTUARINE PLUMES ON THE CONTINENTAL SHELVES

4.7.1 Physical mechanisms

In this section we are concerned with the effect of water from either a river or an estuary pouring out onto a continental shelf, or into a semi-enclosed sea. In some situations where river flow heavily predominates over any tidal effects, the surface outflow onto the continental shelf is mainly of fresh water from the river itself, and is called a river plume. In other situations there is strong penetration of salt water into the river valley to form an estuary, and the outflow onto the continental shelf is of river water mixed with considerable quantities of salt water. This flow constitutes an estuarine plume. Examples of river plumes that have been studied in an integrated biological/physical way include the Mississippi at each of its three passes in its delta, the Amazon River, the Fraser River, and the Connecticut River. Estuarine plumes are much more common, those of the Chesapeake, the St. Lawrence, and the Hudson River being some of the more intensively studied in North America.

When the light upper water from an estuary or a river flows out into the open ocean, it leaves behind a narrow region where the flow is predominantly two-dimensional and enters a less-confined world where the Coriolis force can change the direction of the flow. In the illustration in Fig. 4.15 the water moves down the estuary at a velocity u. Because this motion is relative to the earth there is a Coriolis effect directing water to the right of the flow. The water, however, is not able to change direction because of the side boundaries of the estuary but the water does move to the right causing a slight tilt to the sea surface and a pressure gradient in the direction opposite to the Coriolis force. The flow is, therefore, in geostrophic balance and the slope of the sea surface can be estimated in the way shown in Box 4.05 (see also Gill 1982).

Fig. 4.15 Water moving down the estuary at velocity u is in geostrophic balance. The Coriolis force balances the pressure gradient. In the open ocean the flow is not in geostrophic balance and in the northern hemisphere the Coriolis force causes the flow to turn to the right and flow along the coast.

Box 4.05 The surface slope across an estuary

We first assume a coordinate system in which the x axis points down the estuary. Since there is no flow parallel to the y axis $v = 0$ and the balance of forces in the water is given completely by Eq. 4.03:

$$fu = -\frac{1}{\rho}\frac{dP}{dy}$$

The pressure gradient can be written in terms of the sea surface slope by using the equation that relates pressure and water depth, i.e., the hydrostatic relation

$$P = \rho g h \tag{4.22}$$

where P is the pressure, ρ the density, g the acceleration due to gravity (10 m s^{-2}), and h is the height of the water column. Taking the first derivative of this and inserting $dP = \rho g dh$ in Eq. 4.03 we get

$$fu = -g\frac{dh}{dy} \tag{4.23}$$

which states that the sea surface slope in the y direction (dh/dy) is equal to fu/g. In a narrow channel where the flow may be considered constant over the whole width, the difference in the height of the sea surface from one side to the other will be $dh = fuW/g$, where W is the width of the channel. If u is 0.5 m s^{-1}, $f \simeq 10^{-4} \text{ s}^{-1}$, and $W = 200$ m, then the change in height across the estuary is $\simeq 1$ mm.

When the water flows out of the estuary this surface slope is missing, and cannot provide the pressure force to balance the Coriolis force. Being unopposed, the Coriolis force causes the flow to turn to the right but it comes under the influence of the shoreline again and sets up a situation similar to the one found within the estuary. As the Coriolis force pushes the water to the right, the blocking of the coast causes an opposing pressure gradient in the form of a slight slope in the sea level to be generated against the coast, and the plume of fresher water continues on its way as a coastal current in geostrophic balance parallel to the coast. A laboratory model of this process is described by McClimans (1986).

The Coriolis turning of the water coming out of rivers and estuaries is a common phenomenon and is often observed from aircraft and in satellite photographs. The fresher water in the current is usually distinguishable from the resident water of the ocean because it has more suspended material in it. A northward flowing coastal current driven by the run-off of rivers in British Columbia is described by LeBlond *et al.* (1986).

When the water leaving a river or estuary flows into a region where there is already a current the movement of the fresh water plume is not as simple as in the case described above. The position of the plume in these cases will be the result of the relative strengths of the background flow in the ocean and the new flow entering from the river or estuary. A strong coastal flow will overwhelm a weak outflow and the plume may turn in an unexpected direction. In most cases the flows are tide-dependent and the plume changes position over the tidal cycle. For example, the Connecticut River has a well-developed river plume at its mouth (Bowman and Iverson 1978). Under light wind conditions a layer of fresh water only about 2 m deep spreads out over the coastal waters for a distance of about 10 km from the river mouth. The tendency for the plume to be turned to the right by the Coriolis force is obscured by the local tidal currents, which deflect the plume alternately to the right and to the left. The seaward flank of the plume is marked by a sharp front (see Chapter 6).

Estuarine plumes are normally of much larger volume than river plumes because the freshwater run-off has entrained considerable quantities of sea water within the estuary. For example, the Chesapeake Bay estuary discharges large volumes of water at a salinity of about 16% and the plume is deflected south by the Coriolis force to form a strong coastal current. In many parts of the world estuaries occur sufficiently close together around the main ocean basins that their plumes may overlap and reinforce one another to form a continuous anti-clockwise coastal current in the northern hemisphere or a clockwise current in the southern. In a workshop reported by Skreslet (1986) accounts were given of such currents in Scotland, Norway, Iceland, Greenland, eastern Canada, and the eastern United States, and Drinkwater (1986) suggested that they were part of a nearly continuous flow around the perimeter of the North Atlantic. A modelling study has predicted the existence of a similar anticlockwise flow around the perimeter of the North Sea (Müller-Navara and Mittelstaedt 1985).

Others have pointed out that the plumes are sometimes deflected by wind stress. For example, the Hudson River plume, off New York, took up a position approximately at right angles to the coast after 3 days of southwest winds (Bowman and Iverson 1978). The same river plume also weakened under conditions of low river flow, to the point where it was temporarily obliterated by local tidal currents. More investigations are needed before the strength and variability of buoyancy-driven coastal currents can be assessed.

4.7.2 Biological effects of river and estuarine plumes

The biological effects of freshwater discharge may be considered under three headings: (a) direct effects of the materials carried by the river on biological production in the plume; (b) entrainment and consequent upwelling of nutrient-rich water, which is likely to enhance primary and secondary production; and (c) enhancement of the stability of the water column, which may be expected to enhance productivity at the time of a spring bloom, but which may inhibit vertical mixing and hence reduce primary productivity at other times of the year.

(a) Effects of material carried by rivers

The nutrients carried into coastal waters by river plumes have a marked effect on productivity in a region surrounding the river mouth. The time of peak run-off of the river is normally a time of diatom bloom, and commonly the diatoms are deposited on the sea floor around the river mouth. Riley (1937) found an area of enhanced chlorophyll and phosphorus extending about 50 km to the east and south of the Mississippi delta, and about 125 km westward in the coastal current. Revelante and Gilmartin (1976) found high nutrient uptake by phytoplankton near the mouth of the River Po in northern Italy. Off the mouth of the Amazon peak diatom biomass occurs about halfway across the continental shelf, when the surface salinity is about 5%. It appears that the diatoms produced in the river plume are carried landward in the bottom water and deposited on mud banks (Milliman and Boyle 1975). This localized effect is to be contrasted with other effects, such as lowered salinities and enhanced plankton production, attributed to the presence of Amazon water as far away as Barbados (Kidd and Sander 1979).

The foregoing examples are of distinct river plumes. In estuarine plumes the story is rather different because the nutrient load carried by the river is frequently utilized within the estuary itself. On the other hand, many major estuaries around the world are the sites of dense human settlement, so that the estuaries receive a heavy load of nutrients from land drainage or sewage. Frequently the estuary does not have the capacity to assimilate this material and the estuarine plume is strongly enriched with nutrients. The plume of the Hudson estuary, off New York, is a good example (Malone 1982, 1984). The sewage input into the estuary itself is 1.6×10^5 kg N d^{-1}, which leads to a nitrogen concentration in the water of about $60\,\mu$g-at l^{-1}. Phytoplankton production within the estuary is severely light-limited on account of the high turbidity, so only a small proportion of the nutrients are utilized within the estuary. Most pour out into the estuarine plume, where the amount of chlorophyll in the water column runs at 40–60 mg m^{-3} for most of the year, and may have a monthly mean close to 200 mg m^{-3} during the spring bloom period. In the spring, grazing mortality is low and most of the production sinks to the bottom. Later in the year the zooplankton biomass increases and in summer is thought to consume about 30% of the phytoplankton productivity. Malone (1982) demonstrated that the area of high phytoplankton productivity associated with the sewage enrichment is concentrated within 20 km of the mouth of the estuary. Beyond that, chlorophyll-a concentration decreases as salinity increases. A budget for nitrogen supply and utilization indicates that the area required for the phytoplankton to assimilate all of the dissolved organic nitrogen from the sewage run-off varies from 670 km^2 during summer, when phytoplankton is most active, to 1350 km^2 during winter. In effect, all of the nutrients derived from sewage and land run-off are utilized within that region of the New York bight close to the estuary, known as the Apex.

(b) Effects of entrainment

From the circulation pattern of a partially mixed estuary (section 4.4.3) we can see that nutrients released by decomposition of organic matter on the bottom will be carried to the surface waters when salt water is entrained by the fresh water, and will enhance primary production. In an attempt to determine the importance of this effect relative to other upwelling mechanisms, Sutcliffe (1972) produced a nitrogen budget for St. Margaret's Bay, Nova Scotia. From data on primary production he concluded that there was an average total uptake by the plants of 59 g N m^{-2} y^{-1}. From consideration of the volume of freshwater run-off and its average nutrient content he calculated that a total of 7 g N m^{-2} y^{-1} was supplied from this source. He estimated that 12 g N m^{-2} y^{-1} was supplied by zooplankton excretion and 14 g N m^{-2} y^{-1} from bacterial decomposition of organic matter, leaving 26 g N m^{-2} y^{-1} to be supplied from nitrogen upwelled from the lower layer of the two-layered estuarine circulation. He argued that this new nitrogen could be upwelled either by entrainment in the river plumes or by wind-driven circulation. As a test of whether entrainment processes were important, he proposed to test for correlations between annual freshwater run-off and annual biological production. If the correlations were good, he would conclude that variation in river run-off significantly influenced biological production through the entrainment mechanism. If not, he would assume that the wind-induced upwelling was the major mechanism supplying the required nitrogen.

Unfortunately, information on year-to-year variation in freshwater run-off into St. Margaret's Bay was not available. Undeterred, Sutcliffe turned to the St. Lawrence River, for which excellent data on run-off were available over a long period. He showed (Sutcliffe 1973) that there was an excellent correlation between the discharge of the St. Lawrence River in April and the landing of lobsters in the Gulf of St. Lawrence 9 years later. Since lobsters take about 9 years to grow to marketable size, he concluded that in a year of high river run-off there was good survival of lobster larvae, and that this probability was reflected in good catches of that year-class 9 years later. From this relationship it was concluded that the upwelling of nutrients associated with entrainment in the estuarine plume is a major factor influencing biological production in any given year.

The correlation has held good for all of the intervening 15 years (Drinkwater 1987), probably one of the best correlations of its kind anywhere. However, Sutcliffe thought he was making a simple choice between entrainment and wind-induced upwelling, but in recent years a third possible mechanism has been invoked. As we saw earlier in the chapter, freshwater run-off can lead to earlier stratification of the water column and to greater resistance to vertical mixing. As a result, the surface layer is likely to become warmer in summer, which may easily affect the productivity of the plankton and the survival of lobster larvae. In the Gulf of St. Lawrence it seems that entrainment may be the key factor influencing plankton productivity in one area, while stabilization of the water column is more important in another (Bugden *et al.* 1982; Sinclair *et al.* 1986).

In places where estuaries discharge onto the continental shelf it has repeatedly been shown that deeper waters are drawn toward the mouth of the estuary from a considerable distance. Norcross and Stanley (1967) found that bottom drifters released 70 km seaward of the mouth of Chesapeake Bay (Fig. 4.16) were consistently drawn toward the mouth of the bay, and Pape and Garvine (1982) found a similar circulation off Delaware Bay. It is, therefore, reasonable to assume that nutrients regenerated into this bottom water are carried toward the coast and are eventually entrained into the estuarine plume, increasing biological productivity. Direct evidence of the magnitude of this effect is lacking mainly because of the difficulty of defining the boundaries of a plume that changes its direction under the influence of wind and tides (section 4.7.1).

(c) Effects of fresh water on stability of water column

We have mentioned that addition of fresh water to salt water causes earlier stratification and greater resistance to breakdown of thermal stratification by tidal or wind-induced mixing. This process leads naturally to the idea that in temperate latitudes the spring bloom may start earlier close to the coast, where freshwater influence is most strong.

In the St. Lawrence estuary the presence of a low-salinity surface layer inhibited mixing in the water column and favored phytoplankton production in the spring (section 4.5.1). The effect was particularly dramatic where a plume of fresh water was superimposed in the brackish water in the lower estuary. In this plume were found some of the highest concentrations of chlorophyll in the whole estuary. It is possible that similar effects occur where river or estuary plumes flow out onto continental shelves.

Thordardottir (1986) explained interannual variations in the timing of the onset of the spring bloom of phytoplankton off the coast of Iceland by interactions between freshwater run-off and the wind regime. It is interesting to note that when the primary production is averaged for the years 1958–1982 there are several distinct seasonal patterns. At the station furthest from shore there is little primary production in March and April, a single major bloom in May, and relatively low production for the rest of the season (Fig. 4.17).

Fig. 4.16 The Chesapeake Bay plume, as revealed by salinity contours. The mouth of the bay is near the upper left corner, and the plume is directed south, close to shore. After Garvine (1986).

Close to shore, production begins in March and remains at a relatively constant level through the summer. At intermediate distance from shore there is a pattern with more or less equal blooms in spring and autumn. When the geographical distribution of the rate of primary production was plotted (Fig. 4.18) using data averaged over the years 1958–1980, it was clear that the phytoplankton bloom normally occurred first in a bay on the south-west corner of Iceland known as Flaxafloi. In discussing the factors determining the time of the bloom along different parts of the coast of Iceland, Thordardottir (1986) was able to rule out light and nutrients, both of which were nonlimiting from March onward. As we might expect, the timing of the bloom was clearly related to the formation of a shallow mixed layer. Detailed analysis of the temperature and salinity profiles at various stations showed that the stratification associated with the beginning of bloom conditions was attributable to reduced salinity caused by freshwater run-off, rather than to surface warming. Breakdown of stratification was associated with strong winds. Hence the explanation of the pattern of primary production in time and space shown in Fig. 4.18 is that the bloom normally begins first in Faxafloi, which receives surface run-off and is sheltered from winds on three sides. It then begins close to shore along the south coast, where stratification is facilitated by freshwater run-off, but is delayed somewhat by wind mixing. The

Fig. 4.17 Data on integrated photic zone primary production, pooled for the years 1958–1982 at five stations off the southwest coast of Iceland. Note the tendency for an earlier start to the spring bloom, and higher summer production, at stations close to shore. After Thordardottir (1986).

Fig. 4.18 Seasonal progression of the spring bloom off the southwest of Iceland. Contours show primary production (mg C m^{-3} h^{-1}) at a depth of 10 m. Pooled data 1958–1980. After Thordardottir (1986).

bloom begins last in the open ocean, where there is negligible influence of freshwater run-off, and stratification depends on surface warming.

The general picture for Iceland is that there is a mean residual surface current that flows clockwise around the island, with a speed of 0.06–0.11 m s^{-1} in the area south and west of Iceland. The rate of flow shows considerable variation according to wind conditions and the freshwater run-off. As we saw earlier, the detailed mechanism driving such currents is still under investigation, but it is clear that freshwater run-off is an important factor. It follows that in other parts of the world where coastal currents associated with freshwater out-flows are recognized, it is probable that the additional stability conferred by the lowered salinity is instrumental in bringing about an earlier start to the primary production cycle, at least in temperate and sub-Arctic environments. A systematic investigation of the phenomenon has not yet been made, but the data from the Norwegian Coastal Current in spring, provided by Peinert (1986), indicate that the same phenomenon is to be found there (Fig. 4.19). At the time of the cruise, offshore waters (to the left in the figure) were not yet stratified. The chlorophyll maximum was present at station 113, and according to the author, the bloom in the lowest salinity waters (station 111) had already passed.

We now see that there are two mechanisms that lead to geographic migration of the spring bloom in temperate latitudes. The movement of the spring bloom from south to north in the northern hemisphere has been shown (section 3.3.6) to be associated with surface warming as indicated by the outcrop of the 12 °C isotherm, and it now appears that the presence of river and estuarine plumes in coastal waters leads to a migration of the spring bloom from the coast out to deeper water as the season progresses.

4.7.3 Effects of river and estuarine plumes on secondary production

In section 4.7.2(b) brief reference was made to Sutcliffe's (1973) discovery that there was a good correlation between river run-off into the Gulf of St. Lawrence and the landing of lobsters 9 years later. This finding implies that the St. Lawrence estuarine plume has a profound effect on the survival of lobster larvae, yet in the intervening years little progress has been made in unravelling the mechanism of the interaction. In Iceland, there is also a commercial interest in the effect of freshwater run-off, because the main spawning grounds of the rich Icelandic fish stocks are off the south and southwest coasts, in those areas most affected by fresh water. The early stages of copepods are an important component of the diet

Fig. 4.19 Section through the Norwegian coastal current in spring. For explanation see text. After Peinert (1986).

of first-feeding cod larvae and it has been found that the years in which zooplankton densities were highest were the years in which phytoplankton production started early under favorable conditions provided by freshwater-induced stratification (Thordardottir 1986).

Off the mouth of the Fraser River, in British Columbia, St. John *et al.* (1992b) found that a riverine plume (salinity 0–10 ppt) forms on each ebb tide, and is larger on spring tides than on neap tides. Around the riverine plume is a region of intermediate salinity (10–15 ppt) referred to as the estuarine plume. Entrainment of nutrients at the toe of the salt wedge, together with deep-water entrainment under the riverine plume, leads to enhanced levels of nitrite and nitrate in the plumes during late spring, summer, and fall. This process, in turn, leads to enhanced primary production. A plume front (see section 6.6) containing an aggregation of zooplankton forms along the line where the freshwater plume meets the waters of the Strait of Georgia. Copepods, amphipods, and euphausids were significantly more abundant in the estuarine plume than in either the riverine plume or the surrounding waters of the Strait of Georgia. Herring and salmonids were more abundant in the freshwater and estuarine plumes than in the surrounding water. The highest catches of juvenile salmonids were found at the front between the estuarine plume and the Strait of Georgia. St. John *et al.* suggested that juvenile salmonids migrating out of the Fraser River congregate at the plume fronts to feed on the aggregations of zooplankton and young fish occurring there.

Years of lower-than-average river discharge appear to be associated with above-average production of chinook, coho, and chum salmon, and Pacific herring; the reverse is also true (Beamish *et al.* in press). It is suggested that when discharge is low, the water column has less stability and is more readily mixed by strong winds, thereby enhancing upwelling of nutrients into surface waters.

Seasonal variability in the effects of the Fraser River plume was modelled in terms of nitrate concentration and primary productivity (St. John *et al.* 1992b). In springtime, a large run-off from the Fraser River was found to lead to earlier stabilization of the water column and an earlier spring bloom of phytoplankton. In the summer following a large spring run-off, however, the presence of large amounts of fresh water in the Strait of Georgia (into which the river discharges) has the effect of limiting the vertical movement of nutrients by wind mixing and reducing primary production. This process would explain the inverse correlation between river run-off and fish production.

Another interesting situation of commercial importance is the Zambezi River, which discharges into the waters of the east coast of Africa in the southern hemisphere and has an estuarine plume that turns northward along the coast. The region of this plume is marked by high concentrations of organic matter in the sediments and by very productive fisheries for the shrimp *Penaeus indicus*. Da Silva (1986) has demonstrated that there is a very high correlation between the catch-rate for the shrimps (taken as an index of abundance) and the run-off of the Zambezi from August to March (Fig. 4.20).

The study was further developed by Gammelsrød (1992), using a time series 1974–1988. He confirmed that catch-rate is positively related to river run-off and found that it is also negatively related to effort. In fact, effort contributed more to the interannual variation than run-off. During the years 1980–1988, when variations in effort were small, the correlation between biomass catch-rate and river run-off was 0.983, confirming the importance of this run-off. The author tentatively suggested that the key process is the flooding of the tidal plains near the outlet of the Zambezi River with brackish water, which stimu-

(a)   (b)

Fig. 4.20 (a) Map of part of the Zambezi River and adjacent coastline showing productive shrimp grounds (shaded) and depth contours. (b) Plots of Zambezi run-off (km³) and catch-rate of shrimps. Pooled data, 1974–1983. After da Silva (1986).

lates the recruitment process. A large river outflow leads to lower salinity and flooding of a larger area, both of which are favorable for recruitment.

In general, our understanding of the relationship between river run-off and secondary production remains very much at the correlational and descriptive phase, and much more study is needed to understand the mechanisms at work. Three physical mechanisms are interacting. First, the presence of fresh water in an estuary or coastal area often leads to earlier stratification and, therefore, an earlier spring bloom of phytoplankton. This bloom can be beneficial to organisms feeding in the plankton at that time—for example, cod larvae off Iceland. Second, buoyancy-driven currents can cause upwelling and entrainment of nutrient-rich water, which further enhances primary and secondary production. Finally, strong winds deepen the mixed layer and bring up nutrients that stimulate plankton production. This mechanism is an important determinant of primary and secondary production during the period of summer stratification. If strong buoyancy is present in the mixed layer, it can reduce the effect of wind-mixing.

We see, therefore, that freshwater run-off can enhance plankton production at one time of year and depress it at another time, or it can enhance production in one part of an estuary and depress it in another section. The interactions are complex and the organisms often show matching adaptations. One thing is clear: freshwater run-off has a strong influence on estuarine production, and modification of the magnitude or pattern of run-off can have marked effects on biological production.

4.8 EFFECTS OF MAN-MADE MODIFICATIONS TO RIVER RUN-OFF

The ecological effects of man-made changes in freshwater outflow to coastal waters have been reviewed by Drinkwater (1988). In many parts of the world major rivers have been dammed to provide hydroelectric power generation. Since the economic ideal is to have a year-round constant supply of electricity, the tendency is to store water at times of high flow and release it during times of low natural river flow. This pattern has the effect of modifying the seasonal pattern of river run-off to which the organisms have become adapted over long periods of time, often leading to major disruption of biological events in coastal waters. In warm, dry climates storage of water in reservoirs is associated with considerable loss by evaporation, so that the total annual run-off is considerably diminished. A number of case histories of modification by man are given below.

4.8.1 The (former) USSR southern seas

The effect of water development projects on the biology of the Black Sea, Azov Sea, Caspian Sea, and Aral Sea was reviewed by Rozengurt and Haydock (1991). A well-documented example is that of the Dniester and Dnieper Rivers discharging into the Black Sea (Tolmazin 1985). Major hydroelectric projects were begun in the 1950s and more-or-less completed by the early 1970s. Figure 4.21 shows the seasonal pattern of the run-off of the Dnieper before and after the construction of dams. The loss of a major spring peak, and its replacement by a series of much smaller peaks, is very evident. Accompanying this process was a major loss of water quality, since much of the water from the reservoirs was used in agriculture, in industry, and by municipalities, returning to the system loaded with various contaminants.

The lowering of the amplitude of the peak discharge of the Dnieper River resulted in the low-lying marshes in the Dnieper estuary not being covered during the period of peak fish migration. The average salinity in the Dnieper estuary increased, with an upstream extension of the near-bottom high-salinity layer. These environmental disruptions led to the disappearance of many species of zooplankton and fish from the estuary. Commercial landings of fish from the Dnieper estuary decreased by a factor of five, and in the Dniester River they dropped to close to zero.

Nutrients from agricultural run-off and sewage led to greatly increased phytoplankton production both in the estuaries and in their plumes. The phytoplankton biomass was largely ungrazed and sank to the bottom. With the increased stratification caused by increased summer freshwater discharge, oxygen concentrations in the lower layers were reduced to zero

Fig. 4.21 Run-off of the Dnieper River before and after the construction of a series of hydroelectric dams. Note the major reduction in the spring peak of run-off. After Tolmazin (1985).

and mass mortalities of invertebrates and fish resulted. Oxygen depletion had occurred near the estuarine plumes even before the construction of hydroelectric dams, but the anoxic events became more frequent and widespread, resulting in a sharp decrease in the catches of turbot, flounder, and crab.

A lesson to be learned from this case is that adverse biological effects are not necessarily caused by factors that decrease biological production. The total primary productivity in the estuaries and their plumes increased, but the changed distribution of salinity and oxygen killed the invertebrates and fish that had adapted over long periods to the more natural conditions. The primary production was consumed mainly by bacteria, and the commercial fish landings were severely reduced.

Economic losses in the Black-Azov Sea basins amounted to an estimated $0.8 billion per annum in fin-fisheries, $1.5 billion in molluscs, and $1.5 billion in seaweed harvests (Rozengurt and Haydock 1991). Similar losses have been incurred in the Caspian Sea basin, and the Aral Sea system is now a total disaster (Micklin 1988) as a result of diversion of the inflowing waters to irrigation schemes. The Aral Sea's level has dropped 13 m, and its area decreased by 40%. The commercial catch of fish fell from 48,000 tons in 1957 to zero. Once-flourishing fishing villages are now far from the water's edge and have been abandoned.

4.8.2 San Francisco Bay

Another example quoted by Rozengurt *et al.* (1985) and Drinkwater and Frank (1994) is the extensive modification of the Sacramento–San Joaquin river system, which flows into northern San Francisco Bay. The unmodified discharge of these rivers was about 34 km^3 per year. Of this amount, about 40% is now removed for local consumption, and 24% diverted to central and southern California, leaving only 36% entering the estuary (Nichols *et al.* 1986). Correlated with these changes has been a marked reduction in primary production in northern San Francisco Bay, which is accentuated during summers of abnormally low flow. Two mechanisms for this drop in production have been suggested. Cloern *et al.* (1983) pointed out that stratified and partially mixed estuaries normally have a zone of high turbidity around the region of maximum penetration of the deep saline water, and suggested that reduced freshwater flow allowed the zone of high turbidity to penetrate further upstream than before so that primary production was reduced on account of lower light penetration. Nichols (1985) pointed out that during years of particularly low freshwater flow benthic filter-feeding animals associated with the more saline water penetrated further upstream into the estuary and may have reduced the phytoplankton biomass by their grazing. Whatever the mechanism, reduced river flow into northern San Francisco Bay is correlated with reduced primary production, which is, in turn, correlated with reduced secondary production. The adult population of striped bass has decreased by 75% since the mid-1960s and the chinook salmon population by 70% since the early 1950s (Stevens *et al.* 1985; Kjelson *et al.* 1982). In the summer of 1977, when freshwater run-off was exceptionally low, the phytoplankton biomass dropped to 20% of normal levels, zooplankton and shrimp abundance decreased, and the striped bass population declined severely. The possible mechanisms involved include: (i) a decline in primary production as a direct result of reduced entrainment by the freshwater plume; (ii) degradation of fish spawning areas caused by the upstream migration of the turbidity maximum; and (iii) accumulation of higher concentrations of pollutants, resulting from reduced dilution by the freshwater run-off. At present it is not possible to tell which of these factors is most important. The correlation of reduced run-off with reduced fish production is confirmed by examination of the historical records, which show, for the years before diversion (1915–1944), a strong correlation between mean annual run-off and commercial catches of salmon, striped bass, and shad. Rozengurt *et al.* (1985) suggested that to ensure successful commercial landings of salmon, striped bass, and shad, the run-off should be restored to about twice the present levels.

4.8.3 Nile River

The High Aswan Dam in Egypt was fully operational by 1965. Before this, the Nile River discharged about 40 km^3 y^{-1} into the Mediterranean Sea, with the peak flow in September–October. By 1968, the river discharge had dropped to 10% of the previous value, with 50% of the flow occurring in January and February. This decrease led to increased salinity in the coastal region, a deeper mixed layer, and a sharp decrease in both suspended sediments and nutrients. Before these changes the river plume had been characterized by strong blooms of diatoms and copepods, and an extensive fishery had operated based on the sardines that migrated into the area to exploit the plankton. In addition, a strong shrimp fishery was based on stocks that were adapted to the pattern of river run-off. After the dam was built, the catch of sardines declined by 90% and has remained well below pre-dam values. The catch of shrimp declined by 75% and the stock has been replaced by a smaller, much less productive species.

4.8.4 James Bay

James Bay is in northern Quebec, Canada, and opens into the much larger Hudson Bay. As part of the first phase of a major hydroelectric development, the Eastmain River was diverted northward into the La Grande River system. Prior to this diversion, the mouth of the river discharged about 38 km^3 y^{-1} directly into James Bay, and a brackish plume 1–2 m deep covered about 100 km^2 in the adjacent coastal waters. After the diversion, the flow of the river dropped by about 90%, the tidal amplitude 8 km from the mouth of the river was doubled, salt water penetrated 8 km upstream from its historic position, and a two-layered estuarine circulation was instituted. As expected, this change caused the estuary to act as a nutrient trap and primary and secondary productivity increased markedly. The biota of the affected section changed from predominantly freshwater species to mainly estuarine species. The characteristic freshwater fish, lake whitefish (*Coregonus clupeaformis*) and cisco (*Coregonus artedii*) did not undergo much change in growth abundance or reproduction, but their distribution shifted upstream.

In the La Grande River, which received the diverted Eastmain River, flow was interrupted for about a year after closure of a major power dam. Later, the river's mean flow increased 50–100% over pre-impoundment levels, with up to a 700% increase in flow during winter. The natural spring freshet was eliminated, and summer flow became about the same as before dam construction. The low-salinity plume under the winter ice increased its area by 200% and became deeper. Changes in the population dynamics of lake whitefish and cisco were not important, but the flooding of forest land behind the dam caused a release of mercury compounds, so that fish became loaded with mercury to several times the pre-dam levels. Routine fish consumption by native peoples has resulted in excessive body loadings of mercury.

As these examples show, the physical regime around the mouth of a river has a profound effect on the biology of the organisms, and often on the productivity of populations in the coastal waters. While damming rivers may seem like an attractive, relatively clean option for the generation of electricity, in that it does not involve the burning of fossil fuel or the production of toxic wastes, it is clear that other environmental consequences must be taken into account. Power dams are obstacles to the migrations of fish, water stored in artificial reservoirs or used in irrigation may be lost to the atmosphere if the climate is arid, and land inundated behind dams may release toxic substances into the water.

On the positive side, these man-made experiments have revealed much about the connection between freshwater run-off and the productivity of estuaries and coastal waters. This information may be useful in making predictions about the fluctuations of fish and shellfish stocks resulting from natural variation in run-off. The work on fluctuations in lobster stocks discussed in section 4.7.2(b), and on changes in cod stocks to be dealt with in Chapter 9, are examples of how this information may be put to use.

4.9 THE EFFECT OF WIND ON VERTICAL STRUCTURE

As has been mentioned in earlier sections, wind blowing across the sea surface sets up a stress that causes water to begin moving in the same direction as the wind. The Coriolis effect soon modifies the direction of flow, causing the current to veer to the right in the northern hemisphere, and to the left in the southern. When the situation has persisted for some time the net movement of the water is at right angles to the direction of the wind, an effect known as Ekman transport. When the direction of Ekman transport is away from the coast, surface waters move offshore and their place is taken by deeper water that upwells close to shore (see Chapter 5). In situations where that deeper water is nutrient-enriched, this upwelling serves as a stimulant to primary production.

There are parts of the world where upwelling is a continuous process for many months of the year. These distinctive upwelling systems will be treated in Chapter 5. However, in the context of the present chapter it is instructive to see how coastal waters that are normally stratified in summer as a result of freshwater run-off and seasonal warming can have their vertical structure strongly modified by intermittent wind-driven circulation. In Nova Scotia, Canada, the circulation of water in numerous coastal inlets was studied between 1934 and 1953 and summarized by Platt *et al.* (1972). It was clear that Ekman transport caused major changes in the vertical structure of the water column in every year of study. The following is a typical example: winds blew from the south and southwest, roughly parallel with the coast of Nova Scotia, for about 10 days in early September 1967. The surface temperature in St. Margaret's Bay dropped from 15.6 to 6.9 °C in less than 7 days, while the surface salinity increased from 29.36 to 31.26‰ and the thermocline was destroyed. The nitrate content of the cold water was more than double that of the warm surface water that it had replaced. An increase in primary production occurred at this time. After 11

September, winds blew mainly from the east and northeast, the surface temperature changed back to 14.6 °C, and the thermocline was reestablished.

In this example, which is only one of many that could be given, the effect of the wind is to cause upwelling of nutrient-rich water and stimulate primary production. In section 4.7.2(c) we saw that off the coast of Iceland, wind-driven mixing delays the onset of the spring bloom by preventing the formation of a shallow mixed layer under the influence of freshwater run-off. In the previous chapter we saw that shallow mixed layers formed under the influence of solar warming can be broken down by wind mixing. In coastal waters, then, we find four main factors influencing the vertical structure of the water column: (i) warming and cooling that result from seasonal changes in solar radiation; (ii) freshwater run-off, which tends to produce a low-density surface layer and intensify vertical stratification; (iii) tidal currents that tend to generate turbulent mixing from below, and may prevent the formation of a thermocline or halocline; and (iv) wind-driven currents that may lead to coastal upwelling, or that may generate enough turbulence in the surface layer to break down or prevent stratification. No general statements can be made about the way in which these factors interact. At some times of the year the formation of a shallow stratified layer enhances primary production, while at others it depresses it. Situations in which shallow stratified layers are formed for a time but intermittently broken down are frequently associated with high levels of primary production.

PART B
PROCESSES ON A SCALE OF
1–1000 KILOMETERS

5

Vertical Structure in Coastal Waters: Coastal Upwelling Regions

5.1 INTRODUCTION

A theme running through the two preceding chapters on vertical structure is that the key to high biological productivity is the upwelling of 'new' nutrients from deep waters into the euphotic zone and the retention of phytoplankton in well-lighted waters by stratification of the water column. We have seen that optimum conditions for phytoplankton production can be produced by patterns of alternating convective mixing and thermal stratification, by periods of strong tidal mixing alternating with stratification caused by freshwater run-off, or by some combination of these factors. Wind-induced mixing modifies these patterns by breaking down the stratification, but commonly the stratification is reestablished soon after the wind abates.

In this chapter we consider the special places where wind-induced upwelling is the dominant mechanism for bringing new nutrients to the surface. These upwelling areas have been well studied, partly because they are associated with economically important fish stocks. Cushing (1971), in a review entitled 'Upwelling and the Production of Fish,' estimated the total production of fish and squid in upwelling areas in the late 1960s. He calculated that over 26 million tons, mainly sardines, were produced in the Benguela current system off the southwest coast of Africa and the Canary current system off northwest Africa, over 12 million tons in the Peruvian anchoveta fishery,

139

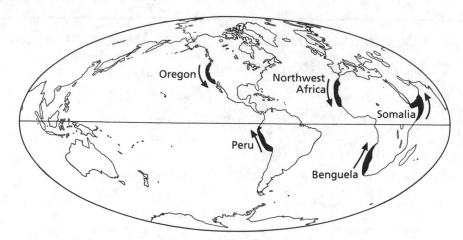

Fig. 5.01 Major coastal upwelling regions of the world, adapted from Thompson (1977). Arrows indicate prevailing winds.

over 5 million tons of anchovy and hake in the California current system, and so on. It is impossible that all of this production could be harvested, but for comparison we may note that this weight of fish is equivalent to more than half of the world's annual commercial fish landings. If reckoned in monetary value, the percentage would be much lower because these species have traditionally been used primarily for the production of fish meal. What is the explanation of this enormous biological production? The subject has been extensively reviewed by Boje and Tomczak (1978), Richards (1981), and Barber and Smith (1981).

The basic concept of wind-induced upwelling was mentioned in section 4.9. Wind blowing across the surface of the sea causes water to begin to move in the same direction. The Coriolis force, resulting from the earth's motion, causes the current to deviate to the right in the northern hemisphere, and to the left in the southern hemisphere. After some time, the net movement of surface water is at right angles to the direction of the wind, a condition known as Ekman transport. When Ekman transport is away from the coast, surface waters moving away from the shore are replaced by deeper water that upwells close to shore. This water is normally nutrient-rich and primary production is stimulated.

As we shall discuss in more detail in Chapter 8, the major ocean basins each have a western boundary current that tends to be fast and deep, and an eastern boundary current that is broad, shallow, and less well defined. Winds of the appropriate strength and direction to cause upwelling are more prevalent on the eastern sides of ocean basins. Nutrient-rich water below the pycnocline is closer to the surface in these areas, and is therefore more readily upwelled, giving a boost to phytoplankton production. Wind-driven upwelling does occur on the western sides of ocean basins, but tends to be of short duration and, because the pycnocline tends to be deeper, often brings up water from above the pycnocline that is not nutrient-rich. Such upwelling is biologically less important.

Worldwide, there are five major coastal currents associated with upwelling areas (Fig. 5.01): the California current (off Oregon and California), the Peru current, the Canary current (off northwest Africa), the Benguela current, and finally a rather anomalous example, the Somali current in the western Indian Ocean. Here the southwest monsoon sets up Ekman transport away from the coast of Somali and the Arab states: this is the only major upwelling area occurring on the western side of an ocean basin. The first four of these currents are plainly visible on our satellite image (see Plate 1).

As we explore the processes occurring in these upwelling areas we shall find that there is great variability. For example, off Peru the upwelling continues more or less year-round but there are years when biological production almost ceases. Off Oregon and Portugal there are large seasonal changes in the strength of upwelling, clearly related to seasonal differences in wind strength and direction. Even within the 6-month season of upwelling off Oregon, there are four or five periods of strong upwelling separated by periods of little or no upwelling. Each of these events gives rise to a burst of productivity equivalent to a spring bloom (Barber and Smith 1981). On still shorter time scales, the diurnal build-up of an onshore seabreeze and the semi-diurnal patterns of tidal current give rise to corresponding fluctuations in biological productivity.

5.2 THE PHYSICS OF COASTAL UPWELLING

5.2.1 The Ekman spiral

When the wind blows over the surface of the water it clearly generates waves but it also drags the water along. This is how wind-driven currents begin. But because the transfer of momentum from the air to the water occurs under the influence of the Coriolis force, the direction of the water movement is not the same as that of the wind. This difference was first pointed out in a theoretical analysis by V. W. Ekman in 1905 and now most phenomena connected with the process have his name attached such as Ekman spiral, Ekman layer, Ekman drift, and Ekman pumping. Our main concern in this chapter is with the Ekman drift, which creates the coastal upwelling zones. We start with the Ekman spiral, which is the name given to the arrangement of the currents generated in the upper layer of the ocean by the wind.

We begin with an elementary physical explanation based on the simple situation outlined in Fig. 5.02. This figure shows the adjustment of the water at the surface to the drag of the wind. Each stage of the adjustment is illustrated on the left with a diagram of the horizontal forces on the water. A separate diagram of the resulting current is on the right. In stage I the wind is blowing parallel to the positive *x* axis and begins to drag the water along in the direction of the positive *x* axis with speed *V*. (Note that the force and velocity scales are not the same.) The friction in the water creates another force, the water drag, opposing this current, and the rotation of the earth generates the Coriolis force directed 90° to the right of the current. These three forces are clearly not balanced, in stage I, as there is no force opposing the Coriolis force and the drag of the wind along the positive *x* axis is greater than the drag of the water along the negative *x* axis.

In the second step the wind and its drag remain in the same direction along the positive *x* axis but the Coriolis force begins to cause the direction of the surface current to rotate around to the right of the wind. The Coriolis force, always perpendicular to the current, also rotates with the current as does the water drag, which is always directly opposed to the current.

Fig. 5.02 The beginning of a wind-driven surface current in three stages showing the forces on the left and the water velocity on the right. In stage I the wind drag creates a flow of water that gives rise to the water drag and the Coriolis force. In stage II the Coriolis force causes the current in the water to rotate around to the right in the northern hemisphere. The force due to the water drag and the Coriolis force rotate with the current in the water. In the final stage the current has rotated the amount required to have the force due to the wind drag balanced by the combined effects of the Coriolis force and the drag of the water.

In stage III the forces have come into equilibrium. The component of the Coriolis force along the negative y axis is balanced by the component of the water drag along the positive y axis. The drag of the wind along the positive x axis is balanced by the sum of the components of the Coriolis force and the water drag along the negative x axis. The balance is attained when the surface water is moving at 45° to the right of the wind. This angle is a consequence of assuming that the Coriolis and the drag forces are the only forces acting on the water and that they are equal in magnitude. Ekman's theoretical analysis is developed more fully in Box 5.01.

The water just beneath the surface is not dragged directly by the wind but by the surface water, which lies directly above and which is moving at 45° to the right of the wind.

Box 5.01 Mathematical derivation of the Ekman spiral

Mathematically the Ekman spiral shown in Fig. 5.02 is derived (Pond and Pickard 1983) by first assuming that the only forces involved in the process are the Coriolis force and the frictional drag force. Forces arising from horizontal pressure gradients are assumed to be negligible. The balances between Coriolis and drag forces, along the x and y axes, are given by the equations

$$fv = -K_v \frac{d^2u}{dz^2} \tag{5.01}$$

$$fu = K_v \frac{d^2v}{dz^2} \tag{5.02}$$

where f is the Coriolis parameter, u and v are the x and y components of velocity, z is the depth, and K_v is the vertical eddy diffusion of momentum, which was introduced in section 2.2.7. The terms on the right of these two equations for the frictional drag are the same ones discussed in the previous chapter in relation to estuarine circulation (Eq. 4.08). Ekman solved these equations for u and v and found that the velocity is described by the spiral illustrated in Fig. 5.02. The equations for the u and v components of velocity are

$$u = V_0 \cos\left(\frac{\pi}{4} + \frac{\pi z}{D_E}\right) \exp\left(-\frac{\pi z}{D_E}\right) \tag{5.03}$$

$$v = V_0 \sin\left(\frac{\pi}{4} + \frac{\pi z}{D_E}\right) \exp\left(-\frac{\pi z}{D_E}\right) \tag{5.04}$$

where V_0 is the water velocity at the surface, D_E is the thickness of the Ekman layer, and z is the depth that is positive downward. At the depth D_E the magnitude of the velocity in the spiral is $V_0 e^{-\pi}$ or $\approx 4\%$ of the surface value. The direction of the current at D_E is directly opposite to the surface current V_0. Pond and Pickard (1983) show that the depth of the Ekman layer can be related to the wind speed by the equation

$$D_E \approx 4.3 \, W/(\sin \phi)^{1/2} \tag{5.05}$$

where W is the wind speed in m s^{-1} and ϕ is the latitude. Thus for a given wind speed the depth of the Ekman layer increases from the pole to the equator where it is infinitely deep. One example in Pond and Pickard (1983) gives an Ekman depth of 100 m at 10° latitude for a 10 m s^{-1} wind. The same wind conditions give a depth of 50 m at 45° latitude.

An adjustment of the forces in this second layer, like the adjustment in the top layer, causes the flow to be slightly to the right of the surface water—that is, at an angle greater than 45° to the right of the wind. This process continues down through the water column until all the momentum that is transferred from the wind to the water is converted to motion in the water. The speed of the water in each layer gets progressively less with depth and turns more to the right in the form of a spiral—the Ekman spiral. A horizontal projection of the currents at 11 equally spaced levels between the surface and the bottom of the Ekman layer is shown in Fig. 5.03. This bottom limit of the Ekman layer (D_E) is also discussed in Box 5.01.

The form of the velocity spiral shown in Fig. 5.03 requires that the friction in the water, K_v, be a constant throughout the depth of the Ekman layer. This requirement is a poor assumption in most regions of the upper ocean because of the vertical stratification that strongly affects K_v. It is also difficult to extract from velocity observations the part that is

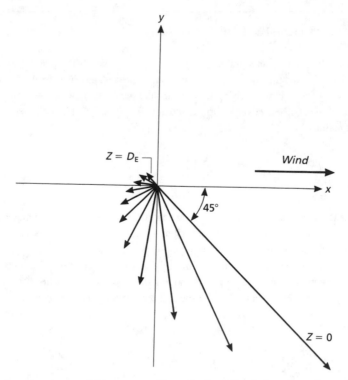

Fig. 5.03 A horizontal projection of the currents at 11 equally spaced levels from the surface to the bottom of the Ekman layer (D_E). The currents are generated by a wind blowing parallel to the positive x axis. For explanation, see text.

due only to the wind. For these reasons the spiral is a difficult phenomenon to observe in the ocean. However, Stacey *et al.* (1986), Richman *et al.* (1987), and Price *et al.* (1987) have been able, through the analyses of long current meter records obtained in the upper layer of the ocean, to demonstrate a spiralling of the flow with increasing depth which is in close agreement with the theory. In an analysis of drogued drifter tracks, Krauss (1993) finds that K_v increases through the upper 25 m of the water column, which leads to an Ekman spiral with a deflection less than the original theory.

5.2.2 Ekman drift and coastal upwelling

One important consequence of the velocity spiral is that the net movement of the wind-driven flow, after averaging over the Ekman layer, is 90° to the right of the wind. This result, which is one of the foundations of the wind-driven circulation in the oceans and of coastal upwelling, is derived mathematically by Pond and Pickard (1983). The result may also be derived, in a less rigorous manner, from the simple schematic diagram given in Fig. 5.04.

Here we assume that at the bottom of the Ekman layer the frictional forces that transfer the momentum of the wind down into the water become very small. This condition follows because the change in the velocity gradient (i.e., the terms on the right of Eqs. 5.01 and 5.02) becomes vanishingly small at this depth. All the wind energy transferred to the water is confined to the Ekman layer. Consequently the whole layer can be

Fig. 5.04 A cube of water in the Ekman layer treated as a layer without frictional coupling to the remainder of the ocean lying below. The force due to the wind drag is assumed to act on the whole cube rather than just at the surface and is balanced by the Coriolis force that is generated by the Ekman drift moving perpendicular to both forces.

treated as a slab that absorbs all the momentum transferred from the wind and that moves without friction over the ocean. This simplification of averaging over the layer of frictional influence (the Ekman layer) allows us to ignore the details of the velocity distribution. We need only consider the force due to the drag of the wind and the Coriolis force, as illustrated in Fig. 5.04.

The wind drag shown in the figure as acting on the whole of the Ekman layer tries to pull the water in the direction of the wind. The Coriolis force, which is the only other force involved, must balance the wind drag and therefore must be equal in magnitude to the wind drag and in the direction opposite to the wind. The flow of water that gives rise to the Coriolis force must be 90° to the right of both the Coriolis force and the wind. This flow is the Ekman drift. The reduction of the Ekman spiral to its net effect of a current perpendicular to the wind allows us to easily calculate the transport of water due to the wind.

The Ekman transport, commonly labeled M_E, is derived by Pond and Pickard (1983) and given by

$$M_E = -\tau/f \tag{5.06}$$

where τ is the wind stress at the surface of the water and f is the Coriolis parameter. A typical wind stress of 0.1 N m^{-2} and Coriolis parameter at a latitude of 45° of $\simeq 10^{-4}$ s^{-1} gives an M_E of 1000 kg m^{-1} s^{-1}—that is, one metric ton of water per second flowing at 90° to the right of the wind for every meter along a line parallel to the wind.

The relationships between the wind, Ekman drift, and coastal upwelling are illustrated in the perspective drawing in Fig. 5.05, which suggests waters off California with an equatorward wind parallel to the coast. The Ekman drift, to the right of the wind, is directed offshore and decreases to zero at the coast. The water that replaces the offshore drift cannot be supplied by horizontal flow because of the coastal boundary. Instead it is upwelled from the deeper layers. The illustration also indicates the commonly observed front in the near surface layers that separates the cooler upwelled water next to the coast from the warmer offshore water.

The upwelling creates a current parallel to the coast in the same direction as the wind. It is generated by the horizontal density gradient that arises because the upwelling water close to shore is denser than the water offshore. This gradient, shown in Fig. 5.06(a) by the rising

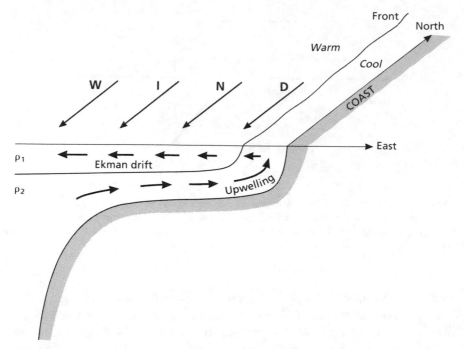

Fig. 5.05 A perspective drawing through an upwelling region illustrating the offshore Ekman drift in the upper layer being replaced near the coast by upward-moving water from the lower layer. The upwelling water, usually cool, is separated from the offshore warm water by a surface front parallel to the coast. The wind blows from north to south.

interface between the two layers of different density, results in a horizontal pressure gradient directed toward the land, which in turn creates a geostrophic current parallel to the shore shown in Fig. 5.06(b). The horizontal pressure tries to push the water back to the shore but the Coriolis force turns the flow into a current parallel to the coast and toward the equator.

In the deeper levels the pressure gradient developed by the upwelling tends to be toward the offshore direction rather than toward the shore, and a poleward current develops. Such equatorward and poleward density-driven flows are predicted by mathematical models (Gill 1982) and are commonly observed in the upwelling areas of the world, as will be illustrated in later sections. The widths of the upwelling zones and the associated poleward and equatorward currents are related to a natural horizontal length scale of the stratified ocean. This Rossby deformation scale is so important in oceanography that we make a slight detour to examine it in detail.

5.2.3 The width of coastal upwelling and the Rossby deformation scale

The vertical section through a coastal upwelling area in Fig. 5.06(a) indicates two horizontal lengths of interest. The first is D, the distance of the front from shore, while R_i is the width of the region where the interface between the upper and lower layer rises up to the sea surface. The distance D is discussed by Csanady (1981), who demonstrates mathematically that after the interface reaches the sea surface the front starts to move offshore. In his example of an upper layer 20 m thick, the interface reaches the sea surface with a 7 m s^{-1} wind blowing parallel to the coast for $\simeq$28 hours. If the wind is greater than 7 m s^{-1} over

(a) *Vertical section*

(b) *Plan view*

Fig. 5.06 (a) A section across an upwelling region in which the upper layer of density ρ_1 and depth H overlies a deep layer of density ρ_2. The interface between the layers rises from the undisturbed depth H_0 to the sea surface in the distance R_i. At the sea surface the interface lies a distance D from the shore. (b) A plan view of the currents parallel to the coast that are generated by the distribution of density in (a).

the 28 hours, the interface first moves to the surface, then moves offshore in proportion to the increase above 7 m s^{-1}.

The width of the region where the interface rises to the sea surface, shown as the length R_i in Fig. 5.06(a), is approximately equal to

$$(g'H)^{1/2}/f \tag{5.07}$$

where g' is the reduced gravity (see Box 5.02) and H is the depth of the upper layer. R_i is called the Rossby internal deformation scale after C. G. Rossby, who first discussed it. Because it is one of the fundamental length scales in oceanography we demonstrate mathematically how it arises in Box 5.02.

5.2.4 Variations in upwelling

The coastal upwelling described so far is a process that is assumed to be constant along a straight coast on the western side of a continent under the influence of a constant equatorward wind parallel to the shore. There are, however, many factors that can produce

Box 5.02 Derivation of the internal Rossby deformation scale

In Eq. 5.07 f is the Coriolis parameter, H is the depth of the upper layer, and g' is called the reduced gravity and is equal to the gravitational acceleration (g) times the fractional increase in density between the layers. Thus,

$$g' = \frac{\Delta\rho}{\rho} g \qquad (5.08)$$

The reduced gravity is, therefore, the gravitational acceleration acting on a parcel of water that has been displaced vertically from its equilibrium position. If no vertical change occurs in density there is no buoyancy force as given by Eq. 3.05 and particles in the water 'feel' no gravitational restoring force when they are moved vertically.

The pressure gradient that drives the current in Fig. 5.06(b) is derived from the horizontal gradient in density where the upper layer rises up to the surface. The shape of the interface between the layers is set by the internal adjustment of the layers and depends on two laws. The first is expressed by the balance of forces in the geostrophic balance. The second is the conservation of angular momentum, which says that if a fat object that is rotating becomes thinner (without losing mass) it will rotate faster. Everyday examples of this rule can be found in the twirling skater who changes her rate of spin by moving legs and arms closer and further away from the center of mass. The water draining in a sink speeding up as it approaches the central drain is another example of this conservation rule.

In oceanography conservation of angular momentum is discussed in terms of vorticity, which is a mathematically convenient term equal to twice the rate of rotation. Physically the two are identical except for the factor 2. In relation to the situation shown in Figs. 5.06(a) and (b) the conservation of vorticity is given roughly by

$$\frac{f + \zeta}{H} = \frac{f}{H_0} = \text{constant} \qquad (5.09)$$

Here again f is the Coriolis parameter representing the vorticity (2 $\times$ rotation) of the earth at the latitude ϕ, which is constant for this discussion. H is the depth of the upper layer with H_0 representing the thickness of the upper layer in the region where there is no current and ζ is the vorticity of the water relative to the earth:

$$\zeta = \frac{dv}{dx} - \frac{du}{dy} \qquad (5.10)$$

Since the v component of velocity is zero in Fig. 5.06(b), we can replace ζ by $-du/dy$, in Eq. 5.09, and the conservation of vorticity is now expressed by

$$f - du/dy = \frac{H}{H_0} f \qquad (5.11)$$

which states that the thickness of the upper layer (H) gets thinner as the horizontal velocity gradient (du/dy) increases. If there is no horizontal velocity gradient, the depth of the upper layer is always equal to H_0.

The equation (Eq. 4.23) for the other law, the geostrophic balance, is

$$fu = -g' \frac{dH}{dy} \qquad (5.12)$$

where g' is the reduced gravity. Now we assume that the thickness of the upper layer decreases exponentially across the current—that is, let

$$H = H_0(1 - e^{-y/L}) \qquad (5.13)$$

and we wish to determine L, the horizontal scale of the deformation of the upper layer, in terms of the other variables. This is easily done by substituting the value for H given in Eq. 5.13 into Eq. 5.11, which gives

$$fu = \frac{g'H_0}{L}e^{-y/L} \qquad (5.14)$$

From this an expression for du/dy may be derived, which when equated to the value for du/dy derived from Eq. 5.11, leads to the value of

$$L = (g'H_0)^{1/2}/f \qquad (5.15)$$

L is then the natural horizontal scale of currents in the ocean and coastal upwelling regions when a geostrophic balance and the conservation of vorticity are the dominant physical laws. The scale is usually represented by the symbol R_i and called the Rossby internal deformation scale. Toward the equator as f goes to zero the Rossby deformation scale becomes very large, which as we shall see in Chapter 9 is an important feature of equatorial dynamics and El Niño. At the mid-latitudes $f \simeq 10^{-4}$. If $g' \simeq 0.02$ a 100 m upper layer will be deformed over $\simeq 20$ km. In other words, the current will be about 20 km wide. The length L is also often called the internal Rossby radius. In this context it refers more to the natural horizontal scale of baroclinic eddies—i.e., those eddies with a vertical velocity gradient.

variations in this picture. Changes in the strength of the wind component parallel to the shore, the vertical structure of the water, variations in the bottom bathymetry, and instabilities in the currents can all create variations in the upwelling process.

The easiest way to view some of the variations in the upwelling is with infrared satellite images that are able to show the horizontal temperature gradient in the surface water generated by the upwelling. Such images usually show the boundary between the upwelling water and the resident water to be very irregular with eddies and plumes of near-shore water moving offshore. The processes that lead to these time-dependent irregularities, some of which have been reviewed by Brink (1983), are difficult to observe in the ocean. Narimousa and Maxworthy (1985, 1987), however, have studied coastal upwelling in a rotating tank in the laboratory and have observed similar irregularities. The authors believe they are created by waves that become unstable through complex interactions in the flow. Petrie *et al.* (1987) attribute the plumes moving offshore of an upwelling band next to Nova Scotia to baroclinic instability, which is a common instability of vertically sheared flows in stratified fluids. It is the process whereby eddies are created from the energy in the mean flow. In the atmosphere these are the passing lows or cyclones of the mid-latitudes that make the weather change from day to day.

Changes in the bathymetry can change the strength of the upwelling. A submarine ridge extending out from the coast, for example, produces conditions more favorable to upwelling than exist in the neighboring regions. Upwelling usually begins first at such ridges and remains stronger at the ridge even when it has developed in the other locations. Numerical models by Peffley and O'Brien (1976) and rotating tank models by Narimousa and Maxworthy (1985, 1987) show that a rise in the bottom is effective in creating preferred conditions for upwelling, whereas a point of land sticking out into the ocean unaccompanied by a rise in the bottom does not significantly alter the rate of upwelling. Mountains close to the coast can also create a preferred location for upwelling by causing the wind to increase in speed as it is deflected by the mountains (Gill 1982; Roden 1961).

All coastal upwelling sites are subject to seasonal and shorter-term variations in the wind strength. The well-studied upwelling west of the continental United States, for example, occurs mainly in the summer but disappears in the winter with the decline of the upwelling-favorable winds (Huyer 1983). The upwelling west of southern Africa in the Benguela current is perpetual at some latitudes but varies with the seasons at the northern and southern limits as the wind gyres that cause the upwelling move north and south (Shannon 1985). Variations in the wind due to storms and the daily heating and cooling are also shown by Shannon (1985) to have significant impacts on the upwelling off southwest Africa.

A change in the depth of the mixed layer in the upwelling zone can result in very large changes in the effects of upwelling. Off Peru, for example, when the upwelling brings nutrient-rich water up into the euphotic zone and stimulates phytoplankton production, the mixed layer is about 20 m deep. In times of El Niño the mixed layer is $\simeq 100$ m deep because of the tilting of the pycnocline across the Pacific (see section 9.2.1), and the upwelled water comes from above the pycnocline where the nutrients have already become depleted. In this case the upwelling can be just as strong as when the mixed layer is shallow but the effects on the biology are drastically different as there is no nutrient-rich water rising up to the euphotic zone.

5.3 THE CANARY CURRENT SYSTEM

Of the various upwelling systems that have been intensively studied, that off the Spanish Sahara of northwest Africa is perhaps the most straightforward to understand, as a basis for comparison with other systems. The results of a cruise in 1974 known as JOINT I have been summarized by Huyer (1976). Progress on a multinational study of this region under the umbrella of the Cooperative Investigation of the Northern Part of the Eastern Central Atlantic (CINPECA) was reviewed in Hempel (1982).

5.3.1 Upwelling and primary production

Figure 5.07 shows that during the JOINT I cruise the alongshore currents at all depths at a mooring in a mid-shelf position about 30 km from shore were well correlated with the wind strength. At a time of weak wind a gentle equatorward flow averaging about 10 cm s^{-1} extended almost 100 km offshore and to a depth of about 200 m on the continental slope, while from 200 to 400 m on the continental slope there was a poleward counter-current with a velocity up to 5 cm s^{-1} (Fig. 5.08a). At a time of strong wind an equatorward coastal jet developed with current speeds greater than 30 cm s^{-1} while the poleward counter-current had slightly increased velocities. During the strong winds a two-layered circulation developed over the shelf. There was offshore transport at rates greater than 15 cm s^{-1} in the upper half of the water column and onshore transport at speeds up to 20 cm s^{-1} in the lower half (Fig. 5.08b).

When isopycnals next to a western coast slope upward we infer that upwelling is in progress. Figure 5.08(c) shows that σ_t contours sloped gently upward toward the coast during the time of weak winds, but turned sharply upward during the strong winds. Barton *et al.* (1977) contoured the surface temperatures from 26 February to 9 April, showing that there were five upwelling events during the JOINT I cruise. The cold water first appeared

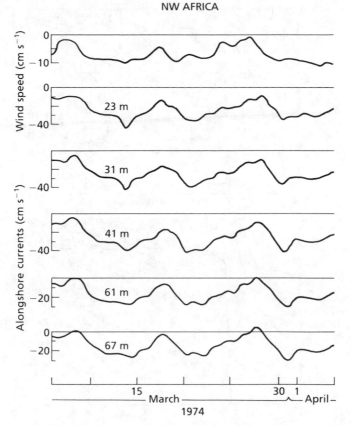

Fig. 5.07 Wind speed and alongshore currents at various depths at a station in the middle of the continental shelf off northwest Africa.

close to shore, but moved out across the shelf as the upwelling event persisted (Fig. 5.09). The distribution of nutrients (Codispoti and Friederich 1978) shows that the nitrate-rich waters followed the same pattern, being brought to the surface first in the inshore area, then carrying the maximum concentrations out to the mid-shelf area (Fig. 5.10).

It would be nice to be able to continue the story by reporting that upwellings of nitrate-rich waters were quickly followed by bursts of primary production. Huntsman and Barber (1977) reported on primary production during the JOINT I cruise, and the averages for the entire cruise show a consistent pattern with highest levels of nutrients in the 10 km zone closest to shore, and with the rate of primary production increasing steadily as they moved offshore, then levelling off to 2–3 g C m^{-2}d^{-1} across the entire shelf. Pulses of high productivity during early April and early May coincided with periods of high and steady wind stress and one might have thought that the wind stress generated upwelling of nutrient-rich water, which in turn led to increased primary production. This explanation is too simplistic. For one thing, there is a time delay between the onset of the wind and the arrival of nutrient-rich water out on the shelf. Second, there has to be a quantity of nutrient-limited phytoplankton cells ready to respond to the availability of nutrients.

Huntsman and Barber (1977) put forward the hypothesis that the observed high productivity results from the alternation of upwelling events and relatively calm periods. The

Fig. 5.08 Results of JOINT I cruise off North Africa, after Huyer (1976). (a) Sections showing the distribution of alongshore flow at times of weak wind and strong wind. Contour intervals are 5 cm s⁻¹. Positive values (shaded) are poleward and negative values equatorward. (b) Sections showing the distribution of onshore and offshore components of the flow during a weak wind and a strong wind. Offshore components (shaded) are negative. (c) Sections showing distribution of σ_t during weak winds and strong winds.

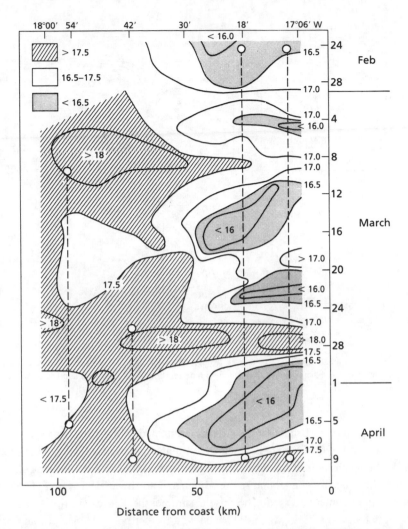

Fig. 5.09 Contours of surface temperatures from 26 February to 9 April, during the JOINT I cruise of northwest Africa. Note that pulses of cold water (shaded) correspond with periods of strong alongshore winds. From Barton *et al.* (1977).

upwelling brings nutrients into the surface waters, but during the calm periods stratification develops and the phytoplankton grows and multiplies while held in the shallow mixed layer. In other words, there is a miniature 'spring bloom' during each calm period.

To test this hypothesis they made numerous measurements of the assimilation number, which is the amount of carbon fixed per unit time, per unit of chlorophyll-a. It is well known that when phytoplankton is growing and dividing rapidly under the stimulation of nutrient enrichment, the assimilation number is high. They found that the assimilation numbers were high during periods of calm and lower during periods of upwelling, thus supporting their hypothesis. Jones and Halpern (1981) published further data supporting this point of view.

Minas *et al.* (1982) took a broad geographical view of the northwest African upwelling region. They showed that the region is divided into two major zones by a front that separates North Atlantic central water (NACW) from South Atlantic central water (SACW), at

Fig. 5.10 Patterns of nitrate distribution (μg-at 1^{-1}): (a) during weak winds, (b) soon after the onset of strong winds, and (c) after persistent strong winds. Reprinted with permission from Codispoti and Friederich (1978), Pergamon Press.

about the latitude of Cap Blanc. (JOINT I was conducted in the Cap Blanc region.) The SACW, to the south of Cap Blanc, is richer in nutrients than the NACW, but the poleward subsurface counter-current of the upwelling region carries SACW well into the Cap Blanc region. Upwelling occurs year-round at Cap Blanc and northward, but south of Cap Blanc it occurs mainly in winter and spring, because in summer the Azores high-pressure cell that drives the equatorward winds (Fig. 5.01) has moved further north. Hence primary production is maximal in the Cap Blanc region, where upwelling is from nutrient-rich SACW and occurs year-round. To the north it is lower because the nutrient content of NACW is less, while to the south it is lower because the upwelling season is limited. In the Cap Blanc region, and probably in several other areas, the annual production is thought to be about 730 g C m^{-2} y^{-1}, or 2 g C m^{-2} d^{-1} on average.

5.3.2 Upwelling and zooplankton

Up to this point we have been able to follow physical and biological processes on the 'event' scale, or meso-scale. Upwelling and primary production follow the onset of a strong wind within a few days. The next step in the food chain, the growth of zooplankton populations, is decoupled from these events because typical zooplankton organisms, such as copepods, require weeks rather than days to complete a life cycle. Trumble *et al.* (1981) have reviewed the available data from JOINT I on seasonal changes in zooplankton biomass. Adult copepods from deep water are upwelled in spring and begin to reproduce. Their offspring thrive on the abundant phytoplankton but tend to be carried offshore during the periods of strong upwelling. The intensity of upwelling falls off in autumn so the zooplankton are able to stay over the shelf and the populations reach their peak density. The annual mean value has been estimated at about 60 g m^{-2} (wet weight), with low values at about 40 and high values at about 120 g m^{-2}. It must be remembered that we are talking about a dynamic situation in which the abundance of phytoplankton is a function of the grazing pressure as well as other environmental variables, and the abundance of the zooplankton is determined partly by the grazing pressure of the fish stocks. The consensus seems to be that, since phytoplankton populations are able to respond rather rapidly to the favorable conditions provided by up-welling areas while the zooplankton respond only slowly, the phytoplankton production normally exceeds the consumption by the zooplankton. In other words, the zooplankton are seldom food-limited.

5.3.3 Upwelling and fish

The yearly cycle for fish species in the Cap Blanc region has also been summarized by Trumble *et al.* (1981). Seventy-five percent of the fish catch comprised only four types. Most abundant were clupeids (*Sardina pilchardus*, the sardine, and *Sardinella aurita*). Next most abundant were jack mackerel, *Trachurus* spp., and *Carynx rhonchus*. Redfish (Spari-dae) and other species followed. *S. pilchardus* tends to dominate in the cooler water in the northern part of the upwelling, while *S. aurita* occupies the warmer, more southern waters. Their ranges tend to be adjusted with the changing seasons, with northward migrations tak-ing place as summer approaches. In a study of the fish eggs and larvae present in the Cap Blanc region in spring, Palomera and Rubies (1982) found 65 different taxa, but 94% of all larvae examined were the sardine *S. pilchardus*. This heavy dominance of clupeids is char-acteristic of upwelling systems worldwide.

There is lack of agreement in the literature on what food is taken by the sardines and related clupeids, yet resolution of this question is needed for any calculation of a bio-logical budget for an upwelling ecosystem. Truly phytophagous fish have a gizzard and a very long, coiled gut. Because larval sardines and anchovies do not have either feature, Cushing suggested that they feed mainly on zooplankton and take phytoplankton inadver-tently. He conceded that first-feeding larvae may depend completely on algae. Muck et al. (1989) found that larval anchoveta in the Peruvian system have smaller mouths and longer guts than sardines, and appear to consume a larger proportion of phytoplankton. He sug-gested that this characteristic might confer a competitive advantage on sardines when phy-toplankton is scarce.

Mathisen *et al.* (1978) assumed that sardines fed mostly on small zooplankton while horse mackerel fed on larger zooplankton. Taking the biomasses of fish estimated acousti-cally during JOINT I, and some reasonable estimates of daily food requirements, these au-thors demonstrated that the estimated supply of food in the zooplankton was always in excess of the requirements of the fish stocks.

We have now followed the process through from the wind stress to the upwelling of nu-trients, to the bursts of phytoplankton production during the periods of slack winds, to the build-up of zooplankton populations and their consumption by fish. The end result, from the point of view of human welfare, is the landing of large quantities of fish. Ansa-Emmin (1982) summarized the fish landings for the northwest Africa upwelling area. The reported catches of all species in 1974 totaled 2.68 million tons. Almost 1 million tons were Clupeidae, 0.67 million tons being sardines. Over 0.5 million tons were Carangidae, the horse mackerels, and 0.2 million tons were squid. The other major catches were true mackerel (*Scomber* spp.), hakes (*Merluccius* spp.), and sparids. The catch was taken by a dozen industrialized countries, the USSR alone taking 287,000 tons of sardines and 55,000 tons of sardinellas, 360,000 tons of horse mackerel, and nearly 200,000 tons of mackerel. Clearly, upwelling areas of the ocean can be very productive indeed.

5.3.4 Regeneration of nutrients

No account of an ecosystem would be complete without reference to the process that closes the loop, nutrient recycling. While upwelling of 'new' nutrients is the distinctive feature of coastal upwelling systems, we should not lose sight of the fact that a proportion of the primary production that takes place makes use of nitrogen, usually in the form of ammonia, regenerated by benthic invertebrates, bacteria, zooplankton, and fish. Two mechanisms are thought to be at work: one in the water column, the other on the bottom. There are two main possibilities for phytoplankton in the surface waters over the shelf. They may be consumed by zooplankton or they may sink to greater depths, even to the bottom. If they are consumed by zooplankton their nitrogen may appear as fecal pellets (which sink) or as excreted ammonia. If the phytoplankters sink to the bottom they may release ammonia during decomposition, or they may be consumed by benthic animals, which in turn excrete ammonia. One way or another, a high proportion of the phytoplankton nitrogen ends up being released in the shoreward-moving lower layer of the water column. As a result, there is an accumulation of ammonia close to shore, which is exactly what has been found during the chemical investigations of JOINT I. Measurements of concentrations do not permit estimations of flux rates but Rowe *et al.* (1977) measured the flux of nutrients out of bottom sediments off Cap Blanc during JOINT I. The average total flux of nitrogen was 410 μg-at N m^{-2} h^{-1}, which at the time of measurement would account for 30–40% of the nitrogen required in the water column for photosynthesis. When we remember that there would be additional regeneration of nitrogen in the water column, it is clear that upwelling of new nitrogen is by no means the only method of stimulating primary production in the area, and regenerated nitrogen may account for more than half of the primary production. The two-layered cross-shelf flow acts as a kind of nutrient trap in much the same way as the two-layered estuarine flow discussed in Chapter 4. When primary production is stimulated by upwelling, much of the nitrogen taken up by the phytoplankton ends up being regenerated in the shoreward-flowing deeper layer and returns to the coastal waters to further stimulate primary production. Barber and Smith (1981) (quoting Whitledge, unpublished data) estimated that on the shelf off Cap Blanc the regenerated nitrogen comprises 72% of the total nitrogen, with 33% coming from zooplankton excretion, 24% from benthos and sediments, and 15% from pelagic fish.

5.4 COMPARISON WITH THE PERUVIAN UPWELLING SYSTEM

The Peruvian upwelling system at 15°S was studied intensively in 1976–1977 in multidisciplinary cruises known as JOINT II (Fig. 5.11). There are three important differences between the upwelling site off Peru and that off northwest Africa. The first is that the shelf off Peru is narrower (20 km versus 50 km) and drops off more steeply (200 m at the shelf break

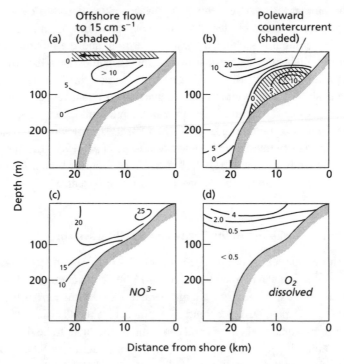

Fig. 5.11 Conditions off Peru, April–May 1977. From Brink *et al.* (1980) and Codispoti *et al.* (1982). (a) Onshore–offshore currents; (b) alongshore currents; (c) nitrate concentration; and (d) dissolved oxygen concentration.

versus 110 m). The second is that the deep water off Peru has higher nutrient concentrations (upwelled water contains 20–25 µg-at 1^{-1} of nitrate compared with 5–10 µg-at 1^{-1} off northwest Africa). Finally, the wind stress is less, and is more constant, averaging 0.79 ± 0.4 dynes cm^{-2} compared with 1.55 ± 1.0 off northwest Africa. The net result is that when the Ekman transport is active, offshore transport occurs mainly in the top 20 m, and shoreward transport is in an intermediate layer over the shelf, at depths of about 30–80 m, instead of in the whole of the water column down to the bottom, as it is off northwest Africa. Off Peru, the water close to the bottom is relatively still, and there is marked accumulation of organic matter to give a reducing sediment.

Two further differences are consequences of those listed above. One is that since the wind stress is less, the wind-induced mixing does not penetrate so deeply. During periods of strong upwelling, phytoplankton is still retained within the euphotic zone, which means that primary production is maintained at a relatively constant level, whether the wind stress is high or low. Second, the poleward counter-current, flowing beneath the equatorward coastal jet, is situated at intermediate depth over the continental shelf off Peru, whereas off northwest Africa it is located on the shelf slope (Fig. 5.08a). Figure 5.11 summarizes the features of the Peruvian upwelling system discussed above.

A detailed analysis of the primary production cycle in a segment of the Peruvian upwelling system at 15° S was published by MacIsaac *et al.* (1985). Because the upwelling is relatively constant it is possible to trace a distinct plume of cold water moving out from the coast and to recognize a number of zones along the axis of the plume. Zone I is the area of intense upwelling within about 7 km of the coast, where nutrients are abundant but phytoplankton biomass is relatively low. The phytoplankton cells are growing and taking up nutrients at rates considerably less than those of which they are inherently capable. In zone II the water column is stabilized by solar warming and the phytoplankton cells are found to increase their rates of nutrient uptake, photosynthesis, and synthesis of macro-molecules, a process known as 'shift-up.' Zone III is characterized by the rapid depletion of nutrients by the 'shifted-up' phytoplankton, so that there is a rapid accumulation of biomass and all processes occur at maximal rates. In zone IV nutrient depletion occurs, so that the cells experience nutrient limitation. In this environment 'shift-down' of physiological rates occurs. Using drogues to track water masses, MacIsaac *et al.* (1985) estimated that phyto-plankton cells moved from zone I to zone IV in 8–10 days, during which time they travelled 30–60 km away from the coast. The chlorophyll maximum occurred about 18 km offshore.

5.4.1 Interannual variability in the Peruvian upwelling system

As Fig. 5.11 shows, the source of upwelled water off Peru in March 1977 was at a depth of 30–60 m. The water was at a temperature of 15.5–16.5 °C, and contained 20–25 μg nitrate. This state of affairs was only part of a larger structure extending across the whole of the South Pacific (see Chapter 9). Trade winds blowing across the Pacific Ocean from east to west drive the south equatorial current and cause the mixed layer to be only 10–20 m deep in the eastern Pacific but up to 80 m deep in the western Pacific. In other words, the ther-mocline is tilted upward at its eastern end. Then the winds blowing from south to north along the coast of Peru cause cool, nutrient-rich water to upwell from below the thermo-cline. This state of affairs is normal.

It is now known that in anomalous years the trade winds weaken or reverse, and the thermocline off the coast of Peru sinks to a depth of about 100 m (see section 9.2.1). Ekman transport along this coast continues, but the water that is upwelled is now much warmer and not rich in nutrients. As a result, there is a sharp reduction in the biomass and productivity of the phytoplankton. The phenomenon, originally recognized from the warm water anom-aly that typically begins about Christmas time, was referred to as El Niño, which means the Christ-child. More recently it has come to be referred to as an El Niño–southern oscillation (ENSO) event. The physics of the southern oscillation will be dealt with in Chapter 9, but we should note that the Peruvian upwelling system seems to be uniquely vulnerable to such drastic changes, so that its interannual variability in productivity is very great. Sea surface temperature anomalies were noted in 1965, 1969, 1972, and 1976 (Barber *et al.* 1985) but in 1982–1983 an ENSO event occurred with a severity that is considered very rare, occur-ring with a periodicity of 100 years or more (Rasmusson and Wallace 1983). Barber *et al.* (1985) showed that at the height of the 1982–1983 anomaly, in May 1983, the upwelling waters were at 29 °C instead of the usual 16–18 °C, and mean primary productivity was only 10 mg C m^{-3} d^{-1}. Two months later conditions had returned to normal and mean pri-mary productivity was 219 mg C m^{-3} d^{-1}.

5.4.2 Total primary production in the Peruvian upwelling system

Many attempts have been made to calculate the total primary productivity of the Peruvian upwelling system, and to compare this figure with the total fish production. A fairly recent attempt is that of Chavez and Barber (1987). They give the references to earlier calculations, which differed widely among themselves. One of the main reasons for the great discrepancies was the lack of agreement about the area of ocean influenced by the upwelling. For example, Cushing (1969) used 479,000 km^2 in his calculation while Ryther (1969) used 60,000 km^2. Chavez and Barber (1987) argued that the offshore dimension of coastal upwelling is limited by the Rossby deformation scale (section 5.2.3). From Eq. 5.15 we see that the formula for this scale is $(g'H_0)^{1/2}/f$. Thus the radius is a function of the Coriolis force f (which varies with latitude), the depth H_0, and the vertical density gradient. Taking these factors into account Chavez and Barber (1987) calculated that the width of the Peruvian upwelling system varied from 270 km at 4° S to 60 km at 18° S. Calculating the width at degree intervals they arrived at an area of 182,000 km^2, intermediate between the two values quoted above. The mean of the large number of determinations of primary production in 1983–1984, after the recovery from the El Niño event, was 2.28 g C m^{-2} d^{-1} or 834 g C m^{-2} y^{-1}. This value converted into a total production of 1.52×10^{14} g C y^{-1}. With increased availability of satellite pictures of chlorophyll distribution, it should be possible to refine the estimate of the average area affected by upwelling, and determine its variability.

5.4.3 Secondary production in the Peruvian upwelling system

Cushing (1971) described the fish and birds of the Peruvian upwelling as follows:

> Anchoveta . . . live in the upper part of the thermocline and presumably migrate towards the surface at night. On the bottom live flatfish in rather shallow water, and near the bottom in deeper water live hake and rosefish. The hake probably migrate upwards at night and feed on euphausids. Between the anchoveta and the hake live the horse mackerel, possibly in the lower parts of the thermocline and below. At the surface live guano birds which are surprisingly abundant. In the Peru current there are 30 million birds, mainly pelicans, boobies and cormorants; during El Niño, the numbers are sharply reduced, because some migrate southwards out of the area when the fish are no longer accessible and some die of starvation. In 1958 during El Niño, the bird populations were sharply reduced but by 1962 the numbers of birds had fully recovered. So they must eat very large numbers of anchoveta.

It is clear that the same types of fish are found in the Peruvian system as in the northwest Africa upwelling system. Cushing (1971) went on to point out that the fish community is a specialized one, and the species are not caught much outside the upwelling areas. The sardines and anchovies usually spawn in the areas of most intense upwelling, close to shore, and Cushing speculated that both the juveniles and the adults make use of the two countercurrents (onshore and poleward) to maintain themselves within the upwelling system. As with the northwest African system, there is lack of agreement about whether the anchovies and sardines are predominantly herbivorous. Walsh *et al.* (1980), summarizing earlier work, concluded that for the Peruvian anchovy, *Engraulis ringens,* larvae of 0.3–0.9 cm consume 'some phytoplankton,' while larvae of 0.9–3.0 cm eat mainly zooplankton. In juveniles from 4.6 to 12 cm 'increasing amounts of algae are found' and adults over 12 cm long are almost entirely phytophagous. It is therefore not surprising that the drastic reductions in

Fig. 5.12 Heavy line, anchoveta catch by Peru; $T°C$, temperature anomaly of the eastern Pacific; Soy, the cost of soya meal ($ ton^{-1}). After Barber *et al.* (1985).

primary production associated with an ENSO event should have a major effect on the biology of anchovies.

Barber *et al.* (1985) showed (Fig. 5.12) that the years of temperature anomalies (El Niño years) were associated with reduced landings of anchoveta. There are several possible explanations: the adults could have starved for lack of phytoplankton food, the fish could have migrated away from the areas where they are usually caught, or the larvae could have failed to survive through lack of both phytoplankton and zooplankton. There is some evidence for each of these hypotheses. Barber *et al.* (1985) report that anchoveta seek out the upwelling areas by exhibiting a preference for water of 16–18 °C. As an El Niño builds, the cool water is found only in certain more persistent centers of upwelling and the anchoveta concentrate there. In 1982–1983 they were found in large numbers in isolated upwelling centers surrounded by warm water on three sides and land on the fourth. When the cool pockets of water were overridden by the anomalously warm water at the peak of the event, the anchoveta died and many were found floating. There were reports of anchoveta having moved to cooler, deeper water at about 100 m, but such water has very little phytoplankton and it seems likely that the fish would not survive there long.

On the question of migration, Valdivia (1978) reported a rapid southward migration of the stocks over about 180 nautical miles in less than 15 days during 1975, at a time when the 1976 El Niño was building. Barber *et al.* (1985) and others have argued that the anchovy stocks had presumably evolved an ability to survive ENSO events, since there is evidence from sediment cores of a dominance of anchoveta off central Peru for more than 2000 years. Migration south to find cooler waters may well have been one of the strategies for survival.

The question of larval survival is a complex one. The anchovies spread their spawning over 7–8 months of the year, presumably as an adaptation to the occurrence of unfavorable conditions at particular times and places. However, there are two peak periods: the austral winter–spring spawning (July–September) and the summer spawning (February–

March) (Valdivia 1978). Walsh *et al.* (1980), investigating the survival of larvae off the northern coast of Peru during winter, found that survival was higher when dinoflagellates in relatively high concentrations were available to the first-feeding larvae. Strong wind events had the effect of dispersing the dinoflagellate concentrations and adversely affected survival. When comparing the northern waters with those further south, they found that the strong wind events were more common in the south and these waters were therefore less favorable to the survival of larvae. Under El Niño conditions, more of the spawning adults were found in the south, and their reproductive success was judged to be lower on this account. Pauly and Tsukayama (1987) conducted a detailed study of historical records of the northern and central stocks of anchoveta, to see whether dispersal of dense layers of food by storms could be used as an explanation for poor recruitment of anchoveta. They concluded that they could not. The relationship between the distribution of food for first-feeding anchovy larvae and their subsequent survival has been investigated in detail by Lasker (1978) and will be discussed in connection with section 5.5 on the California current.

Barber *et al.* (1985) commented on a situation where there was a superabundance of dinoflagellate food for the larvae. At the onset of the 1976 El Niño there was a bloom of the dinoflagellate *Gymnodinium splendens* along 1000 km of the coast from March until the end of May. It was attributable to the stabilization of the water column as the warm water invaded the area. While it provided an excellent feeding environment for the early larvae, it apparently was not a suitable food for the adult fish, which had a reduced fat content, a reduced weight at a given length, and reduced length at sexual maturity. The 1977 recruitment of fish spawned in 1976 was extremely poor and the stock along the Peruvian coast fell to the lowest levels ever observed.

5.4.4 Exploitation of the Peruvian anchoveta stocks

As Fig. 5.12 shows, the ENSO events of 1965 and 1969 led to minor decreases in the landings of anchoveta, but the 1972 event was associated with a drastic decline from which, in fact, the stock has never recovered. The history of the anchoveta fishery in Peru has been reviewed by Glantz (1985). For a century, beginning in the 1840s, the major industrial activity along the Peruvian coast had been the mining of the bird-droppings, guano, from the rocky islands. Large populations of fish-eating birds are characteristic of upwelling populations worldwide, and prominent white accumulations of droppings at their roosting sites are an inevitable concomitant. As Cushing (1971) remarked, it is no accident that upwelling areas commonly have a Cabo Blanco, Cap Blanc, or Cape Blanc. The guano of Peru was mined and exported for fertilizer to many parts of the world. Beginning in the 1950s, a lucrative industry to harvest the anchoveta and convert them to fish meal was developed in Peru. The landings increased rapidly to a peak of about 12 million tons in 1970, then dropped to 2–3 million tons for a few years. Since 1977 the catch has hovered around 1 million tons. In fisheries circles a debate has centered around the question of whether the population crash is part of a natural cycle of events that has occurred many times in the past, or whether it is primarily the result of gross overfishing.

In discussing the issue, Barber *et al.* (1985) pointed out that during the period of low anchoveta abundance there were increases in the abundance of sardine, jack mackerel, and mackerel, suggesting a possible natural shift in species dominance. (We shall see in section 9.3.2 that low offshore Ekman transport appears to favor these species.) However, the sediment records do not support the view that there has been dominance by sardine and mackerel for significant periods in the past. In 1975 conditions for anchoveta reproduction were excellent, the index of recruitment was one of the highest ever recorded, and by early 1976 the anchoveta stocks were estimated at 8.3 million tons. However, 1976 was the year of the *Gymnodinium* bloom mentioned earlier, and it was followed by an ENSO event. Then, from 1977 to 1982 there was an anomaly in the global pressure system that resulted in a weakening of the southeastern Pacific trade wind system and a decrease in the thermocline tilt, which in turn meant poorer nutrient conditions in the Peruvian upwelling system. According to this view, anchoveta went into the 1982–1983 ENSO event at a low population level because of the 1976 event followed by 6 years of unfavorable environmental conditions. On the other hand, Barber *et al.* (1985) pointed out that by 1984 the levels of primary production in the upwelling system were among the highest ever observed, but the anchoveta have not responded by a large increase in stock size.

The explanation for this observation could be the predation on anchoveta by mackerel and horse mackerel. Muck and Sanchez (1987) reviewed historical data on stock size, stomach contents, and energetics of these predators and concluded that their consumption of anchoveta could at times equal or exceed the commercial fish catch. They therefore concluded that years of high abundance of mackerel and horse mackerel could explain, at least in part, years of low abundance of anchoveta.

At this point in the discussion Barber *et al.* (1985) drew attention to the model of Steele and Henderson (1984), which has two components. The first is a population model with multiple equilibrium states, the second is an environmental variable that interacts with the population model. This environmental variable is made to vary stochastically, and the variance increases as the periodicity increases from days to decades, a pattern that conforms with observed and theoretical values. The Steele and Henderson (1984) model predicts rapid changes in abundance of pelagic fish stocks with intervening quiescent periods of about 50 years. Increased fishing pressure leads to an increased frequency of the marked changes in abundance.

While the question is not yet fully resolved, there seems to be both theoretical and empirical evidence for thinking that the low abundance of anchoveta since 1977 has been primarily determined by unfavorable environmental conditions, but that the vulnerability of the stocks to these influences was increased by the heavy fishing pressure to which they were subjected. The failure of the anchoveta to return to their former dominance, even when the conditions became extremely favorable in 1984, may reflect the greater abundance of sardine, jack mackerel, and mackerel in the pelagic system. There have been analogous shifts in relative abundance of pelagic fish in other upwelling systems. A well-known textbook example is the rapid decline of the sardine populations and the rise of anchovy stocks that took place in the California current in the 1940s and 1950s. This event will be discussed in the next section.

5.5 THE CALIFORNIA CURRENT SYSTEM

As we see from Fig. 5.01 there is a northern hemisphere analogue of the Peruvian upwelling system. It is driven by prevailing northerly winds, and upwelling occurs along the Pacific coast of the United States from the Canadian border south to Baja California and beyond. This upwelling is the California current system. Detailed studies of upwelling events have been made, particularly off Oregon, and a comparison of northwest Africa and Oregon was made by Huyer (1976). In several respects the situation off Oregon resembles that off Peru. The poleward undercurrent appears over the shelf as well as the slope, and the shoreward flow is strongest at mid-depths over the shelf. Upwelling events are less strong and of shorter duration off Oregon than they are off northwest Africa.

Bakun (1973) examined records of wind strength over the ocean off northwest United States over a period of 20 years. He calculated the wind stress from the formula

$$\tau = \rho_a C_d v^2$$

where ρ_a is the density of the air, taken as 0.00122 g cm^{-1}, and the drag coefficient C_d was taken as constant at 0.0026. He then computed the Ekman transport, M, from wind stress, τ, and the Coriolis parameter, f, according to the equation

$$M = \tau/f$$

Finally, he integrated the data to give the Bakun upwelling index, a 20-year average of monthly mean Ekman transport for different parts of the coast, expressed as cubic meters per second per 100 m of coastline. The range is from 300 m^3 s^{-1} (offshore direction) to -212 m^{-3} s^{-1} (onshore). Some of the results are displayed in Fig. 5.13. It can be seen that the index indicates year-round upwelling off southern California, with stronger upwelling in summer, but off Oregon and Washington in the north there is strong downwelling in winter, and upwelling is confined to the period April–September.

Figure 5.14 shows that in summer there is a negative temperature anomaly, indicative of upwelling of cold water, all the way along the coast from Oregon in the north to Baja California in the south, with the exception of a warm anomaly off San Diego. The upwelling is obviously most intense between Cape Mendicino and Monterey. Yet when Bernal and McGowan (1981) analyzed a 21-year time series of biological and physical data from the California current, in which the upwelling occurs, they found no significant correlations with upwelling indices. There have been large positive and negative anomalies in zooplankton biomass, but these were best correlated with anomalies in the transport of cold, low-salinity, nutrient-rich water from the north. We shall show in Chapter 9 that these anomalies are correlated with changes in circulation patterns in the whole North Pacific.

Although the temperature anomaly in Fig. 5.14 shows an apparently uniform area of cold water, this area is an artifact of the method of calculation. The situation at any one time, as seen by satellite, is extremely complex, with coastal upwelling systems tending to be centered on topographical features such as capes and canyons, and with plumes of upwelled water extending far out into the California current. As in other coastal upwelling systems, the strength of upwelling is strongly dependent on wind speed and direction, and changes from day to day. Figure 5.15 shows the distribution of temperature and nitrate off Pt. Sur, California, on 9 June 1980, as inferred from satellite imagery supplemented by shipboard observations (Traganza *et al.* 1983). This situation is typical of an early phase of an upwelling event at this site. If the event persists for many days, interaction with the California current may give rise to a cyclonic structure about 100 km in diameter, with high biological production along the associated fronts, or it may extend into a plume up to 250 km long (Traganza *et al.* 1981, 1987). In summertime, patterns of this kind may be

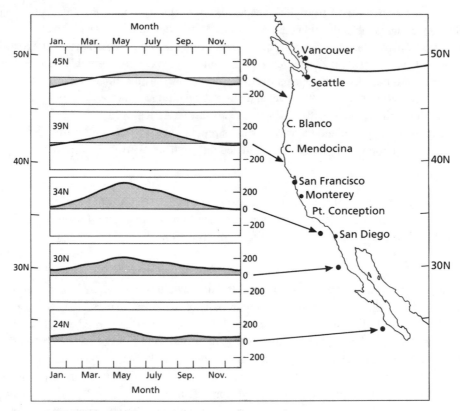

Fig. 5.13 Each plot on the left-hand side of the diagram is the upwelling index for each month of the year, averaged for the 20 years between 1948 and 1967. Each is for a different latitude, as indicated on the map on the right. Note that the strongest upwelling occurs at 34° N. In the southern half of the range the index is positive all year round, but at latitudes 39–45° N the index is negative in winter. Modified from Bakun (1973).

found in various stages of development or decline all the way from Oregon to Baja California. As winter approaches, the upwelling is progressively restricted to the southern portion of the region and in spring the region of upwelling spreads north again.

There seems to be no good synthesis of the biological productivity of this dynamic system. In the southern California bight, where upwelling is not particularly strong, Smith and Eppley (1982) found that the 16-year average for primary production was 0.402 g C m^{-2} d^{-1}, about 150 g C m^{-2} y^{-1}. Using the temperature anomaly at the Scripps pier as an index of upwelling, they found that the highest daily rates were associated with the maximum amount of upwelling, and vice versa. In the California current itself Hayward and Venrick (1982) found wide variation in the biomass and productivity of phytoplankton. Carbon fixation rates varied over an order of magnitude, from about 20 to nearly 200 mg C m^{-2} h^{-1}. These rates might be roughly converted to a range of 0.2–2.0 g C m^{-2} d^{-1}. These differences reflect the heterogeneous nature of the California current, with its admixture of advected and upwelled water. In trying to calculate the carbon flow from primary production to the pelagic fish stocks, Lasker (1988) used a value of 0.5 g C m^{-2} d^{-1} for the southern California bight and adjacent California current, but commented: 'It also seems likely that to the north, off central and northern California, the highly energetic "jets and squirts" which move large quantities of cold, nutrient-enriched water offshore and become entrained

Fig. 5.14 Average coastal temperature anomaly (°C) in summer, calculated as the difference between each location and a smoothed reference temperature for offshore conditions at the same latitude. From Bakun and Parrish (1982).

in the southward moving California current . . . add more primary production to the habitat of the sardine than has been included in the gross overall average.'

5.5.1 Fish production in the California current system

As in other eastern boundary current systems, the most abundant fishes in the California current system are sardines, anchovies, hake, jack mackerel, and mackerel. Sardines, *Sardinops sagax,* were heavily exploited from 1916 to 1967. The peak landings were in 1936–1937 and exceeded 700,000 tons. The catch fell drastically in the 1950s and 1960s and in 1967 the California state legislature imposed a moratorium on the sardine fishery. The next most abundant species is the northern anchovy, *Engraulis mordax,* closely related to *E. ringens* of the Peruvian upwelling. Its numbers began to increase as the

Fig. 5.15 Temperature and nitrate distributions off Pt. Sur, California, from satellite and shipboard operations. Note the difference between actual field data and the averaged long-term trends in the previous figure. From Traganza *et al.* (1983).

Fig. 5.16 The major spawning areas and seasons of the three subpopulations of the northern anchovy off the west coast of North America. From Smith and Lasker (1978).

biomass of sardines declined, and some postulated that the two were in competition so that the decline of sardine stocks released resources for the anchovies. However, Soutar and Isaacs (1969) studied the 1850-year record of fish scales in the anaerobic sediments off California and concluded that northern anchovy scales were present in large numbers throughout the series, while sardine scales appeared intermittently for periods of 20–150 years, with absences that averaged 80 years in duration. They concluded that the two species were not in competition.

The sardines of the California current system are divisible into four stocks. Of these, the largest by far before overfishing was the one that spawned in the southern California bight and migrated to the upwelling areas off northern California to exploit the dense zooplankton stocks that are associated with the coastal upwelling. The anchovies also have several subpopulations (Fig. 5.16). The stock off Oregon spawns at about 44–46° N, mainly in July at the time of the northern upwelling. The central subpopulation spawns principally in the southern California bight. Eggs and larvae can be found throughout the year, but the peak abundance is in the spring, while the minimum is in the autumn. The fish remain in the southern California bight throughout their lives and in recent years this has been the largest stock. There is a southern stock off Baja California, for which peak larval abundance is from January to March.

It thus appears that the largest stocks of both sardine and anchovy spawn in the southern California bight. Upwelling is relatively weak and phytoplankton production is lower than in the California current proper. Bakun and Parrish (1982) have suggested that strong offshore flow associated with Ekman transport is likely to carry eggs and larvae too far offshore, to positions from which they may never return, and that the choice of the southern California bight for spawning area reflects a need to avoid areas of strong upwelling. They also suggested that areas of strong Ekman transport are areas where there are strong winds that may destroy the fine-scale strata of food organisms needed by first-feeding larvae. This idea, attributable to Lasker (1975), will now be examined in more detail.

5.5.2 The survival of first-feeding larvae

Anchovy eggs are most abundant in the southern California bight during February, March, and April (Smith and Lasker 1978). After the absorption of the yolk sac, they must obtain sufficient food to meet their metabolic requirements within 2–3 days or they will die. Hunter (1972) studied the feeding behavior of the larvae and found that the food capture success rate of 3-day-old first-feeding larvae is very low but increases exponentially with age. As a consequence, 3-day-olds require a very high density of food organisms in their immediate environment to survive. He estimated that it was at a minimum of about 1790 dinoflagellates per liter.

Lasker (1975) made a field investigation of the feeding conditions for anchovy larvae in the southern California bight. Eggs were taken from anchovies held in the laboratory and were incubated under controlled conditions. They were then taken to sea so that a ready supply of first-feeding larvae was available for shipboard experiment. In March and April 1974 he found at several stations that there was a marked chlorophyll maximum at a depth of 30–35 m, and that the most abundant organism was the naked dinoflagellate *Gymnodinium splendens* with a diameter of 40–50 μm, occurring at densities of 30–40 particles mL^{-1}. Larvae placed in waters from near the surface took very little food, but larvae placed in water from the chlorophyll maximum fed extensively. The experiments were repeated at stations where other kinds of phytoplankton dominated. It was concluded that spiny or chain-forming diatoms such as *Thalassiosira* or *Chaetoceros* did not stimulate the larvae to feed, nor did any phytoplankton of diameter much less than 40 μm. Finally, the larvae were stimulated to feed only when the phytoplankton was at a population density of at least 20–30 cells mL^{-1}. The latter fact was confirmed in a very striking manner. Toward the end of the study a strong wind caused a deepening of the mixed layer and obliterated the chlorophyll maximum containing the *Gymnodinium*. After that it was impossible to stimulate the anchovy larvae to feed with water taken from any depth in the water column.

Lasker (1978) then investigated the horizontal and vertical extent of areas in the southern California bight suitable for anchovy feeding. Between September and December 1974 suitable food was abundant within 5 km of the shore in a chlorophyll maximum layer. During January 1975, gyral circulation redistributed the water so that adequate concentrations of dinoflagellates could be found at some depth all throughout the bight. In February 1975 northerly winds strengthened and strong upwelling began. The dinoflagellate populations were gradually replaced by small diatoms on which the anchovy larvae would not feed. These conditions persisted through the early spring and summer months.

On the basis of Lasker's (1975, 1978) results we may conclude that physical factors have a strong influence on year-class success of northern anchovy. The first-feeding larvae rely on high concentrations of phytoplankton such as naked dinoflagellates of a size class close to 40 μm. These high concentrations occur mainly at the chlorophyll maximum that forms near the thermocline under relatively calm conditions in the southern California bight. The onset of strong winds at a critical time may mix the water column to such a depth that the chlorophyll maximum is dispersed, or the winds may set up strong Ekman transport so that the water in the southern California bight is replaced by upwelled water containing mostly small diatoms in relatively low concentrations. Either of these situations renders the environment unsuitable for successful feeding, growth, and survival of small anchovy larvae. To put it another way, while the presence of upwelling and the consequent high biological productivity creates optimum conditions for the growth of juvenile and adult sardines, it is the absence of strong upwelling in late winter and early spring in the southern California bight that provides the optimum conditions for survival of first-feeding larvae. The timing of strong winds and the onset of upwelling in this region may well be a major determinant of year-class strength.

To summarize our understanding of factors influencing anchovy and sardine production in the California current, we see that strong wind stress and associated upwelling, which are the most characteristic features of eastern boundary currents, appear in themselves to be detrimental to the success of the larvae. For good survival the larvae require a well-developed horizontal layer of high phytoplankton density, in which dinoflagellates are the dominant form. We have seen in Chapter 2 that diatoms, with their high sinking rate, tend to dominate in highly turbulent or newly upwelled water, while dinoflagellates, which are active swimmers, tend to replace diatoms in more stable water columns with lower nutrient concentrations. Hence, the optimum conditions for success of first-feeding anchovy larvae tend to occur in waters that have been free of strong upwelling and strong wind-mixing for a considerable period. Strong wind-mixing and upwelling at the wrong time can cause heavy mortality. Once past the critical early stages of development, the juveniles migrate toward the areas of intense upwelling and exploit the high biological productivity found there.

Cury and Roy (1989) attempted to quantify the turbulence conditions optimal for recruitment for the Peruvian anchoveta (*Engraulis ringens*), the Pacific sardine (*Sardinops sagax caerulea*), and West African sardines and sardinellas. They found that recruitment increased with upwelling intensity until the year-round average wind speed reached 5–6 m s^{-1}, and decreased for higher values. Thus, up to this value, the advantages of the increased productivity caused by upwelling-favorable winds more than compensated for the disadvantages of having the food layer dispersed by turbulence; above an average of 6 m s^{-1}, the added turbulence outweighed the advantages of the upwelling. The conditions favorable for high recruitment were called an optimal environmental window.

A similar conclusion was reached by Ware and Thomson (1991). Noting a 5-fold decline in the total landings of sardine, anchovy, and hake in the northeast Pacific between 1916 and 1942, and a parallel decrease in the average intensity of wind favoring upwelling, they calculated the average wind speed *during the peak spawning period* at various localities and found the optimum for recruitment to be 7–8 m s^{-1}. This finding is in fair agreement with Cury and Roy's (1989) figure of 5–6 m s^{-1} averaged over the year.

It is interesting to compare this conclusion with that reached in section 5.3.1, where we saw that upwelling provides the supply of nutrients to the euphotic zone but the relatively calm periods between upwelling events permit the development of high phytoplankton biomass—in effect a little 'spring bloom.' On a longer time scale, it is the season of active upwelling that brings nutrients to the surface waters, but it is the season of minimal wind stress and upwelling that provides favorable conditions for first-feeding anchovy larvae.

A second factor to be considered in relation to larval survival is the risk that strong offshore transport will carry the larvae away from the favorable coastal environment. The two mechanisms occur on very different time scales and have quite different characteristics. The destruction of a chlorophyll maximum layer can occur within a few days and is in effect irreversible within the period of development of a cohort of larvae. On the other hand, offshore transport in a system as complex as the California current can take weeks to transport the larvae far enough from shore that their chance of completing their life history is placed in jeopardy. Furthermore, the offshore transport may be part of a gyral circulation that eventually begins to carry the larvae back toward shore, or the larvae may be able to migrate down into a shoreward-moving counter-current. In our present stage of understanding, there is no way to integrate these two mortality factors, beyond pointing out that both sardine and anchovy appear to avoid the regions of strongest upwelling in their choice of spawning location. By spawning downstream of those regions they appear to minimize the hazards while placing their larvae in a location from which, with a short migration, they can benefit from the high productivity associated with coastal upwelling.

5.6 THE BENGUELA UPWELLING SYSTEM

The last of the four major upwelling areas located in eastern boundary currents is the Benguela system, to the west of southern Africa (Fig. 5.01). The physics, chemistry, and biology of this system have been thoroughly reviewed in recent years (Andrews and Hutchings 1980; Chapman and Shannon 1985; Shannon 1985; Payne *et al.* 1987, 1992). The last reference, a symposium volume entitled 'Benguela Trophic Functioning,' contains 26 papers in a section headed 'Physical Influences on Ecological Processes.'

There is a northward-moving current, the Benguela current, occupying a zone about 200 km wide off the west coast of southern Africa from Cape Town north to about latitude 15° S. It is the eastern part of the South Atlantic gyre and the driving force is the wind associated with the South Atlantic anticyclone. Since the system is in the southern hemisphere, a wind from the south blowing parallel with the coast gives rise to offshore Ekman transport. The temperature anomaly produced by the associated upwelling delineates the area occupied by the system, which is seen to be larger in winter than in summer (Fig 5.17). The differences in shape of the SST patterns in the two seasons are largely attributable to the fact that the upwelling centered on Luderitz and on Cape Columbine (near Cape Town) reach their maxima in summer, while the very major upwelling at Cape Frio reaches its maximum in winter.

However, as we saw in connection with the California current, the mean temperature anomaly obscures the complexity of the system. A satellite picture taken at any instant shows a complex pattern of whirls and jets, many of them determined primarily by the bottom topography. In some places the cold upwelled water is carried far out into the Atlantic, while

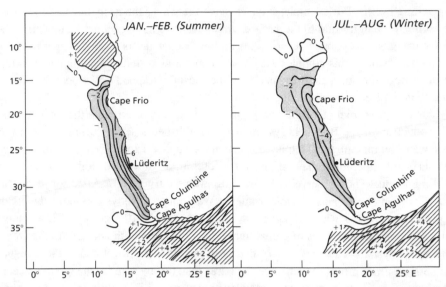

Fig. 5.17 Average sea surface temperature anomalies in the Benguela current system, in summer and winter. From Parrish *et al.* (1983).

at other places the upwelled water is confined to a relatively narrow band close to the coast. The picture is further complicated by variations in wind strength on a scale of a few days.

In the Benguela system it has been confirmed that optimal biological production occurs during an alternation of an upwelling phase, when the supply of nutrients in surface waters is replenished, followed by a quiescent phase during which stratification sets in and phytoplankton biomass increases. A 1–4 day lag occurs between the nutrient input and the blooming of the phytoplankton, and the bloom normally lasts 4–10 days. If the period of upwelling is too short, the phytoplankton will not have time to respond to the nutrients; if the interval between successive upwelling events is too long, the zooplankton will not have a continuous supply of food to permit optimum growth and reproduction. On biological grounds it would seem that a pulsing of upwelling with a duration of less than 5 days would be too short and one of more than 15 days would be too long. Ten days might be the optimum.

Nelson (1992) and Jury and Brundrit (1992) concerned themselves with the causes of variability in wind strength and upwelling intensity. Along the west coast of southern Africa an upwelling event is marked by north-directed upwelling-favorable wind, low sea levels, northward shelf currents, and declining coastal sea surface temperatures. Pulses of upwelling (as indicated by sea-level change) propagate southeastward along the west coast and continue eastward along the south coast, beyond Port Elizabeth, at speeds of 5–8 m s^{-1}. These wave-like features are generated by winds that alternately increase and decrease upwelling. Because the winds are associated with atmospheric high- and low-pressure systems moving eastward past southern Africa, the regions of enhanced upwelling also move around the Cape in concert with the wind systems.

The structure and speed of the upwelling features are similar to coastally trapped waves (Gill 1982). These are internal Kelvin waves which look like the waves pictured in Fig. 7.06, except that the horizontal scales are 30–60 km rather than the 1000 km or so of the waves associated with the tide in the figure. Because of the similarity, coastally trapped waves propagating south along the west coast of southern Africa can enhance or diminish the upwelling events. For example, at the crests of the waves the deep, nutrient-rich water is raised nearer than normal to the surface, which leads to more productive upwelling. At the wave troughs the opposite occurs as the nutrient-rich water is pushed deeper than normal. The net result of this complex system of interactions is that upwelling events with a duration close to the optimal 10-day period of the coastally trapped waves occur with some regularity in the Benguela upwelling system.

Waldron and Probyn (1992) used an innovative method to estimate the total new primary production of the Benguela system. They established a set of correlations between surface water temperature and nitrate nitrogen in the euphotic zone. Using satellite images of sea surface temperature, they estimated the total nitrate upwelled in the course of one 'upwelling event.' They observed 19 upwelling events during 1987. After converting the nitrate to carbon equivalents in photosynthesis, Waldron and Probyn estimated the annual new production of the whole Benguela system to be 4.7×10^{13} gC y^{-1}. This value is 0.6–1.4% of various estimates of global total new production in the ocean. As the area of the Benguela system is about 0.02% of that of the world's oceans, it makes the Benguela system 30–65 times more productive per unit area than the average for the world's oceans.

As in other upwelling systems worldwide, *Sardinops* and *Engraulis* are the most abundant fishes. The history of the fishery off southern Africa has parallels with the California fishery. Beginning in the 1950s, there was an intensive fishery for pilchard, *Sardinops ocellata,* with peak landings of over 1.3 million tons in 1968. The fishery then declined sharply and has not recovered. Attention was then turned to the anchovy, *Engraulis capensis,* which is now the most abundant species in the South African purse-seine fishery.

The stocks may be roughly divided into those breeding in the Cape area and those breeding further north. For the southern stocks, the main spawning ground for both anchovy and pilchards is the inshore water off the southern tip of Africa, centered on cape Agulhas and extending to Cape Town in the west and Cape Infanta to the east. The area is not particularly productive of plankton, but during the breeding season has a layer of warm Indian Ocean water at the surface that confers stability in an otherwise turbulent area. Starting at the western end of this spawning area is a frontal jet, the 'Good Hope Jet,' which runs northward along the shelf edge for more than 250 km (Fig. 5.18). Shelton and Hutchings (1982) showed that anchovy eggs and young larvae are carried in the jet, and appear as recruits in the upwelling system north of Cape Town. Note that this arrangement, where the anchovy spawning ground is 'upstream' of the highly productive upwelling area, is the reverse of that found in southern California where the spawning area is downstream of the intensive upwelling.

In this review of coastal upwelling systems we have seen three categories of fish stocks: (i) stocks poleward of summer upwelling maxima; (ii) stocks equatorward of summer upwelling maxima; and (iii) stocks equatorward of winter upwelling maxima. The anchovy stocks in the Cape Town area, those off the Columbia River in the northern part of the California current system, and the most southern (Talcahuano) stock in the Peru system are poleward of summer upwelling. The anchovy stocks in the southern California bight, various stocks in the Canary current system, and the Arica stock in Peru are equatorward of summer upwelling, their spawning grounds being located in bights with weak offshore transport. It is the most northern stocks off Peru, between San Juan and Chimbote, that are unique in lying equatorward of winter upwelling maxima, and in spawning in a region of strong Ekman offshore transport. Historically, these stocks have been the largest in the

Fig. 5.18 Spawning and recruitment grounds of the South African anchovy, and the jet current that connects them. From Shelton and Hutchings (1982).

world. Various explanations have been offered for the ability of the larvae to flourish in these conditions. The continental shelf is much wider than at other upwelling centers (section 5.4) and, being closer to the equator than the others, the Coriolis force is less, so it has a large Rossby radius and a greater Ekman depth. Hence, it is suggested, for a particular wind stress, that the offshore water velocity is lower than for other upwelling areas, and since the whole system extends further offshore, the larvae have more time to grow to an actively swimming stage before being advected out of the favorable upwelling system. Parrish *et al.* (1983) observed that for stocks poleward of upwelling areas the maximum biomass has been in the range of 0.1–0.4 million tons, for those equatorward of summer upwelling areas it has been 0.8–2.0 million tons, and for those that are equatorward of winter upwelling it has exceeded 10 million tons. It is not clear whether the location of the stocks in relation to upwelling centers is the key determinant of productivity, or whether latitude, with its attendant differences in solar radiation and Coriolis force, may not have more to do with the matter.

5.6.1 A 'Benguela Niño'

We have seen that the phenomenon known as El Niño in the Peruvian upwelling system involves movement of a thick layer of warm, nutrient-poor water into the coastal zone. Consequently, the water that upwells is not productive of phytoplankton, and productivity of the whole food web is impaired by El Niño. As explained in detail in Chapter 9, the immediate cause of El Niño is a weakening of trade winds, which allows warm surface water from the central and western Pacific to move toward the coast of South America. A similar event, in which a thick layer of warm, nutrient-poor water invades the upwelling area, is seen occasionally in the northern part of the Benguela system off Namibia. Taunton-Clark and Shannon (1988) showed that sea surface temperature anomalies had occurred in this region at least once per decade since the turn of the century, and considered that the term 'Benguela Niño' was justified. As Fig. 5.17 shows, the front between warm stratified southward-flowing waters of the Angola current and the cool northward-flowing Benguela current is normally at about 15° S. In a Benguela Niño, the front moves south about 600 km—sometimes as far as 25° S. Coastal surface waters are abnormally warm and saline to a depth of 50 m and to a distance of 150 km offshore. Abnormally heavy rains occur in Namibia, and marked changes usually appear in the abundance of fish, although the details are not clear. Several events have occurred one year after a Pacific El Niño, but the connection between the two is not clear, and the Benguela version seems to be caused by a southward shift in the intertropical convergence zone (see Fig. 3.06) rather than by changes in the strength of trade winds.

5.7 SOME SMALLER-SCALE UPWELLING SYSTEMS

5.7.1 Summer upwelling off Nova Scotia

Upwelling is thought to occur sporadically on the western margins of the major oceans, but few examples have been well documented. A study of summer upwelling off Nova Scotia, Canada, suggests that it is a major mechanism for transporting nutrients onto the shelf from deep water (Petrie *et al.* 1987). First reports of coastal upwelling in this area go back to Hachey (1935). He found that strong winds from the southwest were correlated ($R^2 = -0.7$) with lower than normal sea surface temperatures (SSTs) such that a wind of about 0.9 m s^{-1} would produce a drop in SST of 1°C. The observations of Petrie *et al.* (1987) are derived from infrared satellite images. Thirteen of them were available for the period 24 June to 6 August 1984. On 7 July most of the water over the Scotian shelf was around 16 °C except that there was a narrow band of cold water close to the coast. A week later there was a substantial area of water of 7–8 °C extending about 20 km from the coast on average with plumes extending up to 60 km from the coast. By 21 July the average width of the cold water zone was 20–30 km and plumes extended up to 85 km offshore. Four days later plumes extended up to 175 km from the coast. Between 25 July and 21 August the zone of cooler water disappeared.

The authors examined records of alongshore wind stress and found that the mean for the period 27 June–27 July was 0.03 Pa. They numerically solved the equations of motion for a two-dimensional, two-layer model of the shelf that included time-varying wind stress, representative bathymetry, bottom and interfacial friction, and the effects of varying layer depths. They chose an initial upper layer depth of 20 m, and a density difference of 3 kg m^{-3}. They began with zero wind stress on 27 June, forced the flow by the alongshore wind stress, and set the interfacial stress initially at zero and the horizontal eddy viscosity at 100 m^2 s^{-1}. The model predicted that surface waters would move offshore and the thermocline would reach the surface after 8.1 days. The infrared images were spaced about

7 days apart, but the upwelling first appeared on the image of 7 July, 10 days after the start of the winds. This observation was taken as good agreement with the model.

The pycnocline was found to upwell at a rate of 20 m in 10 days, over a coastal strip about 10 km wide and 500 km long. This finding gives a vertical transport of about $1 \times 10^5 \, m^3 \, s^{-1}$. At a mean nitrogen concentration of 10 μg-at N1^{-1} the authors calculated that the upwelling of nutrients amounted to 10 μg-at N s^{-1} per 1-cm-wide strip of shelf normal to the coast. This value is approximately equal to previously calculated nutrient requirements of the phytoplankton. It was therefore concluded that coastal upwelling is an important mechanism supporting primary production on the Scotian shelf.

The satellite images also showed that nearly stationary wave-like features with scales of 50–75 km appeared to grow rapidly in the upwelled temperature front. Results of modelling exercises suggested that these were caused by baroclinic instability. The properties of fronts will be discussed in detail in Chapter 6.

5.7.2 Summer upwelling on the west coast of Spain

The northwest corner of Spain, near Cape Finisterre, is near the northern limit of the upwelling associated with the Canary current system (Wooster *et al.* 1976). Blanton *et al.* (1984) calculated the Bakun upwelling index off Cape Finisterre and showed that in most years winds from the north predominate, causing strong offshore Ekman transport. The authors argue that offshore Ekman transport lowers the sea level at the mouths of the coastal inlets (rias) setting up a seaward pressure gradient that forces the water above the pycnocline out of the rias. Oceanic water is upwelled onto the continental shelf then enters the ria close to the bottom. Figure 5.19 shows data taken from the Ria de Vigo in 1977. The upper histogram shows monthly averages of upwelling indices. The offshore transport in the summer months (shaded) is clearly seen. When this occurs, cold (<12 °C), nutrient-rich (>5 μm NO$_3$) water is brought into the ria at 20–50 m depth. Blanton *et al.* (1984) also showed that the topography of Cape Finisterre had the effect of intensifying the upwelling.

Tenore *et al.* (1982) have documented the enormous biological productivity of these Spanish rias, especially in the form of cultured mussels. Two thousand 20 $\times$ 20 m rafts in Ria de Arosa produce annually over 100,000 tons total weight of mussels, *Mytilus edulis*. The upwelling and subsequent translocation of nutrient-rich water into the rias is believed to be the secret of this high level of productivity. Blanton *et al.* (1987) used atmospheric pressure maps to calculate the upwelling indices for a point off the western coast of Spain for the period April–September in each of the years 1973–1983. They found that the condition index of the mussels (measured as % solids in the meat) was well correlated with the upwelling index ($R^2 = 0.66$, Fig. 5.20). In the best year, 1977, the meat had 15.8% solids, while in the worst year, 1983, it had only 9.5% solids. The authors suggested that if a prediction could be made in the early spring about the most probable intensity of upwelling in the ensuing summer, the mussel culture could be managed more economically. They pointed out that the winds off Spain are influenced by the relative positions of the high-pressure atmospheric cell off the Azores (the 'Azores high') and the low-pressure cell off Greenland (the 'Labrador' or 'Icelandic low'). Long-range forecasts of synoptic-scale weather patterns could be used to provide the appropriate forecasts of pressure gradients and hence

Fig. 5.19 Physical and chemical data from the mouth of the Ria de Vigo, Spain (1978). From Blanton *et al.* (1987): (a) monthly averages of upwelling index. Positive values (shaded) imply offshore transport; (b) monthly averages of mean sea level; offshore transport corresponds with low sea level; (c) annual pattern of temperature distribution with depth (note strong stratification in June–October); and (d) annual pattern of nutrient distribution with depth, with high surface-nutrient levels during winter mixing.

Fig. 5.20 Correlation between mussel condition index (% solids in meat) and upwelling index. Numbers indicate years. From Blanton *et al.* (1987).

of wind strengths, Ekman transports, and the estimated carrying capacity of the rias for the ensuing season.

It seems highly probable that the productivity patterns of a wide range of coastal habitats are affected by the incidence of wind-induced coastal upwelling. We have evidence that it can occur on the western sides of ocean basins as well as in eastern boundary currents. The extent of the phenomenon appears to be a fruitful line of enquiry.

5.8 CONCLUSIONS

Reviewing this chapter, we see that major anticyclonic gyres in the atmosphere associated with similar gyres in the four major ocean basins—North and South Atlantic, North and South Pacific—give rise to equatorward winds parallel with the coast along the eastern margins of those basins. The combination of wind stress and the Coriolis effect determine that Ekman transport will take place—i.e., surface waters will move away from the coast. Deeper waters upwell along the coast to take their place, but the intensity of vertical transport is not uniform. It appears to be influenced by topographic features such as headlands and canyons, so that satellite images show complex patterns of intense upwelling, which in the course of time give rise to complex plumes, jets, squirts, and gyres that penetrate varying distances from the coast into the main stream of the boundary current.

This complexity is further compounded by temporal variation in wind strength. Off Peru the process may be continuous more or less year-round but off Oregon, for example, it is confined to the summer season. This seasonality is in part explained by the tendency for the major anticyclones to move poleward in summer and equatorward in winter. Superimposed on this seasonality is 'event'-scale variability in which wind stress and the associated upwelling event tend to build and decline over a period of 5–10 days.

Rather surprisingly, it seems that the event-scale variability tends to enhance biological productivity rather than the reverse. There is evidence from several locations that growth of phytoplankton populations (and also feeding by early larval stages of fish) occurs most rapidly when wind stress is low and the water column is well stratified. There seems to be a parallel with the 'Gran effect' discussed in section 3.3.2. In temperate waters it is the alternation between winter mixing and spring stratification that provides conditions required for the spring bloom. In upwelling areas the period of maximum wind stress provides for the upwelling of nutrients and the periods in between provide well-stratified conditions in which the phytoplankton is held in or near the euphotic zone, permitting high photosynthetic rate, rapid uptake of nutrients, cell division, and population growth.

We also saw in Chapter 3 that during well-stratified periods the phytoplankton have a tendency to accumulate in a chlorophyll maximum just above the pycnocline. It seems that these aggregations of phytoplankton cells are necessary for some kinds of zooplankton and for herbivorous fish to feed at their optimal rate. Hence, the times of low wind stress in upwelling areas are the times when strong chlorophyll maxima are formed and when the first-feeding larvae of sardine and anchovy encounter sufficient food for them to make the difficult transition from passive, yolk-absorbing creatures to active herbivores. The timing of their life histories is adapted to the average timing of the upwelling events. Any major disruption of the timetable leads to poor larval survival and hence a poor year-class of fish.

The fish stocks of upwelling areas worldwide are dominated by clupeids such as sardines, pilchards, and anchovies, with mackerel and hakes also being abundant. The clupeids, for the most part, tend to avoid the areas of strong Ekman transport at spawning time. Those in higher latitudes tend to spawn upstream of the major upwelling areas, while those in lower latitudes tend to spawn in sheltered bights with relatively little upwelling, located downstream of the major upwelling areas. Historically, the lower-latitude stocks have had the higher biomasses. The lower-latitude stocks off Peru have had the largest biomass of all, and these appear to be able to spawn right in the plumes of the major upwelling areas. It has been suggested that it is possible because in the Peru upwelling system, relative to the others, the Ekman transport is less intense and less variable, and extends further from shore.

6

Fronts in Coastal Waters

6.1 INTRODUCTION

It is obvious from the preceding two chapters that continental shelves are mosaics of water masses with different properties. The estuaries and estuarine plumes often have lowered salinities as a result of freshwater run-off, while adjacent waters may be fully saline. Some areas are strongly mixed by tidal currents while the next water mass is stratified. At the boundaries between them there are normally sharp horizontal changes in temperature and other properties.

The waters over the continental shelves are usually less saline than the open ocean, but more prone to be cold in winter and warm in summer. The shelf-break front marks the point of transition. Almost all fronts are regions of enhanced plankton production, which leads to higher fish production. Mooers *et al.* (1978) showed that the shelf-break front off eastern North America moves considerable distances under the influence of winds and the meanders of the Gulf Stream. A long-term plot of the positions occupied by the shelf-break front delineates an envelope within which are found most of the highly productive fisheries of the region. An aerial survey of the distribution of fishing vessels on the east coast of Canada showed the highest concentrations in the vicinity of the shelf-break front (Fournier 1978).

In the coastal upwelling situation that we described in the last chapter, upwelled water moves away from the coast by Ekman transport and eventually arrives at a region where the surface water will be pushed back no further, so the upwelled water sinks under it (Fig. 6.10). This frontal region is highly productive and one to which large numbers of tuna are attracted.

There is no agreement about the classification of fronts, but a partial listing of those occurring in coastal waters would be: tidal fronts (also known as shallow-sea fronts), shelf-break fronts, upwelling fronts, estuarine fronts, plume fronts, and fronts associated with geomorphic features such as headlands, islands, and canyons. In this chapter we investigate the distinctive physical and biological properties of each.

6.2 THE PHYSICS OF FRONTS

6.2.1 Definitions, examples, and causes

In the ocean, variables such as temperature, salinity, and density do not vary gradually with horizontal distance. Instead, we find large regions where horizontal variations are small, bounded by narrow regions where horizontal gradients are large. The narrow high-gradient regions are called fronts, following the meteorological custom of naming regions of high horizontal temperature gradient in the atmosphere. Fronts in the ocean come in many sizes. The fronts associated with the large western-boundary currents such as the Gulf Stream, for example, can be easily traced for thousands of kilometers, exhibiting a temperature change of 10 °C in 50 km near the surface and extending with diminishing contrast to the bottom at 4000 m. At the other end of the scale, small transient fronts are generated and destroyed by tidal flow in the upper layers of coastal waters every day. These fronts are typically a few kilometers long, confined to the upper layers and separate waters that differ by only 1–2 °C.

The infrared satellite image of the western North Atlantic Ocean shown in Plate 2 contains a number of examples of fronts. The warmest water in the picture is the orange and yellow water at the latitude of Cuba and Florida that flows north in the Gulf Stream—that is, the long crooked finger of yellow lying close to the coast past Cape Hatteras to the latitude of Chesapeake Bay where it turns eastward. On both sides of the flow, and especially to the west or north, the color changes rapidly to green and blue, indicating a rapid drop in temperature. From shipboard measurements, we know these temperature fronts are intimately tied to similar changes in salinity and density. Thus, a map of one variable is similar to the map of the others.

Across fronts like the one associated with the Gulf Stream, properties are distributed as illustrated in Fig. 6.01(a). Density is used in the illustration because it is the important variable when considering the dynamics of the flow. On either side of the current, in regions A and C, the isopycnals are flat and no horizontal gradient exists. In section B, where the geostrophically balanced current flows perpendicular to the page, the isopycnals rise from one side of the current to the other creating a horizontal gradient in density. Both the vertical and horizontal density gradients diminish with depth as the strength of the current decreases.

In Fig. 6.01(b), for comparison, is an illustration of the idealized density distribution across the typical tidal front discussed in section 4.6.1. In the tidally mixed front the density of the shallow-mixed water is constant with depth and intermediate in value between the surface and bottom density of the stratified side that lies in greater depths. The isopycnals therefore diverge across these fronts—some sloping up and some sloping down.

The front associated with the Gulf Stream is not the only one visible in Plate 2. In the area between the Gulf Stream and Nova Scotia, where the cold red-colored water lies, there are a number of situations where the color jumps from one fairly uniform value to another in a narrow band. Some of these features are tied to permanent currents that are guided over the continental shelf by the bathymetry. Others are signatures of the large-scale turbulent motions that slowly stir the waters of different temperatures being brought together in the region. The density structure in these fronts is similar, except for the size, to that found in the large ocean currents illustrated in Fig. 6.01(a).

Fig. 6.01 (a) Density section through a geostrophically balanced current in which the flow is out of the page (in the northern hemisphere) and decreases with depth. The current is confined to the region B where there exists a horizontal density gradient. In regions A and C, where the isopycnals are flat, there is no current. (b) Density section through a front created by an increase in the turbulent stirring in the shallower water. Region A is vertically stratified while region C is completely mixed by the turbulence. Horizontal density gradients exist in region B, the 'tidal front.'

Images similar to Plate 2 are now available from all parts of the world ocean. From these illustrations, it is clear that fronts are a ubiquitous feature of the sea surface. Some are created, as mentioned, by the large currents and tidal mixing, others by large turbulent eddies, or by the coastal upwelling discussed in Chapter 5. Tidal currents have also been found to cause fronts at the edge of the continental shelf where the flow of water interacts with the rapidly changing bathymetry. The large internal waves that are generated affect the temperature of the surface layer. This type of front, because it is closely associated with the tides, is discussed in the next chapter.

Fronts are important to us because higher biological activity is often associated with them, as has been documented by Pingree *et al.* (1975) and Le Fèvre (1986), who found dense aggregations of phytoplankton and zooplankton along tidal fronts in the English Channel. Our purpose in this section is to explore the physical processes that may account for this increased production. We are looking specifically for processes that can transport nutrients into the stratified euphotic zones where biological activity is fast enough to maintain growth. Our task, however, is not an easy one. The field is new, exciting, and changing rapidly and there is no consensus yet on what the dominant processes are in the different situations. Some of the processes advocated today will be shown to be unimportant in a few months as new observations provide a better understanding. Part of the difficulty is that fronts are relatively narrow features with high gradients. Observing the changes in these fields at the necessary time and length scales requires a large repertoire of highly sophisticated *in situ* instrumentation coupled with accurate navigation and powerful data processing.

6.2.2 Tidal fronts

Tidal fronts, as we discussed in section 4.6.1, sometimes occur over the continental shelves where the stratified water characteristic of the deeper areas lies beside unstratified water characteristic of shallower areas. The strong mixing found on the unstratified side of the front is a result of the high levels of turbulence generated at the bottom by the tidal currents, which tend to be faster in the shallower regions. A useful illustration of a typical tidal front and some of the processes associated with it is shown in Fig. 6.02. In this section the existence of the front is assumed and the problem is to find and explain the mechanisms that transport nutrients into the front to maintain the higher observed levels of primary production.

One mechanism often cited for this purpose and first suggested by Pingree *et al.* (1975) is peculiar to tidal fronts. It arises because the speed of tidal currents varies over the lunar month (28 days) between the spring and neap tides (Chapter 7). As the current speed increases, the energy available for mixing increases. This movement presumably causes the level of turbulence generated at the bottom to increase. The depth of water where the turbulence is energetic enough to break down the stratification will increase. Thus, the boundary between the mixed and stratified water will move toward deeper water, decreasing the area of stratified water. On the other hand, when the tidal currents decrease, the turbulence declines and the front moves toward shallower water, allowing the area of stratified water to increase again.

The newly stratified water behind the advancing front will contain nutrient levels characteristic of the previously mixed water. Observations of nutrients across these features (Simpson and Hunter 1974) show that nutrient concentrations tend to be higher in the mixed region than in the upper stratified layers, where presumably they have been depleted by biological activity earlier in the year. Thus, the newly formed part of the front should contain higher concentrations of nutrient than the water that has been stratified for a long time.

A simple calculation of the nutrient flux into the stratified side of the front by the tidal variation has been used by Loder and Platt (1985). They proposed that if the front moves L km between the neap and spring tides and the depth of the upper mixed layer (D) remains constant throughout the advance, then the amount of nutrient captured by the advancing

Fig. 6.02 Diagram of frontal structure and circulation, after Simpson (1981). Note strong along-front mean flow, convergence and downwelling at the front, upwelling on the well-mixed side, and frontal eddies, some of which close on themselves to form isolated patches.

front is just $L \times D \times \Delta C$, where ΔC is the increase in the nutrient concentration from the stratified to the mixed side of the front. If the front advances in a time T then the rate at which the nutrient is captured per unit length of front and per unit time is

$$F_m = L \times D \times \Delta C / T \tag{6.01}$$

where F_m is in units of mg-at m^{-1} s^{-1}.

The effectiveness of this process critically depends on the length of the frontal excursion (L). This excursion is examined by Simpson and Bowers (1981), who created a model to predict the position of the tidally mixed front at various times throughout the summer season depending on the variations in the current speed. Their first attempt predicted a lateral spring-neap movement three times the observed distance. This discrepancy was thought to be due to the fact that the efficiency with which the turbulence mixes the water changes with the degree of stratification. As the stratification increases it takes more mixing energy to stir up the water. Incorporating this effect decreased the excursion in the model closer to the observed values.

6.2.3 Eddies

The line on the surface of the water that indicates the presence of a front tends to be irregular in shape rather than straight, as is obvious in Plate 2. These irregularities often appear to be eddies that twist the cold and warm waters together in a circular fashion. Such features are thought to be created by instabilities in the front that allow small departures from the geostrophic balance to grow at the expense of the potential and/or kinetic energy in the front. These are the same processes generating the cyclonic lows (eddies) in the mid-latitude atmosphere that provide the variable weather and storms of those latitudes. The frontal eddies are of interest to us because it is often suggested (Pingree 1978; Garrett and Loder 1981; Loder and Platt 1985) that they cause water to be exchanged across the front and thereby contribute a significant flux of nutrients.

In an attempt to examine the potential effects of eddies on the biology of fronts, Pingree (1979) analyzed a number of eddies in the front found at the Celtic Sea shelf break. His analysis showed that the eddies had a diameter of 20–40 km and may have been created by baroclinic instability. His suggestion, which has been taken up by others, is that the eddies transport nutrients across the front and probably contribute to the enhanced biological productivity in the front. Building on the theoretical work of meteorologists, he derived the equation

$$F_E = \gamma (gD\Delta\rho/\rho)^{1/2} D\Delta C \tag{6.02}$$

to estimate the flux, by the eddies, of nutrients across the front. F_E is the flux of the nutrient C that exhibits a difference ΔC across the front, ρ is the density, $\Delta\rho$ is the density difference across the front, D is the depth of the upper layer, g is the acceleration due to gravity, and γ is a constant derived from theoretical considerations and equal to 0.0055.

An alternative method of calculating the nutrient flux due to eddies was suggested by Garrett and Loder (1981). In their scheme the nutrient flux is just the number of eddies times the eddy volume times the difference in the nutrient concentration, ΔC. The radius of the eddies is assumed to be four times the Rossby deformation scale, R_i (section 5.2.3), because this value is the length of the fastest-growing perturbation in baroclinic instability theory. The volume of an eddy of depth D is then $16\pi R_i^2 D$ and the supply of nutrient is $16\pi R_i^2 D \times \Delta C$. If eddies are exchanged across the front every $16R_i$ in time T, the nutrient flux is

$$F_E = \pi R_i \times D \times \Delta C/T \tag{6.03}$$

We shall return to this equation in section 6.3.3, where the relative magnitudes of the various processes are discussed.

6.2.4 Vertical eddy diffusion

The increase in the rate of vertical eddy diffusion from the stratified side to the mixed side of a tidal front may also be an important factor in the higher biological activity found in the front. The increase is caused by the increasing intensity of the turbulence toward the mixed side, which parallels the decrease in the vertical density gradient. Tett (1981), in a mathematical model, used this horizontal change in the vertical eddy diffusivity to explain the increased chlorophyll concentrations in the Ushant front observed by Pingree *et al.* (1975). An outline of the situation is shown in Fig. 6.03.

On the left of the figure, the water is homogeneous because it is kept well mixed by the high vertical diffusion created by the tidal stirring. On the right, the water stays stratified because the vertical diffusion across the pycnocline is low. In the front the vertical diffusion varies between the two extremes and will be higher than where the stratification is fully developed. If the nutrient concentration increases with depth, as observed by Pingree *et al.* (1975), the higher vertical diffusion rate in the front results in a greater upward flux of nutrient there than on the stratified side.

The model produces encouraging results by predicting chlorophyll distributions with many of the observed characteristics. It works because the vertical diffusivity is high enough in the front to supply the required nutrients but it is also low enough to keep the phytoplankton cells in the euphotic layer long enough for them to multiply.

Fig. 6.03 Density section through a tidally mixed front with vertical arrows indicating the increase in the rate of vertical eddy diffusion from the stratified side to the mixed side. At the bottom higher levels of turbulence help to transport nutrients from the cold, high-nutrient layer into the fully mixed region.

Vertical diffusion in the near-bottom waters is also likely to be an important consideration in the nutrient balance near tidally mixed fronts, especially those that have cold nutrient-rich lower layers on the stratified side (Le Fèvre 1986). These layers that are remnants of the winter conditions tend to contain higher nutrient concentrations than in either the fully mixed region or the upper layer on the stratified side. The turbulence created by the high tidal currents is most energetic at the bottom, and at the front is thought to stir the water enough to mix some of the cold high-nutrient layer into the fully mixed region, where it gets distributed throughout the water column (Fig. 6.03). Such a process occurs during each tidal cycle but may increase with the increasing currents of the spring tides. New investigations with instruments capable of measuring levels of turbulence in the water, coupled with detailed nutrient distributions, will be required to quantify this flux and estimate its importance.

6.2.5 Convergences and downwelling

From a ship the first indication of a front is usually a long narrow band of seaweed, plastic garbage bags and other materials on the surface, and feeding birds in the air. The water in the band is also quite smooth compared with the surrounding waters, indicating a concentration of floating organic matter on the surface that damps the small waves. These observations indicate that the surface waters are converging toward the front, bringing the floating material together, before sinking beneath the front. A well-documented visit to a front with a strong convergence is given by Pingree *et al.* (1974).

They described a rather violent tidally mixed front between Guernsey and Jersey in the English Channel, where direct observations both at the surface and underwater showed that the water on one side of the front was flowing toward, then underneath, the water on the other side of the front. The authors concluded from physical measurements that the front was not in geostrophic balance. This conclusion implies that forces in addition to those due to the rotation of the earth and the pressure gradient controlled the flow. This more complicated situation makes it harder to understand the physics of the front without more sophisticated measurements.

6.3 THE BIOLOGY OF FRONTS

6.3.1 Tidal (or shelf-sea) fronts

A concise description of the mechanism underlying the occurrence of shelf-sea fronts was given by Simpson (1981):

> Over the continental shelf tidal flows exert strong frictional stresses on the seabed. In stress terms, these tidal streams are equivalent to hurricane-force winds in the atmosphere blowing regularly twice a day. As well as playing an important part in the tidal dynamics, these stresses serve to produce turbulent kinetic energy which is responsible for vertical mixing.
>
> Variations in the level of tidal stirring divide the shelf seas, during the summer regime, into well mixed and stratified zones separated by high gradient regions called fronts.

The details of tidal waves and currents will be discussed in Chapter 7, but we may note in passing that most of the dissipation of tidal energy occurs on continental shelves, and the extensive shelf seas of northwest Europe are believed to dissipate one-eighth of the world's tidal energy, about 19×10^{10} J s^{-1} (Miller 1966). One of the earliest systematic studies of a shelf-sea front was Simpson's (1971) study of a persistent front in the Irish Sea. His record of the profile is shown in Fig. 6.04. As the diagram shows, there were tidally mixed waters at 13.6 °C on the eastern side of the front and stratified waters (>15 °C at the surface, <13.2 °C near the bottom) on the western side. As one passed over the front there was a change of surface temperature of 1.4 °C within about 2 km. In a later study of the same front, Simpson and Hunter (1974) reported that in August the nitrate-nitrogen content of the tidally mixed water was about 2 μg-at l^{-1} while that of the surface water in the stratified region was less than 1 μg-at l^{-1}. On the other hand, the deep water in the stratified region contained 7 μg-at l^{-1}. They estimated that the chlorophyll content of the mixed water was about 1.5 μg l^{-1}, while that in the surface water of the stratified layer was only 0.5 μg l^{-1}. A moment's reflection on these results (see Fig. 6.04) suggests that if there was a mechanism for mixing the nutrient-rich deep water from the stratified region into the mixed region, or a mechanism for mixing the somewhat nutrient-enriched water from the mixed region into the stratified surface layer, primary production could be expected to increase.

During the early 1970s biologists had been noticing that shelf-sea fronts were often the sites of persistent high densities of phytoplankton and zooplankton. Le Fèvre and Grall (1970), investigating the distribution of *Noctiluca* (a phytoplankter that, in high abundance, imparts a blood-red color to the water, known as a 'red tide'), noted that red tides occurred in a region off western Brittany characterized by convergence of surface waters with tidally mixed waters on the landward side and stratified waters on the seaward side. It was later shown that this frontal structure extends northward toward the coast of England, across the mouth of the English Channel, and is now known as the Ushant front. Pingree *et al.* (1975)

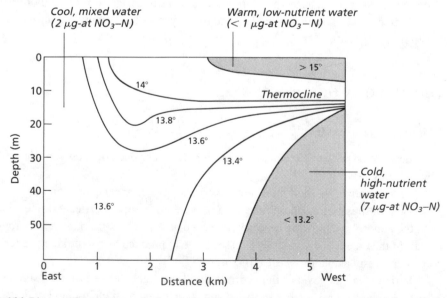

Fig. 6.04 Diagram of the structure of a front in the Irish Sea, based on data in Simpson (1971) and Simpson and Hunter (1974).

reported that there was high phytoplankton biomass along it, and that the most abundant organism in the phytoplankton was *Gymnodinium aureolum,* a species also noted for red tide formation.

Since the northwest European shelf area was the site of these pioneering studies of shelf-sea fronts, we may make this area the subject of our first case study.

6.3.2 Biological production at northwest European shelf-sea fronts

After the work of Le Fèvre and Grall (1970), the next biological study on the Ushant front was that of Pingree *et al.* (1974). They worked on a section of the front between Jersey and Guernsey and drew particular attention to a strong accumulation of floating seaweed, accompanied by numerous crustaceans, fish, and birds, along a line that was obviously a site of convergence and sinking. The authors drew attention to the possibility that high biomass of phytoplankton at a shelf-sea front might be as much due to convergence as to stimulated growth of phytoplankton at the front. The physics of this situation was discussed in section 6.2.5.

On the next cruise Pingree *et al.* (1975) made a series of observations across the Ushant front in the mouth of the English Channel and to the west of Brittany. On each crossing of the front they encountered a chlorophyll peak. A detailed section (Fig. 6.05) showed that there was a persistent chlorophyll maximum from the surface down to the pycnocline on the stratified side of the front, and a subsurface chlorophyll maximum in the pycnocline some distance behind the front. The suggested explanation was that during the spring tide period of maximum tidal currents the front is pushed back into the previously stratified area, so that nutrients are brought to the surface. When the tidal currents relax during the neap tides, the waters close to the front that have previously been tidally mixed become stratified and there is a burst of phytoplankton growth. Simpson and Pingree (1978) obtained similar results. They also obtained records of internal wave activity in the stratified region and explained the subsurface chlorophyll maximum in the pycnocline just behind the front as resulting from vertical transport of nutrients across the pycnocline during the passage of internal waves.

In the earlier study (Pingree *et al.* 1975) measurements had been made of the parameters of primary production. In addition to chlorophyll, the researchers measured the light extinction coefficient, the assimilation index (rate of carbon fixed per milligram chlorophyll-a), and integrated water column primary productivity, which they expressed relative to a value of unity for the vertically mixed regime (Table 6.1). In the stratified waters away from the frontal region they found that primary production below the thermocline was negligible, but primary production within the thermocline accounted for at least 50% of the total carbon fixed. In the frontal region with a weak thermocline, water column production was 6.5 times greater than in the tidally mixed area but it was achieved by having more than 40 times the concentration of chlorophyll, which had a relatively low assimilation index.

Subsequently Tett (1981) constructed a one-dimensional model based on Pingree *et al.* (1975). While acknowledging that several mechanisms for the enhancement of primary production might be at work, he singled out vertical mixing as the dominant factor. He modelled the upper layer of the stratified region as nutrient-limited, and the lower layer

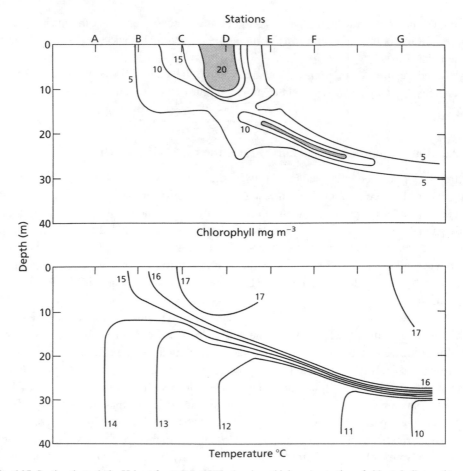

Fig. 6.05 Section through the Ushant front, July 1975, showing a high concentration of chlorophyll near the surface on the stratified side of the front and another high concentration lying just above the thermocline. After Pingree *et al.* (1975).

as light-limited. He considered that in the tidally mixed area the phytoplankton were light-limited because they experienced too low an average level of illumination during their vertical excursions. He considered that in the frontal region, where the pycnocline shallows, the phytoplankton were subject to minimal turbulent transport and were held in a region of moderate illumination from above and moderate rates of nutrient supply by diffusion from below. Model parameters were taken from relevant values in the literature, except that ver-

Table 6.1 Primary production data from Pingree *et al.* (1975)

	Mixed	Frontal	Stratified Wind-mixed layer	Thermocline	Bottom-mixed layer
Chlorophyll-a (mg m^{-3})	0.45	19.0	0.55	9.5	0.66
Extinction coeff. k(m^{-1})	0.12	0.34	0.10	0.1–0.3	0.10
Assimilation index (AI) [mg C h^{-1} (mg chl-a)$^{-1}$]	4.2	1.9	2.2	5.7	0.2
Relative production rate (P)	1.0	6.5	0.6	0.7–1.1	0.05

tical diffusivity was computed from observed vertical temperature gradients and appropriate values for the mean downward heat flux in summer and physical conditions were set to resemble the situation at a front near the Scilly Isles.

The model was used to simulate changes in stratified, mixed, and frontal stations for 40 days, by which time a steady state had been reached and it was found that the distribution of nutrients and chlorophyll closely resembled the natural situation at the front. In discussion, the author emphasized that his aim was only to show that it is *possible* to explain much of the distribution of chlorophyll at a front as a result solely of vertical mixing of nutrients and chlorophyll. He made no claim that this mechanism was the only process at work.

Savidge (1976) studied the Celtic Sea front and showed that as he steamed across it there was a change in surface temperature of about 4 °C within 6–7 km, and that there was a sharp peak of chlorophyll associated with it. He suggested that waters from each side of the front might be deficient in some important ingredient, not necessarily the same one, and that mixing of water types through the frontal structure might lead to enhanced primary production. By means of shipboard experiments he confirmed that, when water from either side was mixed with a relatively small quantity of filtered water from the opposite side, production rate doubled. The hypothesis has been expressed in the phrase 'complementation of waters of inherently different properties induces fertility.' It is rather more complex than the hypothesis of Pingree *et al.* (1975), in which the simple addition of nutrients to stratified water was enough to stimulate phytoplankton production.

On another cruise to the Celtic Sea front, Savidge and Foster (1978) encountered a more complex structure. They passed first through the main front. Just beyond this front was a zone of upwelling marked by a temperature minimum. Further away from the front lay a lens of warm water separated from the main mass of cool, tidally mixed water by minor but distinct thermal gradients. There were three thermal gradients in all, and chlorophyll maxima were associated with each one of them. On a later crossing of the front, strong winds had developed and the pattern had broken up into numerous patches of dense chlorophyll, each about 1.2 km across. As with the earlier cruise (Savidge 1976), Savidge and Foster interpreted the chlorophyll peaks as resulting from 'complementation of waters' by crossfrontal mixing.

Recalling the physical mechanisms discussed in section 6.2.3, we see (Fig. 6.02) that it is now understood that a typical shelf-sea front has a strong mean flow parallel with the front, giving rise to baroclinic eddies that often close in on themselves and break free, forming patches of cool water on the warm side, and vice versa. These patches are readily seen in satellite images. The results of Savidge and Foster (1978) are consistent with their having run a sampling transect through the front and through one of these isolated water masses. If such a water mass, originating from the cool side, eventually mixes with the nutrient-deficient water on the warm side, or vice versa, there will be a net transport of nutrients. This mechanism is one possibility for enhancement of primary production.

In a review of the biological implications of fronts on the northwest European continental shelf, Holligan (1981) pointed out that the distribution of chlorophyll and primary production at the Ushant front varies seasonally (Fig. 6.06). In late April and early May the Ushant front marks the landward limit of the spring bloom taking place in stratified waters offshore. As the season progresses, those diatoms that are not removed by grazing tend to sink and are replaced by flagellates (mostly dinoflagellates) that flourish in the boundary zone between the warm, wind-mixed layer and the cool, tidally mixed water. Finally, by late September, phytoplankton are again most abundant in the surface water on the stratified

Fig. 6.06 Distribution of chlorophyll-a (mg m^{-3}) across the Ushant tidal front on dates given. From Holligan (1981).

side of the front. The highest phytoplankton standing crops of the year are found when the flagellates bloom close to the stratified side of the front in July and August.

Holligan (1981) also pointed out that the depth of the mixed water on the Ushant front is 60–100 m. As a consequence, the growth of phytoplankton is light-limited and inorganic nutrients are never fully depleted. Fronts that form in shallow water have a different biological dynamic, for the phytoplankton in the tidally mixed water may not be light-limited, but more probably nutrient-limited. Pingree *et al.* (1978a) identified shallow fronts in the North Sea off Flamborough Head and in the area between the Scottish Island of Islay and the northern tip of Ireland. The tidally mixed areas appeared to have higher nutrient concentrations than the stratified regions, presumably because nutrients were being regenerated by the benthos. The enhanced chlorophyll concentrations associated with these fronts were once again ascribed to the alternation of mixing and stratified regimes present at certain places as a result of the alternation between neap and spring tides.

After presenting this review, Holligan and his colleagues made several intensive studies of the Ushant front (Holligan *et al.* 1984a, b). In late July and early August 1981, a large interdisciplinary team described the vertical distribution of the plankton at three stations representative of stratified, frontal, and vertically mixed regions. The phytoplankton was most dense in the frontal regions (26.5 g C m^{-2}) and was dominated by the dinoflagellate *Gymnodinium aureolum*. At the stratified station there was a relatively small biomass (0.42 g C m^{-2}) of small naked flagellates that were concentrated at a subsurface chlorophyll maximum. The biomass at the tidally mixed station was intermediate in value (7.91 g C m^{-2}) and consisted mainly of diatoms.

We may note in passing that these results are explicable in terms of properties discussed in earlier chapters. Diatoms, which tend to be larger than flagellates, have a marked tendency to sink, and flourish best under turbulent conditions that bring them back to the surface from time to time. The small flagellates in the stratified waters form a subsurface chlorophyll maximum close to the thermocline, where there is a small but significant steady vertical transport of nutrients, possibly supplemented by bursts of increased vertical transport under the influence of internal waves. It is also possible that the flagellates make use of their locomotory powers to migrate down through the nutricline to take up nutrients and back up again to reach improved illumination (Holligan 1981; see also Kamykowsky *et al.* 1988). Fogg (1985) reported a similar distribution of phytoplankton types in waters adjacent to a front in the western Irish Sea.

When turnover rates were calculated for the phytoplankton, it was found that the small flagellates in the stratified, nutrient-limited surface waters were turning over rapidly, estimated at 1.17 doublings per day, the diatoms in the mixed region were achieving only 0.16 doublings per day, and the flagellates in the frontal region were found to have a negative index of population growth. The authors estimated the vertical diffusion coefficient in the stratified region and the frontal area from observations on heat flux, then applied it to the observed nitrate gradient to determine the nitrate flux through the thermocline. Their results suggested that in the stratified region the small flagellates were obtaining about half their nitrogen needs from upwelled nitrate and the other half from recycling of ammonia. Results in the frontal region suggested that the vertical flux of new nitrogen was sufficiently high to have accounted for the accumulation of high biomass of *Gymnodinium* earlier in the season. They stated that the precise mechanisms that permit the accumulation of large biomasses of dinoflagellates in the frontal region are not yet clear, but suggested that one factor is that they appear to be subjected to low grazing mortality compared with other types of phytoplankton.

At each of the three stations studied, the standing stock of heterotrophs was between 2.3 and 3.2 g C m^{-2}, with bacteria comprising 10–30%. At the stratified station, though not at the others, the biomass of consumers exceeded that of the phytoplankton. Respiration data indicate that the meso-zooplankton were responsible for less than 10% of the oxygen consumption, and micro-heterotrophs were responsible for the greater part of biological energy dissipation.

6.3.3 What causes dense phytoplankton patches on shallow-sea fronts?

From what we have seen so far, there are two possible mechanisms to account for the high densities of phytoplankton commonly found on shallow-sea fronts: advection as a result of the convergent flows that have been observed at these fronts, or *in situ* growth of phyto-plankton populations stimulated by some mechanism or transport of nutrients into the mixed layer of the stratified zone adjacent to the front. The role of advection is still an open question, but the mechanisms of cross-frontal transport of nutrients are becoming better understood.

We have seen that the high biomass of phytoplankton is usually found in surface wa-ters on the stratified side of the front in spring and autumn, but in the thermocline in sum-mer. In Chapters 3 and 4 we saw that a spring bloom in surface waters is normally associated with the shallowing of the mixed layer to a critical depth (the Gran effect). This relationship will account for the bloom formation over the whole of the stratified region of the continental shelf but not for the elevated phytoplankton biomass and productivity adja-cent to the front. To achieve this increase by *in situ* growth requires that the supply of nu-trients be augmented. There are two possible routes for the nutrients to come, either horizontally across the front from the tidally mixed region or vertically from the nutrient-rich water mass below the thermocline. Let us first examine mechanisms of cross-frontal transport.

(a) Spring-neap tidal cycle

The mechanism most often cited by Pingree and his co-workers is the movement of the front during the spring-neap tidal cycle (see section 6.2.2). During spring tides the tidal currents are at their maximum strength, which means that tidal mixing extends to regions with deeper water—i.e., there is an erosion of the stratified region. During this phase, nutrients from below the thermocline will be brought to the surface in the area covered by the excursion of the front. As the tidal currents relax toward the neaps, the area of tidally mixed water decreases, permitting stratification to return in that same area covered by the excursion of the front. Figure 6.07 from Simpson and Pingree (1978) shows the movements of the Ushant front relative to the Brittany coast between neap and spring tides.

Simpson and Bowers (1981) documented the movement of fronts as recorded by satel-lite images. Much of the observed variability in position is due to tidal advection at the semi-diurnal frequency, so this factor was first removed. The residual indicated that the average spring-neap excursion for eight tidal fronts lying west of Great Britain was 4 km.

Loder and Platt (1985) made a rough calculation of the rate at which nitrate might be captured by the upper layer on the stratified side as it advances into the tidally mixed region. They used a variant of Eq. 6.01 and substituted 4 km for the tidal excursion of the front (Simpson and Bowers 1981), 20 m for the depth of the mixed layer, and 20 mg-at m^{-3} for the increase in nutrient concentration between the stratified and the mixed side of the front. From this equation they calculated the average supply of nitrate to the frontal zone as 0.12 mg-at m^{-1} s^{-1}. This value will later be compared with calculations for other methods of cross-frontal transport.

Similar results were obtained by Morin *et al.* (1993) using satellite-derived temperature data and directly observed correlations between temperature and nitrate content of the

Fig. 6.07 Temperature and chlorophyll distribution at the Ushant front. (a),(b) Neap tide conditions, 27–28 July 1976; (c),(d) spring tide conditions, 31 July–2 August 1976. Note that the high concentration of (d) chlorophyll is further west at spring tides. From Simpson and Pingree (1978).

water. They used temperature changes between spring-tide and neap-tide conditions to estimate the vertical transport of nitrate and the corresponding phytoplankton production as the nitrate declined. Their results corresponded well with sea truth data.

(b) Baroclinic eddies

It was shown in section 6.2.3 that the strong net flow parallel with a shelf-sea front gives rise to baroclinic eddies that are known from satellite observations to break free. A cool-water eddy introduced into the stratified side of a front will carry nutrients with it. Loder and Platt (1985) used two methods of calculation to arrive at an estimate of the rate of cross-front transport of nutrients by this mechanism. The first method follows Pingree (1979), who adapted a semi-empirical formula developed for poleward heat transport by baroclinic eddies in the atmosphere (Eq. 6.02). Taking $\Delta\rho/\rho = 6.7 \times 10^{-4}$, $\Delta C = 2$ mg-at m^{-3}, and $\gamma = 0.0055$, the average rate of supply of nitrate by baroclinic eddies, per unit length of front, is given by $F_e = 0.08$ mg-at m^{-1} s^{-1}.

The second method used Eq. 6.03, in which the nutrient flux transported by eddies breaking free from the front is given by the number of eddies times the eddy volume times the difference in nutrient concentration, ΔC. For the two methods to yield the same result, T in Eq. 6.03 would have to be set at 60 days. This period is an unreasonably long interval between the generation of eddies, judging by satellite images. Loder and Platt (1985) therefore suggested that the value of γ used (a constant derived from theoretical considerations) was much too low, and that the value of 0.08 mg-at m^{-1} s^{-1} for cross-frontal transport of nitrate may also be low.

(c) Residual currents

As we saw in section 6.2, the residual currents in a tidal front are primarily along the front, but a weak cross-frontal circulation induced by internal friction is also expected. Garrett and Loder (1981) provided formulae for this circulation and Loder and Platt (1985) evaluated them for the Ushant front. They arrived at a figure for nitrate transport of less than 0.01 mg-at m^{-1} s^{-1}, which is insignificant compared with the transport by the two mechanisms previously described. As we shall see later, work on the tidal front of Georges Bank leads to the view that the role of residual currents in bringing about a cross-frontal flux may have been underestimated.

In addition to the three methods we have just discussed, by which nutrients may pass horizontally across a front to fertilize the stratified low-nutrient water, there remains the possibility of vertical transport of nutrient-rich water from below the thermocline into the mixed layer above. Loder and Platt (1985) attempted a quantification of this term for comparison with the cross-frontal terms.

Vertical transport

Possible mechanisms for vertical transport in the vicinity of a tidal front were described in section 6.2.4. Loder and Platt used the equation

$$F_v = K_v(\Delta C/\Delta z)L_w \qquad (6.04)$$

where K_v is the vertical eddy diffusivity and L_w is the cross-frontal distance along which there is significant vertical transfer. Taking the value of $K_v = 10^{-4}\,m^{-2}\,s^{-1}$, which had been obtained by Pingree and Pennycuick (1975) from a stratified station, and $L_w = 10$ km they calculated the vertical nitrate flux as 0.2 mg-at $m^{-1}\,s^{-1}$.

Summarizing to this point, the Loder and Platt (1985) calculations suggested that reasonable estimates for the various types of transport of nitrogen into the mixed layer of the stratified water adjacent to the Ushant front were: spring-neap tidal cycle 0.12 mg-at N m^{-1} s^{-1}, baroclinic eddies 0.08 mg-at N $m^{-1}\,s^{-1}$, residual currents 0.01 mg-at N $m^{-1}\,s^{-1}$, and vertical transport 0.20 mg-at N $m^{-1}\,s^{-1}$. From this calculation they concluded that all the mechanisms considered might bring 0.40 mg-at N $m^{-1}\,s^{-1}$, which might be approximately equated to a carbon flux of 0.28 g C $m^{-2}\,d^{-1}$ in a 10 km wide frontal zone.

It is instructive to compare this calculation with those made by Holligan *et al.* (1984b). They concerned themselves with the vertical diffusion of nitrate up through the bottom of the thermocline from the mixed layer below. They first assumed that the vertical diffusion coefficient, K_v, was the same for heat and nitrate and constant throughout the thermocline. Thus the flux of heat, F_H, and nitrate, F_N, through the thermocline can be expressed by the equations

$$F_H = -\rho c\, K_v \Delta T/\Delta Z \qquad\qquad (6.05)$$
$$F_N = -K_v \Delta N/\Delta Z \qquad\qquad (6.06)$$

where ρ is the density of the water, c is the heat capacity of the water, and ΔT and ΔN are the differences in temperature and nitrate across the thermocline of thickness ΔZ.

From vertical profiles of temperature obtained at three-month intervals the authors estimated the downward flux of heat (F_H) across the thermocline and derived a value for K_v from Eq. 6.05. Substituting this value of K_v into Eq. 6.06 allowed them to calculate vertical fluxes of nitrate from measurements of the nitrate gradient.

Their main argument is that biological depletion of nitrate in the mixed layer and thermocline causes an increase in the vertical gradient of nitrate at the base of the thermocline— i.e., in the nutricline. This higher gradient must, if the vertical diffusion coefficient stays the same, lead via Eq. 6.06 to a greater upward flux of nitrate.

Holligan and co-workers performed this calculation at a number of stations. At one station a chlorophyll maximum attributable to flagellates was observed in the middle of the thermocline and the gradient of nitrate in the nutricline was not particularly steep. As a result, the calculated vertical transport of nitrate was relatively low, 3.22 mg nitrate-N cm^{-2} d^{-1}. On another occasion there was a chlorophyll maximum attributable to the dinoflagellate *Gymnodinium aureolum* near the base of the thermocline producing a high vertical gradient in nitrate. This higher gradient led to an estimated vertical flux that was much greater than in the first case, 22.4 mg nitrate-N $cm^{-2}\,d^{-1}$.

Similar calculations were carried out for observations in a tidally mixed front that resulted in a maximum value for the vertical flux of nitrate of 39.9 mg nitrate-N $cm^{-2}\,d^{-1}$, equivalent to 'new' carbon production in excess of 3 g C $m^{-2}\,d^{-1}$. Holligan and co-workers concluded that dinoflagellates can increase the vertical nutrient gradient and hence the vertical nutrient flux; however, their flux values are about 10 times those calculated by Loder and Platt (1985). This difference may arise because their estimate of K_v is an average value based on temperature changes in the bottom mixed layer over 90 days while the nitrate gradient was obtained over about 1 hour. Direct measurements (Gargett 1984) indicate that K_v

is not a constant because it must vary continuously with the level of turbulence and stratification. The value of K_v concurrent with the high nutrient gradient observed by Holligan *et al.* (1984b) could have been much lower than the 90 day estimate and unable to provide the high nutrient flux the authors calculate.

We shall now return to the question of the role of residual currents in cross-frontal transport. In particular there have been studies at the Georges Bank front, which is the most extensively studied frontal system after the northwest European shallow-sea fronts.

6.3.4 The Georges Bank frontal system

Georges Bank is a major submarine bank lying off the eastern coast of North America between Cape Cod and Nova Scotia. Its characteristics have been extensively reviewed in Backus and Bourne (1987). Over most of the bank the water depth is 50–60 m and the tidal currents are sufficiently strong to keep the water column vertically mixed throughout the year. Around the bank the combination of freshwater run-off and solar heating causes persistent stratification throughout the spring and summer. As a consequence, a tidal front develops around the central bank, between the mixed and stratified water. (There is, in addition, a clockwise gyre driven primarily by rectification of the strong tidal currents over the sides of the bank, but this condition will be treated in Chapter 7.)

Biological productivity of the waters over the bank is high, and there is intensive fishing of both finfish and shellfish. Since there is a substantial loss of nitrogen associated with removal of commercial catches, it is important to understand the mechanisms for supply of 'new' nitrogen to the top of the bank from deeper water. Work on this topic up to 1988 has been reviewed by Horne *et al.* (1989).

The position of the front is in rough agreement with the prediction made by Loder and Greenberg (1986) using Simpson and Hunter's (1974) formulation (see Chapter 4). From observations made in 1985 it is deduced that the front on the northern and northwestern side of the bank makes a twice-daily excursion of 10–15 km as a result of changes in tidal current velocity. Records from a depth of 33 m at a fixed station in the frontal zone showed semi-diurnal fluctuations not only in current velocity but also in temperature (approximately 7–15 °C on tides of maximum amplitude). From other data sets it was found that there was a good correlation between nitrate concentration and temperature, and it was possible to construct a time series for variations in nitrate concentration with time, indicating semi-diurnal fluctuations between about 1 and 10 mg-at m^{-3} on spring tides at the 33 m-deep fixed station.

The current meters at the fixed 33 m station showed the expected gyral current parallel with the front, but also a cross-frontal current averaging 0.05 m s^{-1} at 13 m depth and 0.03 m s^{-1} at 33 m depth, both directed toward the tidally mixed area on the top of the bank. Use of this information in conjunction with the fluctuating nitrate concentrations enabled Horne *et al.* (1989) to calculate a depth-integrated flux of nitrate through the front toward the center of the bank of $\simeq 12$ mg-at N m^{-1} s^{-1}. We note that this value is about 30 times the flux calculated by Loder and Platt (1985). Horne *et al.* (1989) then went on to consider what physical mechanisms might be responsible for this flux.

They first considered shear-flow dispersion. This explanation would work approximately as follows. A parcel of nutrient-rich water moving onto the bank with the tidal flow and returning half a tidal period later might have been subjected to vertical mixing with nutrient-poor surface water and return with its nutrient concentration reduced. Examination of the data showed that this factor was unlikely to be a major component of the observed flux.

It is instructive to compare the estimated nitrogen flux to the center of the bank (12 mg-at N m^{-1} s^{-1}) with the observed biological requirement for nitrogen. During a cruise in August 1985 *in situ* incubations of phytoplankton were carried out in the presence of either ^{14}C (as HCO^{3-}) or ^{15}N (as NO^{3-} or NH^{4+}), and used to calculate rates of photosynthesis and nitrogen uptake. Measurements were made of nutrients, chlorophyll, and physical characteristics of the water column. It was found that, because of the semi-diurnal movement of the front mentioned earlier, some stations on the bank or in the frontal zone were, in fact, stratified. Grouping stations according to their physical characteristics rather than their position led to the conclusion that the frontal stations had the highest carbon-fixation rates (265 mg C m^{-2} h^{-1}) and nitrate-uptake rates (1.3 mmol N m^{-2} h^{-1}), the stratified stations had the lowest (160 mg C m^{-2} h^{-1} and 0.65 mmol N m^{-2} h^{-1}) while the mixed stations were intermediate. Nitrate-based 'new' production accounted for 60% of the total frontal stations, 41% at the mixed stations, and only 27% at the stratified stations. The nitrate demand of the phytoplankton populations was 0.36 mg-at m^{-2} s^{-1} at the frontal stations, 0.09 mg-at m^{-2} s^{-1} at the mixed stations, and 0.018 mg-at m^{-2} s^{-1} at the stratified stations. At an oceanic station far from the bank the demand was only 0.02 mg-at m^{-2} s^{-1}.

Using reasonable approximations for the area of the mixed zone on the top of the bank, and of those parts of the frontal zone on the bank side of the mooring site, the nitrate requirement of the mixed zone was estimated at 2.5 mg-at N m^{-1} s^{-1} and that of the frontal zone between the bank and the mooring as 2.9 mg-at N m^{-1} s^{-1}. Hence the calculated supply, 12 mg-at N m^{-2} s^{-1}, is roughly twice the total demand by the phytoplankton. Bearing in mind that these calculations are rough approximations, they seem to indicate that at the time of study cross-frontal transport by residual flow was sufficient to meet the observed needs of the phytoplankton, without any requirement for other mechanisms.

In comparing results from Georges Bank with those from the northwest European area, it must be remembered that the front associated with the gyre on Georges Bank may well show very different properties from the shallow-sea fronts off the coast of Europe. Measurements to date suggest that the enhanced primary production at the Georges Bank front is made possible by a strong cross-frontal transport of 'new' nitrogen, and that the principal physical mechanism involved is unknown. At the European fronts it appears that primary production is enhanced by vertical transport of nitrate and by a variety of mechanisms of cross-frontal transport.

The influence of the Georges Bank tidal front on the distribution and abundance of animals is beginning to be studied. Tremblay and Sinclair (1992) concluded that larvae of the scallop *Placopecten magellanicus* were retained on the bank by physical processes associated with the front. On the other hand, when sampling a wider variety of planktonic animals, Perry *et al.* (1990) found that some occurred predominantly on the bank (e.g., small copepods), some predominantly off the bank (e.g., most large zooplankton), and some occurred on both sides of the front (e.g., *Calanus finmarchicus*). It was not clear that the front itself had any influence on their distribution.

6.3.5 Heterotrophic activity at shelf-sea fronts

An integrated multidisciplinary study of several kinds of heterotrophic activity at a shallow-sea tidal-mixing front in the Irish Sea has been reviewed by Fogg (1985). The investigators recognized three water masses as distinct ecosystems: surface stratified water (SSW), bottom stratified water (BSW), and mixed water (MW).

After thermocline formation the SSW system was occupied by an expanding phytoplankton population that gradually used the available nutrients, then reached a plateau of biomass that was maintained throughout the summer, with the greatest concentration of chlorophyll being on the pycnocline. In parallel was development of heterotrophic populations of bacteria and zooplankton that expanded until photosynthesis was balanced by heterotrophic activity. At this stage, nitrogen turnover, as measured by urea turnover rates, was rapid and there was a high degree of interdependence between the food web components that persisted until destabilization of the water column in autumn.

In the bottom stratified water, by contrast, both photosynthesis and heterotrophic activity was at a low level. In the mixed water, where the phytoplankton is light-limited rather than nutrient-limited, bacterial and zooplankton biomasses, heterotrophic activity, and nitrogen turnover rates were lower than in the SSW.

Fogg (1985) reported that, like many others before them, the investigators had found patches of high chlorophyll density in the SSW adjacent to the front and on the pycnocline away from it. However, he said that if chlorophyll concentrations were averaged down to 30 m, the frontal waters were scarcely distinguishable from the rest of the SSW. In other words, the distribution was different, but not the total amount in the water column.

No significant increase in bacterial or zooplankton biomass was found at the front, but a marked increase in physiological activity was observed. In June and July the levels of cellular ATP were about twice those found elsewhere in the SSW, and there were patches of high glucose uptake activity and urea uptake activity adjacent to the front. The tentative explanation put forward by Fogg (1985) is interesting from several points of view. He suggested that eddies formed along the front (see Fig. 6.02) might, by their rotation, induce vertical circulation and bring to the surface concentrations of phytoplankton from the isopycnal region of the SSW. This phytoplankton would be immersed in nutrient-poor water and would be expected to be active photosynthetically but unable to divide. Under these circumstances large quantities of DOM would be liberated, which might explain the high levels of heterotrophic activity. Fogg (1985) also suggested that this mechanism could account for the appearance of high concentrations of chlorophyll at the surface near the front. It is yet another potential mechanism to add to those discussed in the previous two sections.

When the distributions of zooplankton were examined species by species it was found that the front demarcated distinct communities. *Temora longicornis* and the nauplii of *Acartia clausi* were absent from SSW, whereas *Calanus finmarchicus* and *Membranipora membranacea* were absent from MW. *Oithona similis* was concentrated around the pycnocline, while *Microcalanus pusillus* was found only in the BSW.

On the stratified side of the front most of the zooplankton were found above the thermocline and made limited vertical migrations. When zooplankton found themselves in patches of high concentrations of phytoplankton they ceased vertical migration. Fish eggs, larvae, and post-larvae were more abundant in the frontal zone than elsewhere. Birds, such as Manx Shearwaters, were more numerous on the stratified side of the front than else-where, especially in areas of chlorophyll maxima. The general impression that frontal wa-ters are regions of enhanced primary and secondary productivity is sustained, even though the mechanisms responsible are still a matter of debate.

Benthic productivity

It has been observed (Glémarec 1973; Creutzberg 1985) that in the benthic areas corre-sponding with a frontal change from tidally mixed waters to summer stratified waters there is a change in benthic community structure. Creutzberg (1985) showed that on a transect across a front off the Dutch coast in the North Sea (Fig. 4.13) the tidally mixed area to the south was marked by sandy sediments, while the stratified area to the north had muddy sed-iments. The boundary between the two has remained remarkably constant for at least 80 years, and just north of the boundary is a zone about 15 km broad characterized by high or-ganic matter content and a high biomass of benthic animals.

Two hypotheses have been put forward to explain it. The simplest is that the fast-moving tidal streams of the southern region scour the bottom and carry a heavy load of or-ganic and inorganic sediment. Further north, where tidal streams are slower, a point is reached at which the waters deposit their load of organic and inorganic sediment, creating ideal conditions for the growth and reproduction of benthic organisms. A more complex ex-planation invokes the presence of a front with its enhanced phytoplankton biomass in the waters above. Sinking of the phytoplankton leads to the accumulation of organic matter in the sediments below.

Creutzberg (1985) made observations on the position of the front and of its chlorophyll distribution on eight occasions between 1982 and 1984. He discovered that there was a chlorophyll maximum that remained stationary over the enriched benthic zone even when the front itself migrated 50–60 km north. His conclusion was that the first of the two ex-planations was the correct one, namely that the advection of organic matter and its deposi-tion, over a long period, in a position determined by the speed of the tidal currents has led to the development of an enriched benthic community. He suggested that rapid release of nutrients from the enriched benthic community was the cause of the chlorophyll maximum over the top of it. This theory is the converse of the view that holds that deposition of phy-toplankton from chlorophyll maxima associated with the front has led to the enrichment of the benthic community. Creutzberg did not go so far as to suggest that release of nutrients from the benthos was the sole cause of chlorophyll maxima at fronts, but it is worth re-membering that the bottom stratified water below fronts may be receiving unusually large amounts of nutrients across the benthic boundary layer.

Productivity of zooplankton and fish

In addition to the frequent observation that phytoplankton biomass is enhanced at tidal fronts, conflicting observations have been made on the relative abundance of zooplankton and fish at fronts, compared with the waters on either side. In cases where enhanced bio-mass of zooplankton has been observed, it has not been clear whether the effect was due to higher rates of population growth in the frontal zone or to passive advection to the zone of convergence. Uye *et al* (1992) studied the abundance of copepod nauplii at a tidal front in the Inland Sea of Japan. They found that, in response to the higher productivity of phytoplankton, the egg production rate of individual *Paracalanus* was higher at the front than on either side, but the population density of nauplii was lower. They concluded that the mortality of nauplii was higher at the front because of a higher density of predators that had accumulated there by passive advection.

In Delaware Bay, Epifanio (1987) studied the distribution of larvae of the mud crab *Neopanope* in relation to a tidal front and found that a concentrated patch of larvae was maintained on the stratified side of the front. He proposed a model according to which the larvae in surface waters were first advected toward the front, then carried downward with sinking water at the pycnocline. The larvae responded by migrating up toward the light, so that they found themselves in surface water, and the cycle of transport began again, maintaining the integrity of the patch.

Noting that the abundance of larvae of the sprat (*Sprattus*) in the North Sea was highest on the nonstratified side of tidal fronts, Munk (1993) found that larval growth rate was highest in the mixed water some distance from the front, was slightly lower in the area of peak larval abundance, and was much lower in the stratified water. He concluded that the enhanced productivity at the front benefited the larvae, and that the slightly reduced growth rate close to the front could be explained by food limitation in the dense larval population found there.

6.4 SHELF-BREAK FRONTS

Along the outer continental shelf of almost all of eastern North America in winter is to be found a front marking the transition between the colder and fresher shelf water mass and the warmer and more saline slope waters. Data from Flagg and Beardsley (1975) showing the situation south of Rhode Island in early April 1974 (Fig. 6.08) indicate that the shelf water was well mixed at a temperature of 5–10 °C and 32.6–34.4‰ salinity, while the offshore waters were mildly stratified (not shown) with a surface temperature of 14–16 °C and about 36‰ salinity. When the density is considered, these differences tend to compensate for one another so that the σ_t difference across the front was only 0.4 units. In various parts of the eastern seaboard of North America similar changes occur over distances of 7–40 km and a vertical depth ranging from 15 to 60 m (Mooers *et al.* 1978). The front normally straddles the shelf break, intersecting the bottom near the 100 m isobath (60–120 m) and sloping up from the bottom in an offshore direction for about 50 km (0–100 km).

The coming of the spring warming drastically changes the picture. The shelf becomes stratified and below the thermocline, on the deeper part of the shelf, a parcel of water of reduced salinity with a temperature less than about 10 °C becomes isolated from the warmer, shallower waters inshore and the more saline waters offshore. This process was first described in northwest Europe by Vincent and Kurc (1969), who called it the 'cold cushion' (*bourrelet froid*) (Fig. 6.09). The effect is to maintain a weak front near the shelf edge below the thermocline, even through the period of summer stratification.

As was mentioned earlier, Mooers *et al.* (1978) have described how the position of the shelf-break front changes under the influence of wind forcing, by the impact of the Gulf Stream rings (see Chapter 8) and in association with the 'calving' of lens-shaped parcels ('bubbles') of shelf water of scale 10–20 km, which break through the frontal structure and move offshore. The net result is that the position of the shelf-break front off the east coast of North America may vary in distance from shore by as much as 200 km. The movement is

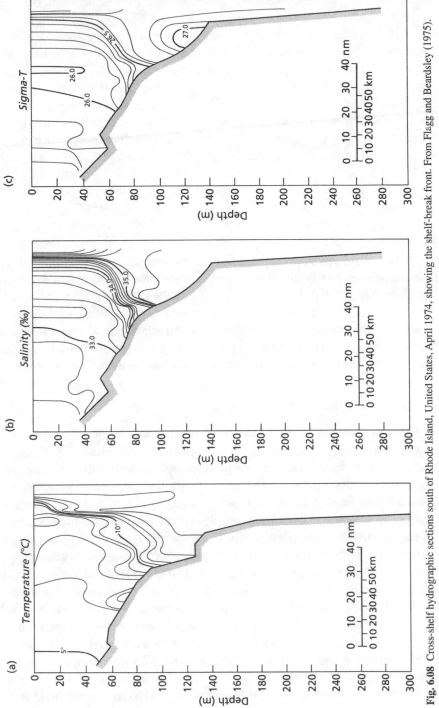

Fig. 6.08 Cross-shelf hydrographic sections south of Rhode Island, United States, April 1974, showing the shelf-break front. From Flagg and Beardsley (1975).

Fig. 6.09 Formation of the 'cold cushion' (*bourrelet froid*) on the shelf off Brittany. The structure is visible in spring and becomes warmer and smaller as the seasons advance. Compiled from the data of Le Magueresse and of Vincent and Kurc, by Le Fèvre (1986).

least off the coast of Florida and Georgia to the south, and reaches a maximum at the latitude of Cape Cod. The envelope in which this front moves corresponds with the position of highly productive fisheries, from the Grand Banks off Newfoundland, south to Cape Hatteras.

6.4.1 Biological production at shelf-break fronts

There have been various observations of nutrient enrichment and/or enhanced phytoplankton and zooplankton biomass associated with shelf-break fronts (Fournier 1978; Pingree and Mardell 1981; Le Fèvre and Frontier 1988). For example, Fournier (1978) reported that the average value for chlorophyll-a across the Scotian Shelf (in March 1977) was less than 1 mg m^{-3}, while over the shelf break chlorophyll reached a maximum of 4.2 mg m^{-3}. As an explanation, he suggested that shallowing of the mixed layer in the vicinity of the front allowed enhanced phytoplankton production, even in winter.

The observations of Herman *et al.* (1981) on the shelf break south of Nova Scotia can be interpreted as supporting the views of Le Fèvre and Frontier (1988). Using an undulating 'batfish' sampler that measured salinity, temperature, depth, chlorophyll-a, and copepods, they found that plant production and copepod abundance were much higher at the front than in surrounding shelf and slope waters. Chlorophyll concentrations were ≈1.5 mg m^{-3} in shelf water, peaked to ≈6 mg m^{-3} in the front, and were 3–4 mg m^{-3} in the slope water. The corresponding abundances of copepods in the shelf, front, and slope waters were ≈30,000 m^{-2}, 70,000 m^{-2}, and 10,000 m^{-2}, respectively. The high concentrations of copepods occurred above the thermocline in the frontal region, and these copepods did not migrate down through the thermocline during the day, while those in the shelf waters often did so. Having considered the various factors that might account for high copepod abundances at the front, Herman *et al.* (1981) concluded that convergence of surface waters at the front was the most plausible mechanism. However, at that time the mechanism of fertilization of surface waters by internal tidal waves (Le Fèvre and Frontier 1988) had not been proposed.

Pingree and Mardell (1981) reported on a series of studies of the Celtic Sea shelf break and pointed out that in summer the edge of the shelf is characterized by a band of water about 1–2 °C cooler than water on either side of it and about 100 km broad. At times the band can be traced for 800 km along the Armorican and Celtic Sea slopes. In it, the levels of inorganic nitrate and chlorophyll are significantly higher than in adjacent water. They noted that there seemed to be an association between the patches of increased chlorophyll and the bottom topography. Although some had invoked upwelling as the mechanism responsible for the presence of cooler, nutrient-rich water at the surface (e.g., Heaps 1980), Pingree and Mardell (1981) were inclined to favor tidally driven internal waves interacting with the bottom topography as the important mechanism.

This idea was developed further by Mazé (1983), Mazé *et al.* (1986), and Pingree *et al.* (1986) to the point where Le Fèvre and Frontier (1988) could write that 'a reasonably safe picture of what is taking place has now emerged.' Details of the physics of internal waves will be given in Chapter 7, but in summary the story goes approximately as follows. The incoming oceanic barotropic tidal wave is rather small, of the order of 2 m. In stratified water the interaction of the tidal wave with the shelf break gives rise to an internal baroclinic tide with an amplitude that reaches a maximum of about 60 m at the very edge of the shelf. From here it propagates in two directions, toward land and away from it, gradually damping out. Propagation toward the open sea is more or less sinusoidal, but propagation onto the shelf is distorted by the barotropic tidal currents.

There has been much discussion about whether an internal tide would produce the observed reduced temperature and elevated nutrient concentrations in surface waters. It is now thought that there is an interaction between the internal waves and the wind mixing of the layer above the thermocline. Assuming some type of equilibrium between the depth of the thermocline, the temperature of the mixed layer, and the prevailing wind conditions, a forced upward displacement of the thermocline leads to a disruption of that equilibrium. Additional energy is injected into the mixed layer, which leads to an increase in the depth of the mixed layer with upward mixing of cool water from below the thermocline. This water carries with it additional nutrients, so that primary production is enhanced. Another way of looking at it is to consider that the depth of the mixed layer is a function of the buoyancy provided by surface warming and the strength of winds causing mixing. If the thermocline is pushed upward, the mixed layer will tend to return toward its original thickness by mixing upward some of the cooler, nutrient-rich water. The frequency of this occurrence is that of the M_2 tide—i.e., twice daily.

Le Fèvre and Frontier (1988) pointed out that augmentation of the nutrients on a daily basis provides a more-or-less continuous increase in phytoplankton productivity to which the copepod population can respond. This condition is in contrast to the situation at a tidal-mixing front, where the cycle of enhanced production follows the fortnightly cycle of spring and neap tides, leading to a short burst of phytoplankton production every two weeks. It seems unlikely that zooplankton populations could increase and decrease in such a way as to exploit this production to the maximum. Hence shelf-break fronts should show that copepod populations increase in parallel with phytoplankton, while tidal-mixing fronts should not. The limited field data available support this view. Results given by Le Fèvre and Frontier (1988) for the Armorican shelf-break front show increased copepod abundance in the frontal region, suggesting that a classical copepod herbivore food chain is functioning. On the Ushant tidal-mixing front the dominant zooplankter was the pteropod mollusc

Limacina, which is known to feed on micro-zooplankton. They suggested that much of the phytoplankton at this front remains ungrazed by copepods and becomes senescent, giving support to a food web involving bacteria and ciliates, which in turn are taken by *Limacina.* This line of reasoning is supported by the high levels of heterotrophic activity observed by Fogg (1985) and others as characteristic of tidal-mixing fronts.

In summary, the most recent view of the mechanism controlling biological productivity at shelf-break fronts is that tidally generated internal waves add energy to the mixed layer, causing a deepening with incorporation of nutrient-rich water from below the nutricline. This process varies in intensity with the M_2 tide so that it produces a twice-daily augmentation of nutrients, a fairly constant elevation of primary production levels, and a corresponding increase in secondary productivity. This mechanism contrasts with that believed to be operative in tidally mixed fronts, where the maximum enrichment of nutrients in primary production coincides with the neap tides at fortnightly intervals. This stimulation of primary productivity is considered to be too infrequent to support an elevated population of macro-zooplankton. Instead, there is a burst of phytoplankton production, followed by a period of decline and decomposition, which supports a dense population of bacteria and micro-zooplankton.

6.5 UPWELLING FRONTS

As we saw in Chapter 5, there are many places around the world where wind-induced offshore transport of surface waters occurs on a major scale, leading to coastal upwelling. While the forcing winds continue to blow, nutrient-rich waters from below the pycnocline continue to rise to the surface and move offshore, displacing the warmer, less nutrient-rich waters that preceded them. The interface between the upwelled waters and the offshore waters constitutes an upwelling front. A section through such a system (Fig. 6.10) reveals that the front is the place where the pycnocline intersects the surface of the ocean.

Fig. 6.10 Conceptual model showing (a) density field and (b) cross-shelf circulation in the vicinity of an upwelling front. From Mooers *et al.* (1978).

During an upwelling event, the front at first forms close to shore, then moves offshore, but at some point an equilibrium is reached at which the front ceases to move offshore and the upwelled waters, as they continue to move offshore, are driven downward beneath the offshore waters (Fig. 6.10). The front lies at a prograde angle. At the surface there is a convergence of waters and a tendency for buoyant biological material to accumulate. When the wind stress decreases and upwelling ceases, the front retreats landward and may disappear.

In spite of the large amount of biological study of upwelling systems (see Chapter 5), there is a surprising lack of detailed information about upwelling fronts. The modelling studies of Thompson (1978) suggested that part of the sinking water is entrained in a shoreward-flowing deeper current, but there may be a one-celled or a two-celled circulation, depending on the wind stress, stratification, current shear, and bottom relief. In any event, the presence of the front is normally marked by a sharp temperature transition from the cold, nutrient-rich upwelled water inshore to the warmer, less nutrient-rich stratified water offshore. Clearly, any mechanism causing a mixing of these waters is likely to stimulate primary productivity.

Pearcy and Keene (1974) studied the upwelling system off Oregon by airborne remote sensing and found abrupt changes in color coincident with changes in temperature that they thought denoted enhanced biological activity at upwelling fronts. Ryther (1967) found large concentrations of a ciliate, *Cyclotrichium meuneri,* which colored the water red, on the inshore edge of an upwelling front off Peru, and Packard *et al.* (1978), after studying the biology of a related ciliate, *Mesodinium rubrum,* in the California upwelling system, showed that large concentrations formed on the inshore edges of upwelling fronts. Peterson *et al.* (1979) studied zooplankton distribution in the Oregon upwelling system and found that very high concentrations were retained shoreward of the front, and were carried below the pycnocline when the upwelling relaxed. Wroblewski (1982) modelled the movement of the various stages of the copepod *Calanus marshallae* in the Oregon upwelling system and showed how vertical migration may interact with the physical oceanography to cause the organisms to move over a spiral path on the continental shelf, moving offshore at the surface, but returning in deeper water after sinking at the front. At the highest trophic level, there is evidence that albacore tuna, which make trans-Pacific migrations to feed close to the California upwelling systems, tend to concentrate in the vicinity of the upwelling fronts (Laurs *et al.* 1977). On the whole, the biological information that we have about upwelling fronts is rather fragmentary, and shows that the study of upwelling fronts is much less advanced than that of tidal-mixing fronts.

One of the more recent field studies of an upwelling front is that carried out by Armstrong *et al.* (1987) across the southern Benguela system north of Cape Town. They emphasized throughout their account that the frontal zone is dynamic and subject to rapid change, especially since they encountered on the outward boundary of the front a strong north-flowing jet current, with speeds up to almost 1 m s^{-1} and extending to a depth greater than 200 m. However, they found during their cruise that the greatest biomass and productivity of the phytoplankton, when integrated over the euphotic layer, was in the frontal region (Table 6.2). Inshore of the front very high concentrations of nutrients and chlorophyll occurred in surface waters, but penetration of light was limited and integrated water-column productivity was correspondingly low. Offshore of the front the nutrient levels in the euphotic zone were extremely low, so that photosynthesis was limited to that level that could be sustained by nutrient recycling. In the frontal region itself, however, relatively high biomasses of

Table 6.2 Station number, water type, sea surface temperature (SST), 1% light depth ($D_{1\%} \simeq$ depth of euphotic zone), integrated phytoplankton production rate $0 - D_{1\%}$, mean phytoplankton production rate $0 - D_{1\%}$, integrated chlorophyll-a and bacterial production rates $0 - D_{1\%}$, and the percentage ratio of bacterial to phytoplankton production. From Armstrong *et al.* (1987).

Station	Water type	SST (°C)	$D_{1\%}$ (m)	Phytoplankton production		Chlorophyll-a (mg m^{-2})	Bacterial production	
				A (mg C m^{-2} h^{-1})	Mean (mg C m^{-3} h^{-1})		B (mg C m^{-2} h^{-1})	B/A (%)
2–03	Inshore	15.3	9	112.8	12.5	17.5	39.2	34.7
4–04	Frontal	16.3	16	80.9	5.1	58.9		
3–03A	Frontal	16.2	19	258.9	13.6	108.5		
4–05A	Frontal	15.7	14	187.0	13.4	73.4	21.4	11.4
2–07	Offshore	18.1	33	65.4	2.0	20.3		
4–10	Offshore	18.5	38	81.0	2.1	15.5	54.5	67.3
1–16	Offshore	18.0	27	27.3	1.0	13.1		

phytoplankton were well distributed through the euphotic zone and production rates were high. It was inferred that there was an enhanced supply of new nitrogen into the euphotic zone of the frontal region. The mechanism suggested was interleaving of water masses of differing salinities, temperatures, and nutrient contents but similar densities, and there was some evidence for this from the CTD data. However, as we have seen, mechanisms of cross-frontal transport are not well understood and are under active investigation.

6.6 RIVER- AND ESTUARINE-PLUME FRONTS

At the edges of river or estuarine plumes (see Chapter 4) where water of reduced salinity meets more saline coastal water, fronts are formed that are conspicuously marked by lines of foam and by sharp change in color of the water. As mentioned in Chapter 4, the less saline, lighter water lens has an elevated surface and is carried by gravity seaward to ride over the top of the more dense coastal water, forming a retrograde front.

An early study of the physics of an estuarine-plume front was that carried out by Garvine and Monk (1974) on the Connecticut River plume in Long Island Sound. They showed that adjacent to the front the layer of brackish water was only about 1 m deep, and that the pycnocline sloped up to intersect the surface over a distance of about 50 m. There was a vigorous convergence of surface waters from both sides of the front and locally intense sinking at the front itself.

There has been little work on the biological significance of plume fronts, but the evidence available suggests that they are sites of enhanced biological activity and that fish tend to aggregate there. Pearcy and Keene (1974), who studied the coastal waters of Oregon by airborne remote sensing, found evidence of increased phytoplankton biomass at the Columbia River plume front, and Owen (1968) found that catches of albacore tuna were higher at that front than at other stations nearby. In Japan, Tsujita (1957) found that the spawning and hatching of Japanese sardine off western Japan was consistently located close to an estuarine-plume front.

In the Chesapeake Bay area, there are several places where the rivers emptying into the bay form plume fronts, and Tyler and Seliger (1978, 1981) have shown that high concentrations of various phytoplankton species, including the red-tide dinoflagellate *Prorocentrum,* are found in these frontal regions especially in winter and early spring.

6.7 OTHER TYPES OF FRONTS

6.7.1 Estuarine fronts

Relatively small-scale fronts commonly appear in estuaries, forming and dispersing with every tide. They are similar in some ways to the tidal-mixing fronts of coastal waters. At certain states of the tide the shallow water at the edges of the estuary is tidally mixed while the deeper water in the middle remains stratified under the influence of the freshwater run-off. Such a front has a marked salinity gradient, as well as a temperature gradient, at the surface. Examples are given by Bowman and Iverson (1978). The influence of these fronts on biological production is not well understood, but striking examples of concentration of pollutants in the surface convergence have been noted.

6.7.2 Effect of geomorphology

When tidal or other currents interact with irregularities in the sea bed, or coastline, it is usual to find consistent patterns of eddies and associated fronts. Pingree *et al.* (1978b) reviewed the formation of fronts at coastal headlands, and Bowman *et al.* (1986) reviewed the interactions of tidal currents with islands and features of the sea floor. There is scattered evidence of enhanced phytoplankton growth—for example, around islands, off headlands, and above sea mounts—but the mechanisms involved, and the role of fronts in these mechanisms, are as yet not at all clear.

Wolanski and Hamner (1988) reviewed the distribution of zooplankton in relation to topographically controlled fronts formed by headlands, islands, and reefs. When a tidal current streams past one of these structures a front (often called the dividing streamline; Fig. 6.11) may be formed that separates the normal tidal flow from the more turbulent eddies on the downstream side of the object. If the water is fairly shallow the turbulent eddies interact with the bottom giving rise to convergence and sinking at the front. As a result, surface-dwelling planktonic organisms tend to aggregate there, especially those that are buoyant. In the vicinity of coral reefs, where many species of coral tend to spawn synchronously, very dense aggregations of buoyant coral eggs are formed at these fronts, which can then be readily identified from the air.

Alldredge and Hamner (1980) documented the formation of dense aggregations of zooplankton in the turbulent area downstream of a headland when tidal streaming was at its

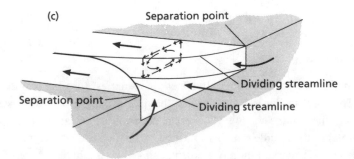

Fig. 6.11 (a) Plan of a commonly occurring flow pattern around a reef in shallow coastal water, showing the front (labelled dividing streamline) separating the area of turbulence in the wake from the less turbulent area beyond. (b) Vertical section showing upwelling produced by the clockwise (southern hemisphere) gyre. (c) Formation of turbulent-mixing region as a tidal stream enters a reef passage. For details see text. From Wolanski and Hamner (1988).

maximum, and Oliver and Willis (1987) reported that high concentrations of coral eggs formed at the topographic fronts at certain stages of the tide remained as coherent linear patches as they drifted away from the places where the fronts had formed. It is clear that calculations of the encounter rates between planktonic organisms around coral reefs, islands, and headlands must not be made on the basis of average density in the water mass, but must take into account the way in which topographically controlled fronts bring about massive aggregations.

It is worth noting in passing that in the situation depicted in Fig. 6.11(c) in which a tidal current streams through a reef passage, nutrient-rich water from depth is upwelled from in front of the reef by Bernoulli suction and stimulates algal production on the reef (Wolanski *et al.* 1988).

6.8 SUMMARY

More effort has gone into the study of shelf-sea (tidal-mixing) fronts than into any other kind of front. It is now clearly understood that tidally induced currents cause mixing from top to bottom of the water column in many shelf areas. The strength of the tidal currents varies with the diurnal tidal cycle, but also with the fortnightly alternation between spring and neap tides. At places where the depth in relation to the tidal currents is too great for mixing to occur from top to bottom of the water column, thermal stratification occurs in summer. Shelf-sea fronts mark the discontinuity between the tidally mixed region and the stratified region and are frequently found to be the sites of enhanced planktonic biomass and productivity.

Early observations on such fronts emphasized the convergence of surface waters and downwelling that occurs there. Any organisms buoyant enough to resist the downwelling would be expected to aggregate at the front. Thus, it was thought that passive advection was one explanation of the concentrations of planktonic organisms found at fronts.

An alternative explanation invoked the concept of enhanced *in situ* production, made possible by particularly favorable conditions of light and nutrients. In the upper mixed layer of stratified water in summer, phytoplankton growth is usually nutrient-limited because the presence of a pycnocline restricts the upward movement of nutrients from the lower regions, where they are more plentiful. Any mechanism that causes a transfer of nutrients to the surface layer on the stratified side of a front is likely to lead to enhanced biological production. A number of mechanisms have been proposed.

1 The spring-neap tidal cycle. At a fixed position relative to the bottom, the water may be tidally mixed at one stage of the tidal cycle and stratified at another. Nutrients are brought up during the mixing phase and utilized in the upper mixed layer during the stratified phase.

2 Cross-frontal transport. A variety of mechanisms have been proposed for transferring nutrients from the tidally mixed side of the front, where phytoplankton is often light-limited but not nutrient-limited, to the stratified side of the front. Baroclinic eddies, in which parcels of nutrient-rich water break through to the stratified side, have been observed by remote sensing.

3 Vertical transport. Some evidence suggests that conditions in the frontal zone are favorable for the vertical transport of nutrients through the front to the stratified water above. This movement could result in enhanced phytoplankton in the immediate vicinity of the front.

The relative importance of these three types of mechanism has yet to be evaluated. It is probable that different mechanisms dominate at various times and places. Only a sophisticated multidisciplinary approach using state-of-the-art technology to yield physical and biological data on a range of scales from microns to kilometers will clarify the mechanisms involved.

Shelf-break fronts occur at the interface between cooler, less saline water on a continental shelf and warmer, more saline water offshore. Production of phytoplankton, zooplankton, and fish is particularly high. The mechanism most widely accepted as causing the enhanced production is the generation of tidally driven internal waves at the shelf edge. The consequent rise and fall of the nutricline leads to vertical transport of nutrients into surface water. The process is seen as being of daily occurrence, thus supporting a relatively constant level of enhanced phytoplankton production on which a food web involving zooplankton and fish can be built. This situation has been contrasted with that found at tidal-mixing fronts. At these, the mechanism of enhanced biological production appears to be linked to the neap-spring tidal cycle, so that pulses of high phytoplankton biomass occur at fortnightly intervals. This period is too long for zooplankton populations to be able to exploit them effectively. As a result, there is accumulation, death, and decay of phytoplankton, with the release of dissolved and particulate organic matter, which is exploited by bacteria, ciliates, and specialized macrozooplankton, rather than copepods and fish (see section 6.4.1).

Upwelling fronts form at the interface between normal shelf water and the cool, nutrient-rich water brought to the surface during wind-driven coastal upwelling. Since upwelling-favorable winds are not constant, but increase and decrease to give recognizable upwelling events, upwelling fronts are also variable in time and space. At the interface, upwelled water is driven down under the warmer shelf water, forming a prograde front. There have been few systematic studies of the biology of upwelling fronts, but the information that is available suggests that planktonic organisms tend to aggregate on the coastal side of these fronts and that tuna and other pelagic fishes occur there in particularly large numbers.

At places where plumes of fresh or brackish waters run into saline coastal waters it is common to find plume fronts, where the less saline waters ride over the top of the saline waters, forming retrograde fronts and where surface convergence and downwelling occur. Once again we find that there are few systematic studies of the properties of such fronts, but there is fragmentary evidence suggesting that aggregations of plankton form at or near the convergence and that good catches of pelagic fish are often obtained there.

Within estuaries, local changes in depth give rise to small-scale variations in the degree of tidal mixing, and to the formation of fronts between mixed and stratified waters. Such fronts are often formed parallel to the shores of estuaries, separating the tidally mixed shallow waters from the deeper waters stratified under the influence of freshwater run-off. Bathymetric features also give rise to a variety of eddies and fronts in coastal waters. Whenever a tidal current impinges on a geomorphic feature such as a submerged canyon, an island, or a headland, transient fronts are found that form and disperse with the daily or twice-daily tidal cycle. Little is known about the biological significance of these small-scale fronts.

7

Tides, Tidal Mixing, and Internal Waves

7.1 INTRODUCTION

Tides are created by the gravitational pull of the moon and the sun and are most familiar as a rise and fall in the level of the sea twice a day. This changing water level leads to interesting patterns of zonation of intertidal organisms. They have been well studied over a long period and we shall not review them here. The tides also generate currents in the water that interact with the bottom to produce turbulence. This bottom-produced turbulence tends to mix the lower layers of the water and if the currents are sufficiently strong the turbulence may prevent any stratification and a whole area may be permanently tidally mixed. It has been shown that the breeding grounds of a number of stocks of herring are each located in a different, discrete area of strong tidal mixing (Iles and Sinclair 1982), and in this chapter we explore the possible significance of this observation.

In situations where tidal mixing is less strong and the water column becomes stratified, the interaction of the tidal currents with the bottom topography may lead to the formation of internal waves on the thermocline. These, in turn, propagate horizontally and cause vertical mixing and the redistribution of nutrients, with important effects on phytoplankton production. The surface of the ocean shows slicks and areas of convergence associated with the passage of internal waves, and these can have a strong influence on the distribution of zooplankton and larval fishes.

As tidal currents move on and off the shallow banks, they interact with bottom topography to generate unidirectional currents that form gyres around those banks. The combination of tidally mixed water over the top of a bank and a gyre around its periphery is thought

211

to provide conditions particularly suitable for the eggs and larvae of fish. In this chapter we shall explore some of these interesting consequences of tidally induced water movement.

7.2 THE PHYSICS OF TIDES

7.2.1 Tide-generating forces and the equilibrium tide

Tides in the ocean result from a slight imbalance between two forces; the first is the gravitational pull of the moon and sun, and the second is the centripetal force that is required to keep the ocean's water moving along with the rest of the earth in a circular path through space. We examine these forces and motions in this section to show how the main features of the tide are generated.

We first examine the effect of the gravitational force of the moon on the world's ocean in Fig. 7.01. Here the earth and the moon are seen from above the north pole of the earth, and both are assumed to be stationary except for the earth's rotation. Also, the earth in this case has no continents but is covered with a uniform layer of water. If the moon was not present, the oceans would cover the earth to a constant depth. The moon, however, exerts a gravitational attraction on each particle of water in the ocean and causes it to pile up under the moon. An observer at the equator at 'O' rotating with the earth through the day will notice that the depth of the ocean rises and falls once a day. The high tide will always be under the moon and the low tide will be on the side opposite to the moon. This elementary model produces a tide but it is not a very good model because most places on the earth usually have two high and two low tides during the day.

One needed improvement in the model is found by considering the rotation of the earth–moon pair once every $\simeq 29\frac{1}{2}$ days. To keep a body moving in a circle, a force must be directed toward the center of rotation; otherwise the body would continue in a straight line according to Newton's first law of motion. This phenomenon is, of course, well known to anyone who has swung a ball on the end of a string and let go. The force that maintains the circular motion is called the centripetal force and in the case of the earth–moon system it is supplied by the gravitational force between the two bodies. If the average gravitational pull over the whole earth was not equal to the required centripetal force, the distance between the earth and moon would have to change until they were equal.

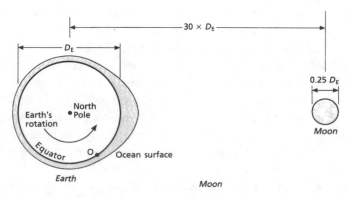

Fig. 7.01 The earth and moon viewed from above the north pole showing the ocean (shaded) pulled by the moon's gravity into a tidal bulge under the moon. An observer is at 'O' on the equator.

If the earth and the moon each had the same mass, they would rotate about a common center of mass that would lie halfway between the two. But because the earth's mass is roughly 80 times that of the moon the common center of rotation is about 1600 km, or about one-quarter of the earth's inner radius. The rotation of the two bodies about this point is illustrated in Fig. 7.02(a). As the moon moves around this center of mass, the earth also moves around it in a circle with a radius equal to the distance between the center of mass of the earth and the center of mass of the two bodies. The curved lines drawn through the center of the earth and the point 'A' illustrate that every point in the earth moves in a circle of similar radius. Because all particles in the earth are moving in circles of identical size, the centripetal force must be the same for every particle. The direction of this force is always toward the moon, which is also toward the center of the circle described by the particle (Fig. 7.02b).

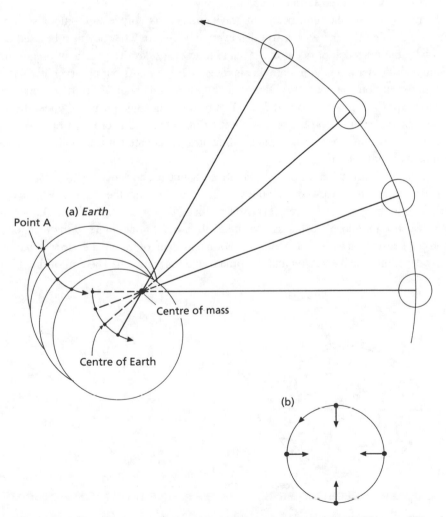

Fig. 7.02 (a) The earth and moon in mutual rotation about the common center of mass. The curves through points 'A' and 'C' illustrate the curves traced out by typical points on the earth. (b) The circular path of a typical particle on the earth through one cycle of the motion illustrated in (a) and the direction of the centripetal force required to keep the particle travelling in the circle.

The moon's gravitational pull on the earth decreases with distance from the moon. Particles on the side closest to the moon experience a greater gravitational pull than do particles on the side away from the moon. The average gravitational pull on particles in the earth is found at the center of the earth and must equal the required centripetal force, but the forces are imbalanced for all other points. This imbalance is illustrated in Fig. 7.03, which shows the earth and the moon again from above the north pole. Along the center line perpendicular to the line joining the earth and moon, the gravitational force on each particle supplies the required centripetal force. On the half closest to the moon, the gravitational force is greater than the centripetal force that causes the ocean to pile up under the moon. On the half of the earth away from the moon, the gravitational force is not strong enough to supply all the centripetal force required to keep the particles moving in the required circle, and the particles tend to move farther away. This process creates the tidal bulge on the side of the earth away from the moon. The observer at 'O' now observes two high tides and two low tides every day as the earth rotates.

So far we see that the centripetal and gravitational forces of the earth–moon system account for the fact that there are two tides every day. Figure 7.03 may also be used to explain why the tides do not occur at the same time every day. The earth and moon rotate around one another in a lunar month that is roughly $29\frac{1}{2}$ days. As our observer at 'O' rotates through exactly one day back to 'O' the moon has moved about 12° further around in its orbit to the position marked 'tomorrow' in Figure 7.03. The high tide that is under the moon will still be under the moon tomorrow, but it will be observed later in the observer's day. Since the earth spins on its axis at about 4 minutes per degree the tides will appear about 50 minutes later each day.

For the next refinement of these qualitative arguments, consider Figure 7.04, in which the earth and moon are viewed from the side. Instead of having the moon directly over the equator, as in the previous figures, it is shown about 25° north of the equator. This situation arises because the moon's orbit around the earth is tilted at an angle to the equator. The moon can be found at various angles to the north and south of the equator up to a maximum of 35° depending on the season and the time of the lunar month. The aim of the diagram

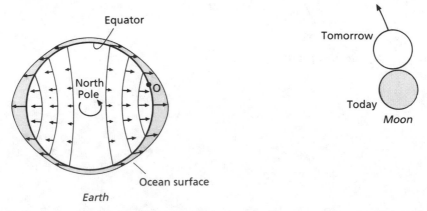

Fig. 7.03 The small arrows on the earth represent the net force due to the imbalance between the gravitational pull of the moon and the centripetal force that leads to 'tidal bulges' on two sides of the earth. The moon is shown at two positions one day apart to illustrate the delay of the tides by ≃50 minutes each day as seen by the observer at 'O.'

Moon

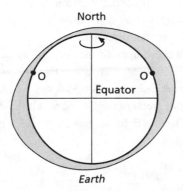

North

Equator

O   O

Earth

Fig. 7.04 Viewed from the side (north up) and with the moon north of the equator, the tidal bulges are asymmetric about the equator resulting in the diurnal inequality in the heights of the high and low tides.

is to show that the tidal bulges are no longer on the equator, as they were imagined in Figs. 7.01 and 7.03, but now one lies north of the equator under the moon while the other lies on the opposite side of the earth, to the south of the equator. Our observer at 'O' now notices that there are two high and two low tides but that the high tides are of unequal height, as are the low tides. This difference is called the diurnal (daily) inequality.

This elementary view of tides on a world with a moon but no sun, without continents, and in an ocean that is always in equilibrium with the forces that are acting is called the equilibrium tide. It cannot predict the tide at any particular location but it does explain some of the main features of the tide such as the diurnal variation, the diurnal inequality in the height of the tide, and the daily delay of $\simeq 50$ minutes in the time of the highs and lows. The next improvement in the model is introduced by considering the effects of the earth–sun system on top of the effects of the earth–moon system.

The mass of the sun is 27×10^6 times the mass of the moon but its distance from the earth is 400 times that of the moon. Because of this great distance, the gravitational attraction on a particle of water on the earth due to the sun is about one-half that due to the moon. So we can construct a diagram like that shown in both Figs. 7.03 and 7.04 for the tide due to the sun, but the height of the tide will be only half that due to the moon. The important effect of the tide due to the sun arises because its tidal bulge moves relative to the moon's tidal bulge throughout the lunar month. When the two tidal bulges coincide they add together to create the extra high tides called the spring tides. When the tidal bulges are opposed their effects tend to cancel one another, creating the neap tides. Thus, the equilibrium tide model can also qualitatively explain the fortnightly inequality in addition to the other effects.

7.2.2 Tides in the real ocean

The equilibrium tide helps us to understand some of the main principles of tides, but when it comes to predicting the tide in the real ocean this theory is of little help because the water that is raised up as the tidal bulge has to move around a world that is spinning on its axis and that is cluttered with continents. It is possible to formulate a rigorous dynamical theory to predict the tides but the way that the tidal wave moves in an ocean basin is very dependent on the shape of the basin, and is not amenable to simple mathematical description. Such theories can be used in simple situations that have elementary geometries—for example, in bays with flat bottoms and straight vertical sides (Bowden 1983)—but generally the tide must be predicted by extrapolating from existing measurements.

This calculation is normally done by measuring the height of the tide for at least a month, then decomposing the record into sinusoidal constituents. There are three main categories of constituents (Pond and Pickard 1983): (i) semi-diurnal, period about 12 hours; (ii) diurnal, period about 24 hours; and (iii) long period, greater than 24 hours. Although 20 or more constituents may be required to predict the tidal height accurately, the four most important constituents are:

the lunar semi-diurnal	M_2	Period = 12.42 h
the solar semi-diurnal	S_2	Period = 12.00 h
the luni-solar diurnal	K_1	Period = 23.93 h
the principal lunar diurnal	O_1	Period = 25.82 h

The M_2 constituent is roughly twice the amplitude of the other three.

The form of the tide, or the pattern of the water's rise and fall, is not the same at all locations around the oceans but varies according to the relative importance of the different constituents. The four main classifications of the form of the tides are illustrated in Fig. 7.05. The top record from Immingham, England, illustrates a tide with two high and two low tides every day. Both of the highs are about the same height and both the lows are about the same height. Such a tide is called a semi-diurnal in form because there are two per day and both are about the same height. In this case the semi-diurnal constituents dominate the diurnal ones. This fact is often quantified with the ratio $F = (K_1 + O_1)/(M_2 + S_2)$, where each letter stands for the amplitude of the constituent. If F is small (0.11), as in the upper record, the sum of the amplitudes of the diurnal constituents $(K_1 + O_1)$ is small relative to the sum of the semi-diurnal ones $(M_2 + S_2)$.

The four records in Fig. 7.05 show a marked decrease from top to bottom in the amplitude of the semi-diurnal tidal oscillation relative to the diurnal oscillation, which is confirmed by the increase in F from 0.11 to $\simeq 19$. At San Francisco there are always two tides per day but they are usually of unequal amplitude. At Manila there are two tides per day during the neap tides but only one rise and fall per day during spring tides. At Do-Son there is only one rise and fall of the tide per day throughout the month, which is a purely diurnal form of tidal oscillation. This occurrence is the rarest form of tide. The changing form of the tide between locations is due partly to the shape of the ocean basin in which the tidal wave is contained and partly to the latitude (Hendershott 1981).

7.2.3 Moving the tidal bulge over the earth: Kelvin waves

The tidal bulge shown in Fig. 7.04 travels over the surface of the rotating earth as a very long wave. And because the length of the wave is very much greater than the depth of the water it is called a 'shallow water' wave, similar to the waves that can be set up in a bathtub by making the water rock back and forth. The other kind of wave, the 'deep water' wave,

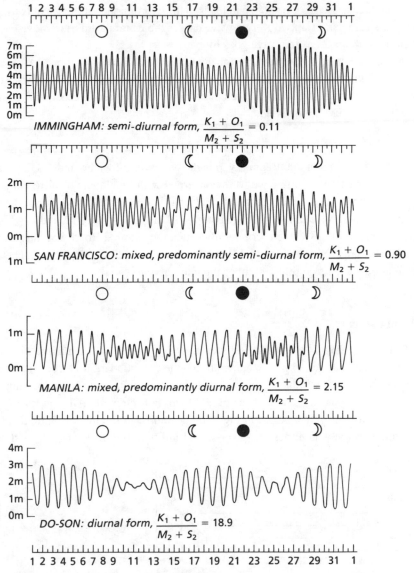

Fig. 7.05 Tidal records through March 1936 at four coastal stations illustrating variations in the amplitudes of the semi-diurnal ($M_2 + S_2$) and diurnal constituents ($K_1 + O_1$). Adapted from Defant (1958).

is the one normally seen generated by the wind on the ocean's surface. One important feature of shallow water waves that sets them apart from the deep water waves is that the velocity of the wave motion is constant throughout the depth of water while the motion in the wind waves dies out a few meters below the surface. Thus the velocity in the tidal wave is approximately constant throughout the depth of the ocean. Such flow is sometimes referred to as a barotropic wave—more specifically a barotropic Kelvin wave.

Because the tidal waves cause the water to move relative to the earth for a long time the Coriolis force is an important feature of the motion. The effect of the Coriolis force is to push the water to the right in the northern hemisphere. Figure 7.06 shows a Kelvin wave travelling south, with the coast to the west of it. The Coriolis effect causes the amplitude of the wave to increase toward the shore and leads to the expression that the wave is 'trapped' against the shore. Such a trapped Kelvin wave causes the water particles to move back and forth parallel to the coast as the wave goes by.

The horizontal scale of the wave, or the approximate distance from the coast to where the sea level is undisturbed by the wave, is estimated by the Rossby radius. We used the same scale in estimating the width of the coastal upwelling regions in section 5.2.3 but in the present case we use the formula

$$R_e = (gh)^{1/2}/f \tag{7.01}$$

where g is the acceleration due to gravity, h is the depth of the water, and f is the Coriolis parameter. In our earlier example we were concerned with two layers of different density and g was replaced with the reduced gravity (g') and h represented the depth of the upper layer. For the Kelvin wave in Fig. 7.06 we put $g = 10\ \text{ms}^{-2}$, $h = 4000$ m, and $f = 10^{-4}\ \text{s}^{-1}$ to get $R_e = 2000$ km. This value is commonly called the external Rossby radius, or deformation scale, as opposed to the internal Rossby radius derived in section 5.2.3. The external deformation scale is roughly 100 times the internal one.

7.2.4 Tidal currents

We see from Fig. 7.06 that a coastally trapped Kelvin wave will retain its integrity only if it is moving in a counter-clockwise manner around the coastal margin of a basin. The Coriolis force ensures that it will have a high amplitude close to the coast and a low amplitude far from the coast, typically at a distance of 2000 km. For the North Atlantic Basin, therefore, a pattern similar to that shown in Fig. 7.07 develops. In the mid-Atlantic there is an amphidromic point at which there is zero rise and fall with the tides, and the tidal ampli-

Fig. 7.06 A Kelvin wave travelling from the upper right to the lower right with the coastal wall to the right of its direction of travel. Note that the height of the wave decreases with increasing distance from the coast.

tude increases as one moves toward the coast. High tide appears successively at points along the coast, proceeding in a counter-clockwise direction. Similar patterns are found within smaller basins, such as the North Sea.

In the deep ocean the vertical range of the tide, as suggested by the illustration in Fig. 7.07, is only a few centimeters and tidal currents tend to be only a few centimeters per second. Over the continental shelves, however, the currents can be in the meters per second range as they are associated with much higher tidal amplitudes.

The direction of tidal currents varies greatly and depends on the way in which the tidal wave propagates in the local area. Along a straight coast or in a confined channel the currents tend to be parallel to the shore. In open areas the tidal wave is not constrained to be rectilinear, and currents in general will have both north–south (v) and east–west (u) components. Each of these components may be decomposed into the same sinusoidal tidal constituents that were found in the vertical excursion of the tide at the coast.

One method of presenting these data is shown in Fig. 7.08(a), which shows an ellipse representing the currents over one cycle due to the M_2 tidal constituent. The ellipse is constructed by plotting the u component against the v component and is commonly called a hodograph. The variations of the u and v components through time are further illustrated by the curves in the upper left and lower right of the figure. The curve in the upper left represents the flow along the east–west axis while the one in the lower right is the v component. It is quite common to specify the M_2 tidal flow by the length of the major and minor axes of the ellipse plus the orientation of the ellipse relative to north. If this method is used the M_2 tidal ellipse in Fig. 7.08(a) has amplitudes along the major and minor axes of ≈ 0.4 m s^{-1} and 0.1 m s^{-1}, respectively, and the major axis lies along 145 °T.

By noting the velocity of water along two axes at right angles at a fixed point throughout a tidal cycle, it is possible to calculate the total distance that a parcel of water has moved—the tidal excursion. Movement is usually highly variable and even oscillatory, so the displacement over a tidal cycle is calculated by dividing the records of u and v into segments so short that each can be assumed constant, then summing them. A shorter method, shown in Box 7.01, gives the same result.

Fig. 7.07 The amplitude (dotted) and phase (solid) of the M_2 tide in the North Atlantic Ocean. The amplitude is in meters and the phase in hours after the passage of the moon over the prime meridian. Adapted from Defant (1958).

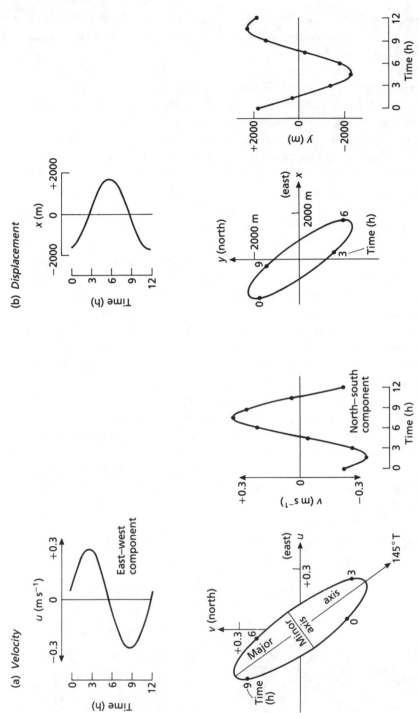

Fig. 7.08 (a) A tidal ellipse for the M_2 constituent calculated from a current meter record obtained near the bottom in 150 m of water at 53.8° N 55.2° W from August 1986 to August 1987. The numbers around the ellipse refer to the time in hours after the passage of the theoretical equilibrium tide. Separate plots of the u and v components versus time are in the upper left and lower right, respectively. (b) Plots of the particle displacement due to the M_2 tidal currents shown in (a).

Box 7.01 Estimating tidal displacements mathematically

We start with the u and v components of velocity displayed in Fig. 7.08(a). These are single-frequency oscillations and can be expressed by sine functions, so mathematically the components can be written in the form

$$u = A_u \sin(\omega t + \phi_u)$$
$$v = A_v \sin(\omega t + \phi_v)$$

(7.02)

where A_u and A_v are the amplitudes of the oscillations in meters per hour, ω is the frequency of the oscillation in radians per hour, which for the ≈ 12 h M_2 constituent is $2\pi/12 = 0.52$ rad h^{-1} (or 30° h^{-1}), t is the time in hours, and ϕ_u and ϕ_v are the phase angles to indicate where the zero of the sine function is relative to the time zero. For the u and v components shown in Fig. 7.08(a) the amplitudes (A) are 0.25 m s^{-1} and 0.34 m s^{-1}, respectively. In meters per hour these numbers become 900 m h^{-1} and 1224 m h^{-1}. The phases are 10°, or 0.17 rad, and 220°, or 3.84 rad, respectively, and we can write the u and v components as

$$u = 900 \sin (30t + 10)$$

and

(7.03)

$$v = 1224 \sin (30t + 220)$$

where we have used degrees to indicate the phase inside the brackets because it is a more familiar unit than radians.

The displacement, D_x, due to the oscillating velocity component, u, is given by the integral of the velocity as

$$D_x = \int A_u \sin (\omega t + \phi_u) dt = - A_u/\omega \cos(\omega t + \phi_u)$$

(7.04)

where the symbols are the same as in Eq. 7.02. The amplitudes of the displacements due to the u and v velocity components are $A_u/\omega = 900/0.52 = 1719$ m and $A_v/\omega = 1224/0.52 = 2338$ m, respectively. The displacement functions in the x and y direction are therefore

$$D_x = -1719 \cos (30t + 10)$$

and

(7.05)

$$D_y = - 2338 \cos (30t + 220)$$

where again the phase of the argument is in degrees. Outside the brackets ω must be in radian measure. These functions are plotted in the upper left and lower right of Fig. 7.08(b). By plotting the displacement in the x direction at a given time against the displacement in the y direction the ellipse in the lower left results. This plot is the same type as the velocity hodograph and results in a similar ellipse except that it describes the time sequence of a particle displacement rather than the time sequence of its velocity. Comparing the ellipses in Figs. 7.08(a) and (b) shows both to have the same shape and orientation but the phase is different in the two that we already knew because sine (velocity) and cosine (displacement) functions with the same arguments are 90° out of phase.

It should be remembered that the estimation of displacement developed in Box 7.01 does not use velocity observations obtained at the location of the moving particle but uses data obtained at one location, which is sometimes far from the moving particle. The calculation assumes, for lack of better information, that there are no horizontal variations in the velocity field. This assumption is not a bad one over short distances in regions where horizontal uniformity is likely, but near the shore and over rough shallow bathymetry the velocity field can change significantly in a few kilometers and extrapolating particle excursions from velocities at one position should be done only with care.

The current and displacement ellipses in Figs. 7.08(a) and (b) specify only the currents and displacements associated with the M_2 constituent and each of the other constituents may be plotted in the same manner. The other constituents tend to be of smaller amplitude and the main features of the tidal flow can usually be approximated by the M_2 tidal ellipse alone. At some locations the tidal ellipse appears more than those shown in Fig. 7.08, while at others the ellipses collapse into a straight line. These variations are due to the differences in the amplitudes and phases between the u and v components of velocity. These can be easily demonstrated by plotting figures similar to Fig. 7.08(a) after changing the values of the amplitudes and phases in the functions for u and v in Eq. 7.03.

7.2.5 Internal waves and internal bores

Tidal currents often cause internal waves to be generated on the pycnocline. These waves were mentioned in section 3.2.4, where it was shown that when the pycnocline is displaced vertically a restoring force is generated that pushes the pycnocline back toward its undisturbed position. This force leads to the possibility of a vertical oscillation and of waves being propagated on the interface. Some features of such a wave on a sharp density interface are illustrated in Fig. 7.09.

The interface in the figure has been placed close enough to the sea surface so that some effects of the internal wave motions are felt at the surface. The lower layer is assumed to be infinitely deep. The wave is propagating from left to right and causes the water particles to move along the particle streamlines. These lines indicate the instantaneous paths of the particles while the arrows within the streamlines suggest the magnitudes of the particles' velocities, which diminish with distance away from the interface. As the wave progresses, the pattern of streamlines moves along with it but the individual particles of water remain in their original locations moving in wave-induced circular orbits. The examples of these orbits in Fig. 7.09(b) illustrate that the orbit diameters, like the particle velocities, decrease with distance from the interface.

Interesting biological consequences of the internal wave motion are created by regions of convergent and divergent flow in the upper layer. Above the pycnocline the thickness of the upper layer clearly varies along the length of the wave. As the wave advances, water must flow from the thin regions above the wave crests to the thicker regions above the wave troughs. The water must, therefore, converge behind the wave crests and diverge behind the wave troughs as noted in Fig. 7.09(a). These regions of convergence and divergence exist up through the water column to the sea surface with diminishing intensity but their strength is enough to create a wave in the sea surface. This surface wave has the same length as the interface wave but it is 180° out of phase with the internal wave and has a much smaller amplitude.

On the sea surface the convergent zones cause floating organic matter to accumulate in bands that are ahead of, but parallel to, the wave crests. The organic material increases the surface tension of the surface layer, which reduces the amplitudes of the smaller surface ripples. This change gives the appearance of a band of water that is smoother than the water over the divergent zones where the small waves continue to exist. These smooth patches are

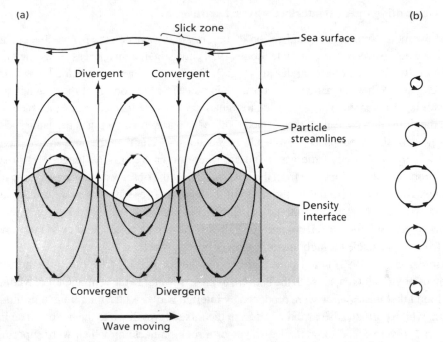

Fig. 7.09 (a) A wave progressing from left to right on the density interface showing the instantaneous particle streamlines, the resultant surface wave, the zones of convergence and divergence, and the region where slicks are likely to form on the surface. (b) The particle orbits over one cycle due to the passing internal wave. The diameter decreases away from the interface and the direction of the trajectory reverses across the interface as indicated by the double arrows on the central orbit.

often called 'slicks.' The contrast in the roughness of the sea surface can also be detected by a special instrument called the synthetic aperture radar. When such instruments are carried in aircraft or satellites they allow mapping of the sea surface features over large areas, as demonstrated by New (1988) for an area near the Celtic Sea shelf break.

Below the pycnocline the particle trajectories (Fig. 7.09b) due to the internal wave are opposite in direction to those above the interface. In addition, the regions of divergence and convergence below the interface are displaced by one-half of a wave length from the ones above the interface. These features make sense because the thickness of the lower layer varies along the wave in the opposite sense to the thickness variations above the interface. The reversal in the particle orbits across the interface leads to a shear across the interface that is especially strong at the wave crests and troughs. Halfway between the crests and troughs the motion due to the wave is predominantly vertical, as indicated by the vertical streamlines in Fig. 7.09(a).

The interface waves of Fig. 7.09 are a special form of internal wave. Throughout most of the ocean's depth the density of water increases gently and smoothly rather than rapidly across thin layers. The larger-scale density stratification also supports internal waves but they tend to be oscillations of the whole water column rather than waves confined to the region near an interface. For a description of these large scale oscillations and a more thorough examination of the interface waves introduced here, see Pond and Pickard (1983), Roberts (1975), New (1988), or Apel (1987). More information of the biological significance of internal waves is given in section 7.4.1.

7.2.6 Generating internal interface waves: lee waves

One of the more common mechanisms producing internal waves on the pycnocline occurs when the water flows over an obstacle. Long (1954) investigated this process in a long tank filled with two layers of unequal density (Fig. 7.10). The denser layer fills the lower third of the channel. When a small smooth obstacle is placed in the bottom of the channel and the lower layer is forced to flow from left to right, lee waves are formed on the interface between the fluids downstream of the obstacle. These lee waves are moving at the same speed but in the opposite direction as the water. Waves that are shorter than the ones shown travel too slowly for the speed of the water and are swept downstream. The longer and faster waves propagate upstream away from the obstacle. Thus, the observed lee waves are the interface waves that are selected by the speed of the stream. These lee waves are usually associated with steady flows but often tidal flows generate such waves on the lee side of shallow ridges. When the tidal stream slows down, the waves continue to exist but move away from the obstacle through the more slowly moving water.

Witman *et al.* (1993) noted pulses of warm water containing 2–3 times the normal amount of phytoplankton, close to the bottom, in 29 m of water in the central Gulf of Maine. They found that these pulses were produced by internal waves with a maximum amplitude of 27 m, which periodically carried water from the mixed layer down to the bottom. The internal waves were found to be the lee waves of a rock pinnacle on which were superimposed solitons propagated from Georges Bank (see below). Hence, tidal currents passing over features of the geomorphology probably serve both to mix nutrients up into the surface waters and to carry phytoplankton down to benthic filter-feeding animals.

Another type of lee wave generated by the same mechanism but over the period of the flood or ebb tide is sometimes created at the edge of continental shelves. An elementary illustration of its generation is given in Figs. 7.11(a) and (b). We first assume that the tidal flow through section A in the deep water is the same as that through section B in the shallow water and that far away from the shelf edge the currents are the same from top to bottom—that is, the currents are barotropic. On the flood tide, when the water is streaming onto the continental shelf, there must be more water passing through the upper layer in section B than through the upper layer in section A. This condition occurs because the layer is the same thickness in each location and the water through section B is going faster than through A. In the lower layer there is more water passing through section A than through B. The deficiency in the upper layer coupled with the surplus in the lower layer causes the interface between the two to rise in the vicinity of the shelf break. On the ebb flow the opposite occurs and the interface between the layers is forced down, as indicated in Fig. 7.11(b).

The raising and lowering of the pycnocline can be tens of meters and represents a large fraction of the depth of the upper layer. These large undulations propagate as waves both

Fig. 7.10 The interface in a two-layer fluid in which the lower layer is moving from left to right over a smooth obstacle whose height is less than the thickness of the lower layer. Adapted from Long (1954).

Fig. 7.11 (a) A section through the ocean in the vicinity of a shelf break illustrating the higher currents of the flood tide over the shallow water at B than in the deep ocean at A and the resultant rise in the pycnocline at the shelf break. (b) The same situation as in (a) except the tide is ebbing and the pycnocline is depressed at the shelf break.

into deep water and over the shelf but because of their large amplitudes nonlinear effects tend to dominate their existence. The fate of the wave that is called the internal or baroclinic tide is outlined in Fig. 7.12. First, the wave is in the form shown in stage 1 and travels toward the left over the shallower water. The front of the wave (to the left) begins to steepen because the troughs of the waves travel faster than the peaks. As the waves get steeper (stage 3) shorter waves begin to form that grow with time until only high amplitude but short waves remain (stage 5). These short waves, because they are quite far apart relative to their wavelength, are sometimes called solitary waves or solitons. They usually travel in groups, however, and are not really solitary but they may be treated mathematically as isolated phenomena.

As these packets of solitary waves propagate further onto the shelf their amplitude decreases as the high shears within the waves lead to Kelvin–Helmholtz instabilities (section 4.4.2) and turbulence (stage 6). The waves are dissipated within about 30 km of the generation region and it is the conversion to turbulence in this 30 km strip of wave energy that is thought to provide the mixing that increases the nutrient flux up into the euphotic zone and the subsequent biological enhancement (Sandstrom and Elliott 1984; Pingree and Mardell 1985).

A similar process has been observed over the northern edge of Georges Bank in early summer by Loder *et al.* (1992) and Brickman and Loder (1993). They found a depression in the thermocline over the edge of the bank being generated regularly on the ebbing tide. The length of the feature was 2–3 km, rather than the 40 km indicated in stage 1 of Fig. 7.12 for the internal tide. The authors suggest the depression they observed was an internal hydraulic jump.

Hydraulic jumps are commonly seen on the downstream side of dams where water flowing rapidly down a steep incline reaches a level where the incline is less steep. At this point, the flow slows a bit and since more water reaches this point than can be carried away the depth of the water increases sharply. The phenomenon is also often observed in water flowing rapidly over rocks in fast-moving rivers and streams. Brickman and Loder (1993) think the rapid tidal flow off the shallow part of Georges Bank creates a hydraulic jump when the flow reaches the deeper water at the edge of the bank where it slows down. Because the flow occurs in the interior of the water column, they call it an internal hydraulic jump.

The authors also suggest that the observed depression may be partly created by internal waves generated at the edge of the bank as part of the internal tide. On the ebbing tide these waves are unable to propagate away in the direction of the shallow bank against the rapid off-bank tidal flow. They therefore tend to congregate to produce the internal hydraulic jump.

The biological importance of the depressions comes when the tide turns and they are partly carried and partly propagated over the shallow part of the bank where they eventually break as in stage 6 of Fig. 7.12. Brickman and Loder (1993) calculate that the dissipation of these features supplies 0.42 μmol m^{-2} s^{-1} of nitrate into the upper layer. As this amount is roughly equal to the estimated nitrate demand of 0.36 μmol m^{-2} s^{-1}, the process may be a major contributor to the increased productivity observed in the region.

Another phenomenon associated with internal tides comprises internal bores. These conditions are occasionally observed when the wave of the internal tide runs into shallow water before it is dissipated in breaking solitary waves. In the familiar case of surface waves meeting the shore, the waves are fairly short and the water level rises and falls rapidly as they break. With the long wave associated with the tide, the water level rises and falls slowly over a very large region. We do not normally see the tide wave break. In some gently sloping river estuaries, however, the outflowing water of the river impedes the advancing tide. Eventually the rising tide overcomes the retardation by forming a steep turbulent front that leads the tide or tidal bore up the river. The advancing wave of the internal tide can end up in similar circumstances and generate an internal bore. In these cases the lower—i.e.,

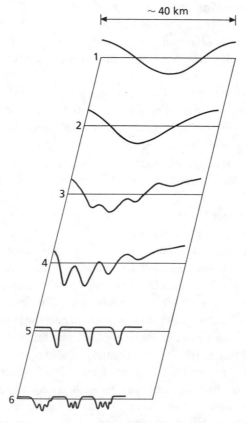

Fig. 7.12 Six stages in the life of the internal tide as it progresses to the left over the shelf, first becoming steeper at the leading edge, then forming shorter waves that become solitary waves that decay into turbulence.

denser—layer advances shoreward behind a front in which the depth of the layer increases dramatically.

7.2.7 Tidal rectification

The bathymetry of continental shelves is often uneven, being marked by shallow banks and basins. An example of this changing bathymetry in the Gulf of Maine is shown in Fig. 4.12. Oscillating tidal currents washing back and forth over the banks sometimes generate mean currents at the edge of the bank where the depth is increasing rapidly. This process of extracting energy from an oscillating tidal current to produce a mean unidirectional flow is called tidal rectification. Two places where it is thought to play an important role are around Georges Bank in the Gulf of Maine (Loder 1980) and around large sandbanks in the North Sea (Huthnance 1973).

The rectification of the oscillating flow depends on the fact that the tidal excursion is larger over the bank than in the deep water. This situation is illustrated in Fig. 7.13(a) in a cross section of the ocean through the edge of the bank, similar to Fig. 7.11 but without density stratification. The tidal flow through the vertical section 'A' is the same from top to bottom. If the same amount of water is to pass through section 'B' as section 'A' the speed of the flow in the shallow water must be larger than in the deep water. The same phenomenon is illustrated in Fig. 7.13(b) by the comparison of the M_2 tidal ellipses over the bank and in the deep water.

The change in speed of the tidal flow between the deep and shallow water occurs in a narrow band over the sloping edges of the bank and creates nonlinear forces associated with the acceleration and deceleration of the water. The rapidly changing speed also affects the magnitude of the forces due to the Coriolis effect and bottom friction, both of which depend on the speed of the flow. It turns out that in this region of rapidly changing forces the balance between them is disturbed in such a way as to produce the clockwise mean flow around the bank as illustrated in Fig. 7.13(b). The details of the mechanisms are not easily described outside the elegant analytic formulations of Loder (1980) and Huthnance (1973) and numerical models of Greenberg (1983).

Fig. 7.13 (a) A section through the ocean across the edge of a bank such as Georges Bank in the Gulf of Maine illustrating that the tidal current is higher in the shallow water than in the deep water. The shaded region over the sloping bottom indicates where the currents are accelerated and decelerated leading to 'tidal rectification' that generates the mean flow around the bank. (b) A view down on the bank shows the large tidal ellipse over the shallow water relative to the ellipse in the deep water and the tidally generated clockwise flow around the bank.

7.3 TIDAL MIXING IN THE WATER COLUMN

We saw in section 4.6 that tidally driven vertical mixing changes the seasonal pattern of phytoplankton production. There is also evidence that tidally mixed areas are favored as spawning grounds by herring. This section explores the evidence for these ideas.

7.3.1 Tidal mixing and plankton production

If tidal currents are strong enough to mix the water column all year, there is a continuous supply of nutrients from near-bottom waters up to the euphotic zone, which permits phytoplankton production to continue at a good level throughout the summer. This sequence of events is in contrast to the situation in stratified waters, where the supply of nutrients tends to become depleted after the spring bloom. As a result, tidally mixed areas like Georges Bank, near the Gulf of Maine, and the Dogger Bank in the North Sea have levels of primary production considerably higher than adjacent stratified areas of the shelf. It has been found that these sites are often selected as breeding grounds by commercially important fish stocks, and it is of interest to explore the connection between tidal mixing and the early life histories of fish.

7.3.2 Tidally mixed areas and the spawning of fish

Fisheries biologists have long been puzzled by the ability of a single species such as herring to divide into a number of discrete breeding stocks, each with a characteristic place and time of breeding, and often with recognizable small differences in body structure. The feeding adults of the various stocks are often found mixed together in the same place but at breeding time they segregate and return to the place where they were hatched. In this way they maintain genetic differentiation of the stocks. How do they recognize those breeding grounds, and what defines their limits? For some stocks, the breeding ground is readily identified as a group of fjords (e.g., the Norwegian coast) or estuaries (as on the Maine coast), but in other cases there is no clear physical boundary to the spawning ground or the nursery area in which the young fish develop. Iles and Sinclair (1982) noticed that the breeding grounds of a number of herring stocks are located in areas of vigorous tidal mixing, bordered in summer by a tidal front. They proposed that tidally mixed areas act to define and delimit the breeding and nursery areas of a number of stocks of herring on both sides of the North Atlantic.

The early work on predicting the occurrence of tidally mixed areas was done in waters around the United Kingdom (section 4.6.1). When tidally driven currents moving across the sea floor are sufficiently strong to generate turbulence that breaks down any stratification in the water column above, that area is said to be tidally mixed. As we showed in earlier chapters, Simpson and Hunter (1974) and Pingree *et al.* (1978a) showed how to predict the occurrence of tidally mixed areas on Georges Bank, Nantucket Shoals, in the mouth of the Bay of Fundy, and off the southwest coast of Nova Scotia.

Iles and Sinclair (1982) showed that there was a remarkable similarity between the distribution of larval herring and the occurrence of tidal mixing, on both east and west coasts of the North Atlantic. In the Gulf of Maine area larval herring are found on Nantucket

Shoals, on Georges Bank, and in the tidally mixed areas off New Brunswick and Nova Scotia. In the North Sea, the spawning grounds of the 'Downs,' 'Banks,' and 'Buchan-Shetlands' stocks of herring are in tidally mixed areas. Even in the Gulf of St. Lawrence, where the occurrence of tidally mixed areas had been predicted by Pingree and Griffiths (1980), five of six major herring spawning grounds are in tidally mixed areas.

The question then arises: what is it about these tidally mixed areas that makes them suitable as herring spawning grounds? Is the food-production process particularly favorable or could it be that the strong tidal currents remove fine particles and leave behind a gravel particularly suited to herring spawning? Some of these questions were addressed by Lough *et al.* (1989) during a series of dives in a manned submersible on the northeastern edge of Georges Bank. Cod (*Gadus morhua*) and haddock (*Melanogrammus aeglefinus*) are pelagic in the water column for the first few weeks after hatching in spring, then become bottom-dwellers about July. They were observed to live and feed primarily on a large pebble–gravel deposit (the gravel pavement) at 70–100 m depth where their coloration matched that of the substrate. In September the fish began to make night-time forays into the water column to feed on invertebrates. The clockwise gyre (section 7.2.7) then distributed them more widely over the perimeter of Georges Bank. Parts of the gravel pavement were also found to be the preferred spawning habitat of herring (Valentine and Lough 1991) and a good habitat for scallops. The gravel pavement is produced by the winnowing action of strong currents, which removes fine particles from the material originally deposited by glaciation. This interaction between bottom deposits and tidal currents may prove an important factor in the ecology of fishes and good management could potentially include a concern that the structure of the gravel pavement community is not destroyed by trawling.

In general (see Chapter 4), tidally mixed areas have a relatively uniform biomass of phytoplankton throughout the growing season, and the zooplankton biomass tends to peak late in the year. Since the herring in these areas may be spring spawners or autumn spawners, there appears to be no critical pattern of food production to which they are adapted. In fact, off southwest Nova Scotia there is an autumn-spawning stock that occupies the tidally mixed area principally in winter, when food production is minimal. There remains the possibility that the young herring make particular use of the high productivity associated with tidal fronts, but this suggestion has not been verified in the field.

Iles and Sinclair (1982) suggested that the important feature of these tidally mixed areas is that the boundaries act as natural barriers to the dispersion of the larvae during early development, thus ensuring the geographical discreteness of the stock. They pointed out that as tidal currents flow over the edges of banks like Georges Bank and the waters surrounding Grand Manan in the Gulf of Maine, the changing topography causes rectification of the currents and anticyclonic gyres are formed (Loder 1980; Loder and Wright 1985; Smith 1989; and see section 7.2.7). These persist even in winter when the front between stratified and tidally mixed waters disappears, and might well be the physical barrier that reduces dispersal of the young fish. In general, the idea that tidally mixed areas assist in the retention of fish in young stages of development, thus defining the geographical limits of the breeding grounds of the stock, has come to be known as the larval-retention hypothesis. An important aspect of the hypothesis is the suggestion that the size of the retention area sets an upper limit on the size of the stock that breeds there. Where the breeding ground is in a fjord or an estuary, the land forms a natural boundary of the breeding area. Otherwise, the boundaries of the tidally mixed area serve the same purpose. At one site in the Gulf of

St. Lawrence that is an important herring breeding ground but is not tidally mixed, there is a geographically fixed gyre that might well serve as a retention area.

Both here and in section 9.5.2 it is very noticeable that the argument does not proceed in a rigorous 'hard science' manner from one testable hypothesis to another, but consists of fragmented observations on different aspects of a very complex system. Ecologists who choose to investigate the properties of large ecosystems face a dilemma. They may isolate a small part of the system, find a 'manageable' problem, and formulate a rigorously testable hypothesis. But the total system, functioning in an integrated manner, is much more than the sum of its parts and at the present stage of development of ecosystem science one often has to resort to the process of collecting evidence that appears to support or not support a particular hypothesis.

7.4 THE BIOLOGICAL SIGNIFICANCE OF INTERNAL WAVES

7.4.1 Internal waves as nutrient pumps

We may recall what was said about shelf-break fronts in section 6.4.1. It has often been observed that the shelf break is an area of enhanced biological productivity. Fish are particularly abundant there (Fournier 1978) and it is very common to find greater amounts of phytoplankton and zooplankton than in adjacent areas. Fournier (1978) reported that in March 1977 the average value for chlorophyll-a across the Scotian Shelf was less than 1 mg m^{-3}, while over the shelf break it reached a value of 4.2 mg m^{-3}. Herman *et al.* (1981) found that chlorophyll-a was ≃1.5 mg m^{-3} over the shelf, rose to ≃6 mg m^{-3} at the front, and was 3–4 mg m^{-3} in the slope water. The corresponding abundances of copepods in the water columns of shelf, front, and slope waters were 30,000, 70,000, and 10,000 m^{-2}, respectively. The high concentrations of copepods occurred above the thermocline in the frontal region, and these copepods did not migrate down through the thermocline during the day, while those in shelf waters often did so.

Pingree and Mardell (1981) reported on a series of studies of the shelf break adjacent to the Celtic Sea in the eastern North Atlantic. In summer this region is characterized by a band of water about 100 km broad that is 1–2 °C cooler than water on either side of it. At times the band can be traced for about 800 km south from the Celtic Sea. The levels of inorganic nitrate and chlorophyll are significantly higher in this band than in adjacent water. Pingree and Mardell suggested that the interaction of tidally driven internal waves with the bottom topography was the cause of the phenomenon. During 1983 and 1984 Pingree and his colleagues continued to investigate the phenomenon, in collaboration with French oceanographers (Mazé 1983; Pingree *et al.* 1986; Mazé *et al.* 1986). Using both fixed current meters and an undulating towed sampler, they showed that during the off-shelf streaming phase of the barotropic tide the isotherms in the upper shelf region were depressed. When the tide slackened, the depression separated into on-shelf and off-shelf propagating internal tides. At large spring tides the crest-to-trough height of the internal tide was 50–60 m at the point of origin. The off-shelf waves had a wavelength of about 46 km and propagated at about 1.03 m s^{-1} while the on-shelf waves had a wavelength of 31 km and propagated at 0.7 m s^{-1}. The latter had a more rapid decay rate, indicating that they caused more vigorous mixing in shelf waters than offshore. As the field data show (Fig. 7.14) the large amplitude waves had relatively high-frequency internal waves associated with them

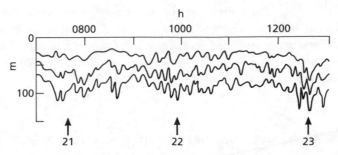

Fig. 7.14 Isopycnals along a line between points 50 and 150 km off the shelf break of the Bay of Biscay. Isopycnals are plotted at densities of 1026, 1026.5, and 1027 kg m^{-3}. Numbers 21–23 indicate tidal troughs with packets of short-period waves. From New (1988).

and concentrated in the troughs. This situation is a clear example of the phenomena described in section 7.2.5 and Fig. 7.12.

According to New (1988) these short waves produce a roughening of the sea surface that is often visible as a series of 'rips,' many kilometers long and spaced about 1 kilometer apart. As mentioned earlier, these are detectable from space by synthetic aperture radar imaging and illustration was given of the occurrence of several groups of these waves over the shelf edge in the Bay of Biscay. Each group of waves indicates the position of a depression of the thermocline in one of the long-period waves.

New (1988) also investigated the properties of such a system by means of a linear numerical model that described the topographic generation of the internal tides. He was able to show that there was good agreement between his model and field observations, and then went on to investigate variations in the Richardson number (Eq. 4.09). Since a steady, stratified shear flow is thought to become unstable to small perturbations (subcritical) when $R_i < 0.25$, New (1988) was able to calculate the position and duration of regions of potential instability and the probability that these would cause upward mixing of nutrient-rich water through the thermocline. He found that at spring tides during summer stratification, mixing was highly probable on the shelf and beyond the shelf break. At neap tides the Richardson number was not critical at any location, so no mixing was predicted.

The effect of internal tidal waves when they approach close to shore was investigated at the head of Monterey Canyon in Monterey Bay, California, by Shea and Broenkow (1982). They identified a tidal bore that, at the peak of the internal wave, forced a 20 m lens of cold, nutrient-rich water out of the canyon and onto the continental shelf. They considered this tidal bore to be a major contributor to coastal productivity when upwelling was not in progress.

7.4.2 Internal waves and phytoplankton production

We saw in Chapter 3 that a chlorophyll maximum is often close to the pycnocline in stratified waters, and net primary production in this region often constitutes a considerable fraction of the total for the water column. Phytoplankton in the upper part of the pycnocline are often nutrient-limited, while those near the base of the pycnocline receive nutrients by turbulent transport from below, but tend to be light-limited. Lande and Yentsch (1988) pointed out that internal waves travelling along the pycnocline are likely both to increase turbulent transport of nutrients and to cause the phytoplankton to oscillate in depth, thereby increasing the average light intensity experienced by them. This condition should be taken into account when making *in situ* measurements of primary production. Vandevelde *et al.* (1987), working in the central Gulf of St. Lawrence, produced a 3-day time series in which the chlorophyll maximum oscillated vertically with a frequency of about 16 hours, corresponding to the frequency of the internal waves, and nutrients in the photic zone were renewed by vertical turbulence.

7.4.3 Internal waves concentrate and transport planktonic organisms

It has been observed many times (e.g., Halpern 1971, Chereskin 1983) that as the tidal current generated by an ebbing tide flows over the edge of the continental shelf, or over the edge of a reef or bank, one or more depressions of the thermocline are formed on the downstream side. These depressions are known as lee waves. When the ebbing tide slackens off, the lee waves generate a train of internal waves that propagate shoreward (see section 7.2.6).

The water movement associated with these internal waves, when they are close enough to the surface, is as shown in Fig. 7.09. There are alternating zones of upwelling with divergences and downwelling with convergences. Ewing (1950) drew attention to the fact that convergences may be visible at the surface as smoother areas, or slicks, on a lightly rippled sea. He pointed out that a surface film of organic matter is naturally present on biologically productive waters and that in the convergence zones this film becomes thicker and is able to damp out small surface ripples. The internal waves are quite long, so that slicks may be hundreds of meters apart, and they move slowly shoreward (≈ 20 cm s^{-1}) with the internal waves. There is a tendency for buoyant material such as floating seaweed or amorphous organic matter to aggregate at the convergences, and for planktonic organisms to become associated with the aggregations. This action suggests that the shoreward transport by these slicks may have significance in the adaptations of organisms inhabiting them.

Shanks (1983) made daily collections of the pelagic megalopa larvae of the crab *Pachygrapsus crassipes* from the end of the 320-m-long pier at the Scripps Institution of Oceanography, California. He found that the daily abundance fluctuated with fortnightly maxima that occurred about 5 days before the spring tide. Noting the occurrence of surface slicks, and their relationship to the fortnightly tidal cycles, he used styrofoam cups weighted with sand as small drogues and deployed 50 cups 20 m apart on a line at right angles to the lines of the slicks. On three of the five occasions that he performed this experiment, the cups were not transported by the slicks, but on the other two occasions the cups were caught by the slicks and carried shoreward. At the end of the experiment over 90% were concentrated in two slicks and had been carried 1–2 km shoreward. Associated with these slicks were dense accumulations of flotsam.

In parallel with these physical experiments standard neuston net tows were taken in the slicks and in the rippled water between the slicks to sample the larvae of invertebrates and fishes. On the days when the drogues were concentrated, but not on the other days, the larvae of invertebrates and fishes were concentrated 6–40 times more densely in the slicks than in the water between them. While the author had no data on the differences between days when concentration and shoreward transport occurred and days when it did not, he suggested that the slicks might be a common method of onshore transport for larvae.

If larvae are to make effective use of slicks induced by internal waves, they must have behavioral adaptations that will take them to the surface and enable them to stay there in downwelling conditions. Shanks (1985) found that when megalopa larvae of *Pachygrapsus crassipes* were individually released by a diver at 1–5 m depth, they all swam to the surface. Laboratory experiments showed that they swam toward a light source, away from the direction of gravity, and in the direction of decreasing pressure. They also showed a tendency to cling to objects in their vicinity. Under a variety of experimental conditions their average swimming speed was 9.5 cm s^{-1}, which was considered more than enough to enable them to swim against the downwelling observed in the vicinity of slicks. Having arrived at the surface, they could then hold their position by clinging to floating objects.

A New Zealand crab, *Munida gregaria* has bright red larvae that aggregate near the surface of the sea in daylight. Zeldis and Jillett (1982) carried out aerial surveys and published beautiful photographs showing the larvae concentrated on what appeared to be mid-shelf internal wave slicks, as well as on river plume fronts and headland fronts. While their report lacked the experimental evidence provided by Shanks, it appears to be strong confirmatory evidence for the role of these physical mechanisms in transporting crab larvae shoreward.

Satellite observations show that tidally generated internal waves are refracted by bottom topography, and that some coastlines are more suitable for the production of internal waves than others (Apel *et al.* 1975). Shanks and Wright (1987) thought that the differential pattern of shoreward transport of slicks might account for some of the patchiness in barnacle distribution on rocky shores. They repeated the experiments of Shanks (1983) using styrofoam cups as drogues and found that barnacle larvae were indeed concentrated in slicks associated with the internal waves, and that settlement of barnacle larvae was more abundant in places where the drogues tended to accumulate. In addition to the barnacle larvae, they found that two kinds of crab larvae and gammarid amphipods were at times concentrated in the slicks.

Pineda (1991) placed a different interpretation on some of the earlier work off the Scripps pier. He drew attention to the work of Cairns (1967) showing that internal waves approaching shore in summer develop steep leading edges and greatly increased amplitude. If these internal waves break, they transport water shoreward, so that they may be regarded as internal tidal bores. At large amplitude they cause cold bottom water to move to the surface, which appears in the temperature records as abnormally cold surface water.

Analyzing the historical temperature records from the Scripps pier, Pineda showed that the low-temperature anomalies tended to occur most frequently in summer and 6 days and 19 days after the new moon, indicating a tidal mechanism. He also found a strong correlation between relatively low water temperatures and settlement of barnacle larvae, so he postulated that internal bores—not surface slicks—were the mechanism transporting the larvae onshore. If surface slicks were the main mechanism, we should expect that the larvae would be most abundant above the thermocline, whereas Barnett and Jahn (1987) found them to be most abundant near the bottom, supporting the tidal bore hypothesis.

In a later paper Pineda (1994) investigated events that follow the arrival of a parcel of cold water at the shoreline. He found that this water reverses direction and moves off shore close to the bottom, generating a gravity current. This movement has the effect of drawing warm surface water into the shore. In other words, an upwelling of cold water occurs as the internal bore (section 7.2.6) arrives at the shoreline, and downwelling appears as the cold water mass retreats along the bottom (Figs. 7.15 and 7.16). The onshore movement of warm water was found to be associated with a distinct warm water front in which crab larvae and fish larvae were concentrated. Pineda suggested that a moving front associated with an

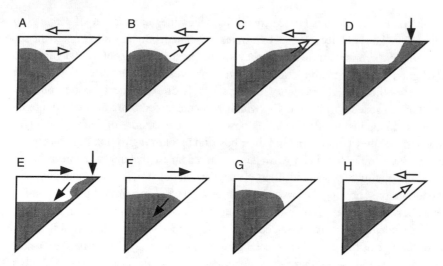

Fig. 7.15 Diagram showing onshore advection of dense water (shaded, A–C), followed by down slope progression and corresponding onshore advection of warm water (unshaded, E–H). Vertical arrows in D and E mark warm fronts. Other arrows represent currents. From Pineda (1994).

internal bore is more likely to transport a water mass and its contained organisms than is a slick associated with an unbroken internal wave.

Full resolution of the differences in interpretations of Pineda (1991, 1994) and of Shanks (1983, 1985) await further investigation. Kingsford and Choat (1986), working with the Shanks hypothesis, paid particular attention to internal waves and larval and juvenile fish at a site off the coast of New Zealand. They consistently found greater abundances in the slicks than in the rippled areas. They also found that fish were more abundant in the

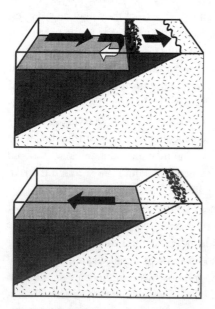

Fig. 7.16 Diagram showing the delivery of debris and larvae to the shoreline by a warm front, and its stranding as the cold water recedes. From Pineda (1994).

vicinity of drift algae. When the internal waves and their slicks were not present, the algae and the fish were in scattered patches at different distances from shore. When the slicks formed, the algae and the fish were arranged in lines and carried shoreward.

The implications of these findings are that quantitative studies of the larval abundances of shallow-water and intertidal organisms must take into account the nonrandom distribution of those larvae in coastal waters and of the physical mechanisms that may carry them to the coastal shallows from far out on the continental shelf. Since the occurrence of internal waves and slicks is dependent on the tides, it will not be surprising if there is a fortnightly tidal rhythm in inshore transport and settlement. Further studies of the relationship between internal wave packets and bottom topography will be expected to throw light on the patchy distribution of many coastal organisms.

7.5 TIDAL CURRENTS AND TOPOGRAPHY

7.5.1 Tidal currents and island stirring

Ever since Doty and Oguri (1956) described enhanced phytoplankton biomass and production in the vicinity of Hawaii and coined the phrase 'island mass effect,' it has been suspected that the disturbance of flow caused by the presence of an island may lead to upwelling of water from below the thermocline and hence nutrient enrichment of surface waters. However, there has been a lack of simultaneous physical and biological measurements to give firm support to the concept. Looking at this question more closely, Simpson and Tett (1986) drew attention to the evidence of enhanced phytoplankton biomass around two groups of islands on the continental shelf adjacent to the British Isles. They suggested that the cause was an interaction of the tidal currents with the islands.

Both the Scilly Isles (off the southwest tip of England) and St. Kilda (west of the Hebrides, western Scotland) have surface temperatures around the islands about 3 °C lower than the surface layer of the summer stratified region offshore. In part this variation may be explained by tidal mixing in the shallower water close to shore, but other effects are at work. For example, as tidal currents flow in a curved path around the island mass they are accelerated to a velocity about twice that in the far field, which leads to centrifugal displacement of surface waters and a compensatory upwelling of nutrient-rich deeper water. The biological productivity stimulated by these nutrients gives rise to increased phytoplankton biomass that is advected away from the islands by the residual currents. To the west of the Scilly Isles a plume of chlorophyll exceeding 4 mg chlorophyll-a m^{-3} sometimes extends for about 50 km into water that otherwise contains less than 0.5 mg chlorophyll-a m^{-3}. At St. Kilda, where the tidal currents are less strong, the zone of enhanced chlorophyll-a is at a rather lower concentration, 2–3 mg m^{-3}, but it extends over a much larger area, about 5000 km^{-2}. A numerical model based on the interaction of tidal currents with the Scilly Isles gave results that were in good agreement with the field observations.

Both island groups are nesting grounds for large numbers of sea birds (about one million on the Scilly Isles alone), and the enhancement of primary production by the interaction of the island mass with the tidal currents is thought to be a factor in their success. That is not to say that this mechanism is the only one operating. There may be a positive feedback in which the enhanced phytoplankton production gives rise to a richer population of planktonic and benthic animals close to the islands, which in turn stimulates the recycling of nutrients so that total productivity is increased both by the 'new' nitrogen upwelled from depth and by the nitrogen regenerated locally.

Sander (1981) considered internal tidal waves to be an important component of the island mass effect in Barbados. He demonstrated the presence of internal waves of up to 30 m amplitude and showed that the water beneath the thermocline was enriched with nitrate and phosphate. When he compared the potential for fertilizing surface waters from the internal waves with other potential sources of nutrients, he concluded that the internal wave mechanism was the most important. In other situations, increase in biomass of plankton and fish around islands has been attributed to convergence of water masses in the lee of the islands, to the fertilizing effects of land run-off, or to the normal processes associated with shallow nearshore waters, such as tidal or wind-induced mixing. Hernandez-Leon (1991) gives further references.

7.5.2 Tidal currents and coastal or bottom topography

The flow of tidal currents past coastal headlands often results in vertical transport of nutrients and stimulation of biological production. In a study of these processes around a promontory on an island in British Columbia (St. John and Pond 1992, St. John *et al.* 1992a), it was found that significant vertical mixing occurred in a plume downstream of the promontory during the ebb tide when the mean velocity through the water column was over 12.7 cm s^{-1}. At this velocity the authors suggested that the Richardson number (section 4.4.2) was low enough to allow vertical mixing events to occur. As a result, the nutrient levels of the upper mixed layer of the plume significantly increased. Calculations indicated these higher levels could lead to an increase in new primary production of between 19 and 314 tons in June and July.

At the southwest tip of Nova Scotia, Canada, lies Cape Sable. The waters have long been known to have anomalously low temperatures and high nutrient concentrations and to support high productivity of lobsters, groundfish, shellfish, and herring. Upwelling obviously occurs, and an early explanation (Garrett and Loucks 1976) was that it was driven by centrifugal forces associated with the strong tidal currents along the convex coastline. Tee *et al.* (1993) proposed a new mechanism involving bottom topography, offering field observations and a model in support of their proposal. Upwelling is generated by residual currents flowing from deep to shallow water across a submarine ridge. These upwelled parcels of water are transported by a longshore current away from the upwelling region and into a region of strong tidal mixing. The combination of topographic upwelling and strong tidal mixing leads to the observed cold-water anomaly and its associated high biological productivity off Cape Sable.

7.6 SUMMARY

In this chapter we have chosen not to review the enormous body of literature on life between tidemarks. The diurnal or semi-diurnal rise and fall of the tide creates on the shore a vertical pattern of physical zonation in which the lower regions have less exposure to the air, less tendency to dry out, and a more equable temperature, while the upper regions have converse properties. This knowledge creates a delightfully interesting stage on which organisms enact complex biological plays that are only now beginning to be understood. Since this zone is accessible for study by both naturalists and experimental scientists, there is copious literature on the subject that would require a separate volume to do it justice.

Instead, we have concentrated here on the effects of the tides on biological processes remote from shore. We have seen that the vertical mixing in the water column, caused by tidal currents, creates conditions in which nutrients are supplied to the phytoplankton all summer long, creating conditions favorable for the growth of larval and juvenile fishes. It has been noted that a very large proportion of the known herring spawning grounds are in areas that are tidally mixed all year round. Many of these are on offshore banks, where the water shallows enough to permit tidal mixing throughout the water column.

The interaction between tidal currents and offshore banks also tends to create a strong circular flow around the perimeter of a bank, and this feature is thought by some to create a barrier to the dispersal of fish larvae, retaining them in the biologically productive waters over the banks long enough for them to learn to recognize the bank as a site to which they should return for spawning. It has been suggested that this mechanism is important in the evolution of distinct stocks within a single species of fish.

Tidal currents moving over the bottom in stratified water also tend to generate internal waves at the pycnocline. Particularly large waves, 50–70 m in amplitude and of the order of 30 km wavelength, are generated where strong tidal currents flow over the shelf break. These waves are thought to mix nutrient-rich water up into the nutrient-depleted mixed layer, stimulating biological production near the shelf edge. Details of the food webs have not been worked out, but we know that the shelf-break region of the North Atlantic is particularly productive of fish.

On a smaller scale, as internal waves approach shore they create convergences at the surface in which organic films accumulate and produce visible slicks. These convergences also accumulate planktonic organisms and appear to be important mechanisms for carrying aggregations of planktonic larvae toward shore. It is thought that various crabs, fish, and corals may rely on this mechanism for the completion of their life histories. An alternative view is that as the internal waves approach shore they break and become internal bores, which are more effective than slicks in transporting small organisms.

Finally, we have seen that as tidal currents interact with islands they cause vertical mixing that brings nutrients to the surface and stimulates biological productivity around the islands. It has been suggested that this mechanism accounts for the large populations of sea birds that find various islands the best places for rearing their young.

PART C
PROCESSES ON A SCALE OF THOUSANDS OF KILOMETERS

8

Ocean Basin Circulation: The Biology of Major Currents, Gyres, Rings, and Eddies

8.1 INTRODUCTION

In this chapter we begin to consider the ocean basins in their entirety. We look at the major gyres that circulate anticyclonically in the subtropical regions of the North and South Atlantic and Pacific basins. We note that they are driven by the major global winds, which in turn are driven by the inequalities in the solar energy flux between the equator and the poles. North of the subtropical gyres in the North Atlantic and North Pacific are subpolar gyres that rotate in the opposite sense (i.e., cyclonically). Some of the most productive waters in the world are in the southern halves of these subpolar gyres, but the seasonal patterns of production are quite different in the two ocean basins.

Along the western margins of all these gyres are particularly intense currents—for example, the Gulf Stream and the Kuroshio current in the northern subtropical gyres, and the Labrador current and the Oyashio current in the sub-Arctic gyres. Organisms within these currents are transported long distances very rapidly, and we find that a variety of commercially important marine organisms, such as salmon, eels, and squid, migrate in these currents in the course of completing their life histories.

There is a tendency among biological oceanographers to take these intense western boundary currents as given natural phenomena, without delving too deeply into the mechanisms that drive them and cause them to be different from eastern boundary currents. However, as we enter an age in which human activity appears to be capable of altering the climate of the planet, and with it the circulation of the ocean, it is appropriate to try to understand the underlying mechanisms in order to try to predict the consequences of climate change. In section 8.2.1 the theory of what drives boundary currents is explored. The story is not a simple one, and physical oceanographers do not have all the answers, but a study of the theory sheds new light on the functioning of whole ocean basins, and on the way in which they may respond to climate change.

Satellite images of boundary currents show that along some parts of their length they begin to meander. Sometimes the meanders grow so large that the ends are cut off to form isolated rings of water, of the order of 100–300 km in diameter. These rotating water masses, which are particular forms of eddies, retain their distinct identity for several months and have their own special biological characteristics. If they are formed between the boundary currents and the coast, they may impinge on the continental shelf. Their enormous kinetic energy has the effect of drawing large volumes of water off the shelf on one side of the ring, while adding water on the other. The biological consequences for juvenile fish populations inhabiting the shelf water are often quite drastic.

It is now known that eddies are present throughout the ocean, although their energy often appears to decrease with distance from major ocean currents. It seems possible that there is a cascading of energy from the rings associated with the major currents, down to eddies remote from those currents. Since cyclonic eddies have the potential to cause upwelling in their centers, they may well be important in the global primary production budget.

Finally, in this chapter we note that the centers of the major subtropical gyres are permanently stratified and are thought to be biologically relatively unproductive. However, recent work suggests that these areas may have transient episodes of upwelling of nutrients that stimulate bursts of primary production. If this is true, we need to know more about them, for the subtropical gyres occupy a large fraction of the world's ocean, and their fixation of carbon dioxide is an important factor in the budget for carbon dioxide in the atmosphere.

8.2 THE WINDS AND THE WIND-DRIVEN CIRCULATION

All the major surface currents in the oceans are created by the drag of the wind on the surface of the water. The winds, in turn, are created because the earth's surface is heated unevenly by the sun, making the tropical regions warm and the polar regions cold. If the system were simple, the cold and dense polar air would flow under the warm and light tropical air while the tropical air moved north over the top of the cold polar air. But the fact that it all happens on a spinning sphere complicates the patterns. A very simplified diagram of the two major wind systems, the trades and the westerlies, is given in Fig. 8.01.

The trade winds arise because the warm air in the equatorial regions rises and is replaced by air flowing toward the equator in both the northern and southern hemisphere. The Coriolis force deflects the equatorward flows to the west giving rise to the northeast and southeast trades (winds are named after the direction from which they come, while ocean currents are named after the direction in which they are going). The air in the trade winds comes from the air that rises above the equator and then flows to higher latitudes before descending back to the earth's surface in the subtropical highs. These circular convection cells are known as the Hadley cells and the zone near the equator where the surface winds converge is called the intertropical convergence zone (ITCZ). The ITCZ moves north and south with the seasons, being furthest north in August and furthest south in February, but is always north of the equator (Gill 1982).

The Hadley cells tend to mix the air between the equator and 30° latitude, resulting in low horizontal temperature gradients between 30° S and 30° N throughout the lower 10–20

Fig. 8.01 Major wind systems of the world.

km of the atmosphere (the troposphere). Poleward of the Hadley cells the temperature at all levels in the troposphere decreases rapidly. This gives rise to a strong horizontal pressure gradient between $\simeq 30°$ latitude and the pole and to strong westerly winds. These winds are perpendicular to the pressure gradient because of the Coriolis effect and are in near geostrophic balance. The westerlies are continuous bands of flow right around the earth and are sometimes called circumpolar vortices. The speed of the flow is not constant at all levels but increases with height and reaches a maximum in the so-called jet stream at $\simeq 12$ km over $\simeq 30°$ latitude as illustrated in Fig. 8.02. This figure emphasizes the dominance of the westerly air flow compared with the small easterly flows of the trade winds near the equator and the polar easterlies. In the summer the jet stream is weaker and moves poleward about $10°$ in latitude.

The westerlies do not stay at a constant latitude right around the globe but meander north and south in waves that are about 10,000 km long. At any one time there are three to five complete waves around the globe but averaged over a month only three are usually visible. These waves are called planetary or Rossby waves and they cause the westerlies at a given longitude to move north and south over periods of a month or so. Since the movement of the weather systems is guided by the westerlies, the shifts in position also alter the

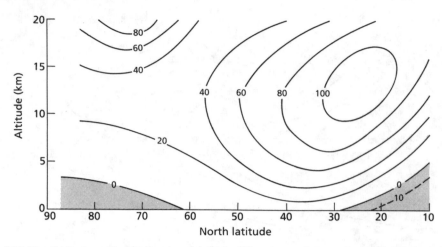

Fig. 8.02 The distribution with altitude and north latitude of the average winter zonal wind speed in km h^{-1}. Winds are toward the east except where shaded. Adapted from Pettersen (1969), with permission of McGraw-Hill, Inc.

paths of the weather systems, and is the process that causes the weather in the mid-latitudes to be different from the average for weeks at a time. As we shall discuss in the next chapter, the position of the westerlies can be shifted in special locations such as the North Atlantic Ocean for much longer periods than Rossby wave fluctuations. The longer term changes can last for years and are known as oscillations.

The winds shown in Fig. 8.01 drive the ocean currents shown in Fig. 8.03. However, the Coriolis force and its variation with latitude make the link between the wind stress and the water motion less than straightforward. At the equator, where the Coriolis force is zero, the situation is simple and the water moves in the same direction as the wind. The equatorial currents in the Atlantic and Pacific, for example, are mainly parallel to the equator. At other latitudes the Coriolis force is not zero and causes the moving water to be deflected to the right of the wind in the northern hemisphere and to the left in the southern hemisphere. The net direct effect of the wind is a flow perpendicular to the wind in the upper layer of the ocean. This is the Ekman transport, which begins the series of processes that generate the currents seen in the upper layers of subtropical and subpolar oceans.

As is evident in Fig. 8.03 the currents in the subtropics, between latitudes 15 and 45°, form large gyres in which the poleward-flowing currents on the western sides of the oceans are strong and narrow while the currents throughout the remainder of the oceans are broad and slow. In the North Atlantic Ocean, for example, the subtropical gyre is made up of the broad, slow North Atlantic, Canary, and North Equatorial currents flowing east, south, and west, respectively. The northward-flowing current that completes the gyre is the Gulf Stream, which is narrow and rapid. Similar subtropical gyres of broad, slow currents with strong narrow poleward flows like the Gulf Stream are found in all the other major oceans except the South Pacific. In the subpolar regions of the North Atlantic and North Pacific, gyres are observed in which the sense of the flow is opposite to that in the subtropical gyres—that is, counter-clockwise. These gyres, like the subtropical ones, contain broad, slow flows throughout most of the ocean with a strong narrow flow on the western side. The strong narrow flows, the Labrador and Oyashio currents, however, flow toward the equator instead of toward the poles. In the higher latitudes of the southern hemisphere,

Fig. 8.03 The principal surface currents of the world's oceans. Western boundary currents such as the Gulf Stream are emphasized. Positions of ocean weather ships B (Bravo), 56.5° N 51.0° W; I (India), 59.0° N 19° W; J (Juliett), 52.5° N 20° W, and P (Papa), 50° N 145° W are indicated by circled dots with letters.

Fig. 8.04 An idealized rectangular ocean (on the right) extending from 20° N to 40° N and from 20° W to 80° W. The coordinate origin is at 30° N 80° W and y is positive to the north; x is positive to the east. The wind stress (on the left) is purely zonal and decreases sinusoidally from 0.1 N m^{-2} at 40° N to −0.1 N m^{-2} at 20° N.

where there are no obstructing continents to force the formation of gyres, the main current is the Antarctic circumpolar current, which circles the globe at about 50° S.

8.2.1 Theory of the wind-driven circulation

The asymmetry of the currents in the large ocean gyres has been known for centuries but the reason for it remained obscure until Stommel (1948) showed mathematically that the westward intensification of the currents is due to the change in the Coriolis force with latitude. In the following short description we try to outline the main physical processes creating the current, but it is clear from the reviews of Fofonoff (1981), Rhines (1986), Stommel (1965), and Pedlosky (1990) that the subject is vast, complicated, and dynamic and we can only 'scratch the surface.' Our account concentrates on the North Atlantic Ocean and begins with the large-scale wind patterns and the Ekman transport.

The important parameter associated with the wind is the force per unit area or stress of the wind drag on the water surface. For the North Atlantic Ocean the yearly average of the strength and direction of the wind stress is shown by Isemer and Hasse (1987) to be similar to the distribution illustrated in Fig. 8.01. There is a maximum eastward stress between 40 and 50° N associated with the westerlies, a region of relative calm at 30° N associated with the subtropical highs, and the maximum westward stress associated with the northeast trade winds at about 15° N. Analyses by Hellerman and Rosenstein (1983) demonstrate that this north–south variation in wind stress is symmetrical across the equator.

To simplify our discussion we reduce the ocean to an idealized rectangle with idealized east–west (zonal) wind stress (Fig. 8.04). Our model ocean extends from 20 to 40° N and from 80 to 20° W with a wind stress that varies sinusoidally from a maximum of 0.1 N m^{-2} toward the east at 40° N to 0.0 N m^{-2} at 30° N, then increases again to 0.1 N m^{-2} to the west at 20° N. This idealized wind ignores any mean winds in the north–south direction (meridional winds), which are small relative to the zonal winds. Mathematically the idealized wind stress can be written

$$\tau_x = 0.1 \sin \frac{\pi}{2L} y \tag{8.01}$$

Fig. 8.05 A perspective drawing of a 1-m-wide slice of the ocean between 20° N and 40° N and from the ocean's surface to the pycnocline at 1000 m depth. The wind stress, changing from eastward at the north end to westward at the south end, is shown at the top of the picture. The top 100 m of the ocean is the Ekman layer, which converges toward the center (30° N) under the influence of the wind. Ekman pumping is indicated at the base of the Ekman layer and the southward Sverdrup transport is indicated in the layer above the pycnocline.

where τ_x represents the component of the wind stress in the x direction in units of N m^{-2} and y represents the distance north and south of 30° N, which varies between $+L$ and $-L$.

Referring back to section 5.2.2, we recall that the net effect of a wind stress on the water surface is to produce a transport in the upper layer of the ocean to the right of the wind known as the Ekman drift. The mass transport, M, of this flow is calculated by dividing the wind stress τ by the Coriolis parameter f as in Eq. 5.06:

$$M = -\tau/f \tag{8.02}$$

If we represent the wind stress with Eq. 8.01, the Ekman transport is found by substitution of Eq. 8.01 into Eq. 8.02 to give

$$M_{yE} = -\frac{0.1}{f} \sin(\pi y/2L) \tag{8.03}$$

where M_{yE} is the meridional Ekman mass transport through a 1-m-wide strip of the ocean in units of kg m^{-1} s^{-1}. This transport is illustrated at the top of Fig. 8.05, which shows a

1-m-wide slice of the ocean between 20 and 40° N from the surface to the main pycnocline at 1000 m. It is a perspective view looking toward the northeast. At the very top, above the surface of the water, is a representation of the wind field with a maximum stress to the east at 40° N and a maximum toward the west at 20° N. The Ekman layer takes up the top 100 m. North of 30° N the westward wind creates the southward Ekman transport shown by the shaded blocks of water with the arrows to indicate movement to the south. South of 30° N the east wind creates a northward transport in the Ekman layer. These oppositely directed transports, therefore, create a convergence in the Ekman layer that increases from zero at the extremities to a maximum in the center of the ocean at 30° N. Because the sea level cannot rise enough to accommodate the convergence it generates a downward velocity in the upper layer, which is indicated in the figure by the small arrows at the base of the Ekman layer. The downward flow is referred to as Ekman pumping. (If the velocity at the base of the Ekman layer is upward because of a divergence in the Ekman layer, it is called Ekman suction.)

An expression for the vertical velocity associated with the Ekman pumping can be derived from Eq. 8.03 by calculating the derivative of the transport with respect to the horizontal coordinate, which in this case is y. The result,

$$dM_{yE}/dy = -\frac{0.1\pi}{2Lf}\cos(\pi y/2L) \tag{8.04}$$

gives the amount of water that is piling up at latitude y. The maximum of this function is found at 30° N where the cosine is 1.0. At this latitude with $f = 0.73 \times 10^{-4}\ \text{s}^{-1}$ and with $L = 1.1 \times 10^6\ \text{m}$ (10° of latitude) the convergence works out to $2.0 \times 10^{-3}\ \text{kg m}^{-2}\ \text{s}^{-1}$ or 169 kg m^{-2} d^{-1}, which converts to a downward velocity at the base of the Ekman layer of $\approx 0.17\ \text{m d}^{-1}$. This velocity decreases to the north and south along with the cosine function and reaches 0.0 m d^{-1} at 20° N and 40° N.

The ocean beneath the wind-driven layer responds to the downward push of the Ekman pumping with a horizontal flow toward the equator that is many times larger than the Ekman transport. This derived flow, called the Sverdrup transport, is dictated by the fact that the Coriolis force varies with latitude. To understand this response we briefly review the consequences of the law of conservation of angular momentum or vorticity, which we first introduced in Box 5.02.

In oceanography the law is usually expressed by Eq. 5.09,

$$\frac{f + \zeta}{H} = \text{constant} \tag{8.05}$$

where f is the Coriolis parameter or twice the rate at which the earth rotates about the local vertical relative to the stars, ζ is twice the rate that the water is rotating relative to the earth about the local vertical, and H is the height of the column of water. In discussions of vorticity, f is called the planetary vorticity, meaning that it is the vorticity given to the water by virtue of its being on the rotating planet at the latitude ϕ. The term ζ is called the relative vorticity representing the vorticity of the water relative to the earth.

The two parts (f and ζ) are illustrated in Fig. 8.06. The cylinder of water on the surface of the earth at latitude ϕ moves with the earth as it rotates about its axis at the angular velocity Ω. The cylinder, therefore, rotates in a counter-clockwise sense relative to the stars

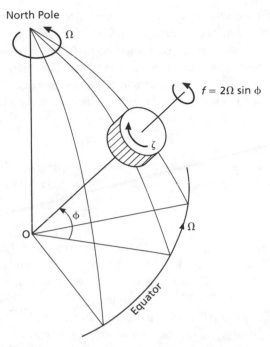

North Pole

Ω

$f = 2\Omega \sin \phi$

ζ

ϕ

O

Ω

Equator

Fig. 8.06 A cylinder of water sits on the surface of the earth at latitude ϕ north of the equator. The earth rotates at an angular velocity Ω causing the cylinder to rotate about its local vertical relative to the stars at $\Omega \sin \phi$, giving the cylinder a positive planetary vorticity $2 \Omega \sin \phi$. The cylinder is also shown to be rotating relative to the earth to give it a negative relative vorticity ζ.

at the rate $\Omega \sin \phi$, as was shown in section 4.2. The planetary vorticity f is twice this rate of rotation, or $f = 2\Omega \sin \phi$, which is counted positive in the northern hemisphere. If the cylinder in the figure is given a clockwise rotation relative to the earth, we call this its relative vorticity ζ and clockwise is taken as negative.

The consequences of the conservation law are illustrated in Fig. 8.07. Figure 8.07(a) shows the cylinder of water illustrated in Fig. 8.06 in cross-section. We assume that the cylinder is not rotating relative to the earth ($\zeta = 0$) and possesses only planetary vorticity f, which stays constant. If the cylinder is squashed so that H decreases, the water spreads out as shown in Fig. 8.07(b). The outward horizontal flow generates a Coriolis force to the right and gives the water enough clockwise rotation or negative relative vorticity to keep the ratio of Eq. 8.05 constant. If the column of water is stretched instead of squashed, depth H increases and the water in the cylinder flows inward. The Coriolis force generates a

(a)   (b)   (c)

H

$-$ve

$+$ve

Fig. 8.07 (a) A cross-section of the cylinder in Fig. 8.06 before and after squashing. (b) Overview of the cylinder spreading out as the height diminishes. (c) The same as (b) for a cylinder increasing in height and diminishing in radius.

counter-clockwise rotation or positive vorticity to the water, as shown in Fig. 8.07(c), which again maintains the constancy of the ratio in Eq. 8.05.

In the open ocean away from the coasts and intense currents, squashing or stretching water columns does not tend to produce relative vorticity. The water instead moves north or south (the Sverdrup transport) to seek a value of f that keeps Eq. 8.05 true without generating relative vorticity. For this part of the ocean, where ζ is always close to zero, we can rewrite Eq. 8.05 as

$$f/H = \text{constant} \tag{8.06}$$

Since we wish to know the consequences of a change in the height (dH) of the column of water we calculate the derivative of Eq. 8.06 with respect to the north–south coordinate y, to get

$$\frac{df}{dy} = \frac{f}{H}\frac{dH}{dy} \tag{8.07}$$

The term df/dy, the rate of change of the Coriolis parameter with latitude, is commonly called β. It exists only because there is a change in the vertical component of the earth's rotation with latitude, and the effects that depend on its existence are referred to as 'β effects.' If the Coriolis force did not change with latitude there would be no tendency for the water to change latitudes to satisfy Eq. 8.07 and Sverdrup transport would not exist. Since $f = 2\Omega\sin\phi$ we can calculate β as a function of latitude, $\beta = df/dy = 2\Omega\cos\phi \, d\phi/dy$. The derivative dy can be written as $Rd\phi$, where R is the radius of the earth (6.4×10^6 m), which results in the formula

$$\beta = 2\Omega\cos\phi/R \tag{8.08}$$

The angular velocity of the earth, Ω, is 2π rad d^{-1} or 7.27×10^{-5} rad s^{-1}. Therefore, at 30° N, where $\cos\phi$ is 0.87, β is $\simeq 2.0 \times 10^{-11}$m^{-1} s^{-1}. Using β in place of df/dy we rewrite Eq. 8.07 as

$$dy = \frac{fdH}{\beta H} \tag{8.09}$$

and by dividing both sides by the time interval dt the ratios dy/dt and dH/dt are velocities and the equation becomes

$$V_S = \frac{f}{\beta H}W_E \tag{8.10}$$

where V_S represents the Sverdrup velocity and W_E represents the velocity of the Ekman pumping. This equation can be used to calculate the north–south movement of the squashed cylinders of water at the base of the Ekman layer.

The downward push in the Ekman layer in our idealized model ocean is $\simeq 0.17$ m d^{-1}, which is W_E in Eq. 8.10. If $f = 0.73 \times 10^{-4}$ s^{-1}, $\beta = 2.0 \times 10^{-11}$, and $H = 1000$ m, Eq. 8.10 gives a value for V_S of 620 m d^{-1}. This value is based on the maximum value of W_E and is the maximum Sverdrup flow. Since the column is about 1000 m deep the flow represents a transport across 30° N of 6.2×10^5 m^3 d^{-1} or 7.2 m^3 s^{-1} through each 1-m-wide meridional strip across the ocean. If the ocean is 5000 km wide the total Sverdrup transport southward across 30° N is $\simeq 36 \times 10^6$ m^3 s^{-1} or 36 Sv, where 1 Sv or 1 Sverdrup equals 10^6 m^3 s^{-1} in honor of H. U. Sverdrup, the man who contributed so much to understanding the oceans.

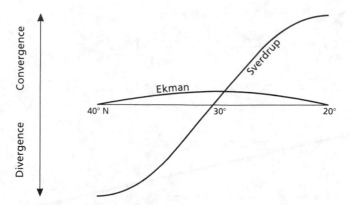

Fig. 8.08 The convergences and divergences due to the Ekman and Sverdrup transports between 20° N and 40° N.

This flow is illustrated in Fig. 8.05 by the shaded blocks of water between the Ekman layer and the pycnocline that are moving toward the equator. The volume of the flow varies from a maximum at 30° N to zero at the northern and southern extremes. The diagram also illustrates the fact that the Sverdrup flow in this example is always toward the south even though the winds and the Ekman drifts are in opposite directions in the northern and southern halves of the ocean. The important variable is, of course, the vertical velocity in the Ekman layer, which is downward at all positions across the ocean.

It is a curious fact, evident in Fig. 8.05, that the maximum Sverdrup transport at these latitudes is much larger than the maximum Ekman transport. Intuitively one might expect that the convergence in the Ekman layer would create an equal and opposite compensating divergent flow to keep the sea level and the depth of the Ekman layer constant in time. The Sverdrup flow, however, does nothing to compensate for the Ekman convergence but creates a much larger convergence and divergence in the water columns in the northern and southern halves of the idealized ocean as shown in Fig. 8.08. The Ekman convergence is a maximum at 30° N, while the divergence created by the Sverdrup flow is a maximum along the northern edge of the basin, and the convergence of the Sverdrup flow is a maximum along the southern edge. The Sverdrup flow is clearly not providing a compensation for the converging Ekman layer. If the Sverdrup flow was able to operate alone, the sea level at 40° N would decrease about 1 m per day while the sea level at 20° N would increase by the same amount. Such a change in sea level does not occur because large compensating zonal flows are generated that provide the water for the regions of divergence and remove the water from the regions of convergence.

Because a flow toward the east or west does not involve a change in the Coriolis force and hence a change in the planetary vorticity, it is easier for water to move in these directions than north or south. This explains why the compensating flows are east and west. Rhines (1986), when discussing this point, says the ocean is 'more rigid' to meridional flows than to zonal ones. An illustration of the zonal flows and the Sverdrup transport is given in Fig. 8.09. The arrows to the left represent the zonal flows that compensate for the convergence and divergence in the Sverdrup flow. The flow from north to south is shown on the right. The thickness of each tube is proportional to the volume of the water being transported through a 1-m-wide slice in the ocean and the small darker arrows indicate the direction of the flow.

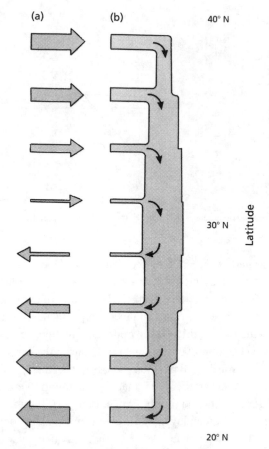

Fig. 8.09 (a) The zonal flows required to compensate for the convergences and divergences in the meridional Sverdrup transport. (b) The Sverdrup flow through a 1-m-wide north–south slice in the ocean with compensating zonal flows. The width of the tubes is proportional to the volume transport.

Fig. 8.10 The north–south Sverdrup flow across 30° N is constant at each meridian but the compensating east–west flows at each latitude (+y and −y) must increase to the west.

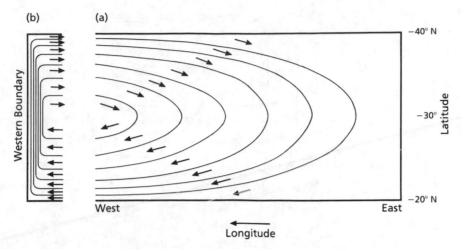

Fig. 8.11 (a) The total wind-driven circulation in the idealized ocean without a western boundary including the Sverdrup flow and the compensating zonal flows. The transport between contour lines is constant. (b) The presence of the western boundary forces the westward-flowing water of the gyre into a narrow northward current.

Each 1-m-wide north–south slice across the ocean requires the same zonal flows to compensate for the convergences and divergences in the Sverdrup flow. By placing a series of diagrams like the one in Fig. 8.09 side by side to represent successive north–south slices it is clear that the zonal compensating flows must increase toward the west. This increase is illustrated in Fig. 8.10, where each thin line represents the flow of water crossing 30° N through a 1-m-wide section that comes from the west along latitude $+y$ and returns to the west along latitude $-y$. The broad arrows on the left indicate the zonal flows at latitudes $+y$ and $-y$.

The diagram in Fig. 8.10 represents the flow delivered from one parallel of latitude north of the center line to a symmetric parallel to the south of the center line. A similar picture, from Stommel (1948), is shown on the right in Fig. 8.11 except that the contour lines represent the flow in the whole of the idealized ocean except near the western boundary of the ocean. The contour lines define the paths of the water particles and between two contours the volume transport stays constant. As the lines get closer together, the speed of the flow increases. The diagram incorporates both the meridional Sverdrup flow, which is constant across each parallel of latitude, and the compensating zonal flows, which increase toward the west. The pattern of currents illustrated in this part of the diagram will continue toward the west for as long as the ocean continues in that direction. There is no mechanism in what we have discussed so far to close the gyre with a flow toward the north. The reason is the fact that water is much easier to move east or west than it is to move north or south.

At the western side of the ocean some force must be generated to overcome the resistance of the water to go north to complete the gyre. The required force was shown by Charney (1955) to derive from the inertia of the westward-flowing water, which must slow down as it approaches the coast. Any acceleration or deceleration of a mass requires a force. The magnitude of such inertial forces is usually insignificant in the ocean, where normally the only forces of consequence are the Coriolis and pressure gradient forces involved in the geostrophic balance. The inertial force is dominant in only a narrow region

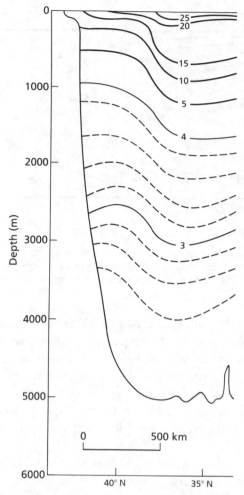

Fig. 8.12 A north–south cross-section of temperature (°C) between Bermuda and Nova Scotia showing the rapid depth change of the isotherms across the Gulf Stream. Adapted from Fuglister (1960).

near the coast but its generation causes the deflection of the westward-flowing current toward the north and causes the flow to be squeezed into a narrow rapid flow as shown on the left in Fig. 8.11.

The narrow current forces the density distribution in the ocean to change to satisfy the geostrophic balance discussed in section 4.3. The result is that the pycnocline along with the thermocline and the halocline rises up across the current. This is illustrated in Fig. 8.12, which shows the temperature distribution between Nova Scotia and Bermuda with the Gulf Stream between 38 and 39° N evident by the sudden decrease in the depth of the isotherms between the Sargasso Sea and the north side of the stream. The 10° C isotherm, for instance, rises from ≈800 m to ≈200 m, and this sudden rise is often used to map the position of the stream. The rapid rise in the isotherms means that on a level surface there is a rapid drop in temperature when crossing the stream; it is often called the 'cold wall.'

Because both a western boundary and inertial forces are instrumental in creating the Gulf Stream, it has been called a western boundary current or an inertial boundary flow. In

Fig. 8.13 The wind-driven circulation in the North Atlantic subtropical gyre including the Gulf Stream and its branches with some estimates of volume transport in Sverdrups. The rapid narrow currents such as the Gulf Stream are indicated by heavy lines. The slow broad flows are shown by light lines.

our idealized ocean this northward flow completes the gyre of the subtropical wind-driven circulation, but how does the model compare with the observed circulation?

8.2.2 The observed circulation

The near-surface circulation of the North Atlantic subtropical gyre presented in Fig. 8.13 is based on the works of many authors. The general path of the Gulf Stream is similar to the original diagram of Sverdrup *et al.* (1942) (page 684) but the northern and southern recirculation gyres are shown with greater intensity according to Hogg *et al.* (1986) and Worthington (1976), respectively. The maximum Gulf Stream transport of ≃150 Sv occurs at about 65° W. The branching and path of the stream near the southeast Newfoundland ridge are according to Mann (1967) and Clarke *et al.* (1980), while the branches at ≃45° N and ≃51° N to the east of Newfoundland are discussed by Krauss (1986). The continuation of the stream's branches in the North Atlantic current to the mid-Atlantic ridge is detailed by Harvey and Arhan (1988) and Sy (1988), and the circulation to the east of the ridge is adapted from Saunders (1982) and Klein and Siedler (1989). The circulation east of the Bahamas is based on Olson *et al.* (1984) and Stommel *et al.* (1978). Some of the flowlines in the eastern basin are not continued south of 35° N to indicate that much of this flow is thought to descend below the surface layers of the southern part of the gyre.

It is clear from Fig. 8.13 that the circulation pattern of our idealized ocean does not reproduce all the features of the real ocean. For example, the location of the maximum flow in the Gulf Stream is not where the model predicts it to be and the maximum flow of the

Fig. 8.14 A schematic illustration of meanders in the Gulf Stream and the formation of rings based on Richardson *et al.* (1978) and Parker (1971).

stream is almost four times the Sverdrup transport, which we estimated in the previous section to be $\simeq 36 \times 10^6$ m^3 s^{-1}. In the model the maximum flow occurs at the latitude where the wind stress is zero and changing from eastward to westward. This point actually occurs at about Cape Hatteras; however, the Gulf Stream continues to increase in intensity north of this latitude. The increase in the flow above the predicted amount is thought to have nothing to do with the Sverdrup transport but is associated with the recirculation gyre between Cape Hatteras and the Grand Banks, which is not well understood (see Fofonoff 1981).

One other feature of the Gulf Stream and its continuation, the North Atlantic current, that is not in the model is the fact that the strong narrow current continues toward the east well past the region where it would disperse in the model. Also the branching of the flow that occurs to the east of the Grand Banks of Newfoundland is not part of the model. That the stream does split into branches was first postulated by Iselin (1936) and has been confirmed by various authors since, as mentioned above, but the reason for the splitting is not yet understood. It is obviously associated with the decay mechanism of the current, which cannot spread out evenly but only in smaller and smaller well-defined narrow streams.

Another curious feature of the western boundary currents that does not seem to have a satisfactory explanation and that is not part of our model is the fact that they turn away from the western boundary and flow out into deep water. The Gulf Stream, for example, leaves the continental shelf at Cape Hatteras and travels northeast over deep water rather than continue over the continental shelf. This behavior is also seen in the other western boundary currents in Fig. 8.03. The Agulhas current is the most impressive case, for when it leaves

Fig. 8.15 (a) Plan view of Gulf Stream meander propagating in the direction of the broad arrows (i.e., east). The circle indicates the cyclonic flow generated by the convergence associated with the moving front and the dotted line suggests the disturbed particle path deflected to the right at B and to the left at C. (b) The depth profile across the current of an isopycnal in the pycnocline. The region of cyclonic motion is indicated at A by the curved arrow. Arrows on the isopycnal indicate the disturbed motion as it appears along the isopycnal.

the coast it turns almost completely around to flow in the opposite direction. It is said to retro-reflect. The fact that the current is travelling along a western boundary that is trending toward the west as it moves poleward, rather than toward the east and the coast of North America, emphasizes the turning of the current. A number of suggestions to explain this phenomenon have been reviewed by Fofonoff (1981) but he concludes that no satisfactory answer has yet been found.

8.2.3 Meanders, rings, eddies, and gyres

(a) Meanders

After the Gulf Stream leaves the coast, it continues eastward as a strong narrow stream. At about 65° W, it becomes unstable and begins to develop large north–south oscillations, or meanders (Fig. 8.14). Recent experimental and theoretical studies (Lindstrom *et al.*, submitted, Flierl and Davis 1993) have shown that vertical motion is generated within the current when these loops are translated eastward. The motion is such that nutrient-rich water from below the euphotic zone rises into the euphotic layer on the western side of the north-facing loops, which may lead to increased production.

The key elements of the process are illustrated in Fig. 8.15. The horizontal map in Fig. 8.15(a) indicates an idealized Gulf Stream meander moving east. The diagram in Fig. 8.15(b) shows the depth change of the pycnocline across the current. For simplicity we assume the upper- and lower-layer waters are homogeneous but of different densities.

As the meander translates eastward, upper-layer water must move out of the way of the approaching tilted pycnocline and deep-layer water must move in behind the front. The latter flow constitutes a convergence of water in the neighborhood of A. Such a convergence, as shown in Fig. 8.07(c), creates a counter-clockwise or cyclonic flow because of the conservation of angular momentum.

The net result of the cyclonic motion is the addition of a small perpendicular component of flow that deflects water particles to the right at B and to the left at C—that is, along the dotted line in Fig. 8.15(a). Because the pycnocline slopes down to the right, the water deflected to the right moves along isopycnal surfaces to greater depths and the water deflected to the left moves to shallower levels, as suggested by the arrows on the isopycnal in Fig. 8.15(b).

The biological signal is derived from the leftward deflection that pushes water to shallower levels. This motion replaces some of the water in the pycnocline, and possibly the mixed layer on the left side of the current, with water of higher nutrient levels. Flierl and Davis (1993) analyze this process and its effects with a numerical model and calculate a 10–20% enhancement of phytoplankton production in these areas.

(b) Rings

The meanders in the Gulf Stream often grow too large to continue as meanders. The ends then separate into isolated rings of water, as illustrated in Fig. 8.14. The rings to the south of the Gulf Stream rotate counter-clockwise and enclose colder water from the north side of the current; they are known as cold-core rings. The clockwise-rotating rings to the north of the stream are called the warm-core rings because they contain the warmer water from the Sargasso Sea and are surrounded by the cooler waters found north of the stream. Both kinds of rings are baroclinic phenomena—that is, they are confined to the thermocline or the upper $\simeq 1000$ m.

Each year there are about 10 cold-core rings formed. Each exists for about a year before it is either reabsorbed into the Gulf Stream or loses its unique properties to diffusion (Ring Group 1981). At any one time the cold-core rings represent about 10% of the surface area of the Sargasso Sea. According to Richardson (1983) the number of warm-core rings formed each year is about half that of cold-core rings. The production rate, however, is quite variable. Sometimes none can be identified in the region north of the Gulf Stream, while at other times the region is almost filled with rings. The warm-core rings also tend to be smaller than the cold-core rings with diameters of $\simeq 100$ km as opposed to 100–300 km for the cold-core ones.

The rings appear as pools of anomalous conditions because they contain water from the opposite side of the stream having the other side's physical, chemical, and biological properties. As the anomalous conditions change over the approximately 1 year ring's life, through mixing with the ambient water, the biology changes in ways that are unique to the rings, as we discuss in sections 8.4.3 and 8.4.4.

(c) Eddies

Although rings are a form of eddy, the term eddy is usually reserved for the lower-amplitude current variations that are found throughout the ocean. These oscillations, described by Robinson (1983), tend to be barotropic (constant with depth) as opposed to baroclinic, and because the ocean is full of them they interact in a way that is reminiscent of turbulent eddies. For this reason the properties in the eddies change much more rapidly than the properties within the rings, so the biology of an eddy is less individualistic than that of a ring.

The time scale of one rotation of an eddy is usually 10–30 days, and the horizontal scale between 10 and 100 km, close to the internal Rossby radius (section 5.2.3). Two terms often used to describe eddies are 'synoptic' and 'meso-scale.' Synoptic is borrowed from meteorology, where it is used to indicate the eddies in mid-latitudes associated with cyclonic storms. It implies a horizontal scale that can be sampled more or less simultaneously. Meso-scale is an ill-defined term indicating an eddy smaller than ocean basin scale, of the order of 100 km rather than 1000 km.

The energy level of eddies tends to be highest in those close to narrow boundary flows like the Gulf Stream. Currents in these eddies may be about 1 m s^{-1}, while currents in eddies remote from major currents (e.g., in the southeast Pacific) are nearer 0.01 m s^{-1}. It has been suggested (Robinson 1983) that most open-ocean eddies are formed indirectly from strong currents, with a transfer of energy through the eddy field in a way resembling the cascade process described in section 2.2.3. However, neither the location nor the mechanisms of such processes are yet known.

Two other types of eddy are recognized. One type is clearly formed by the interaction between flows and bathymetric irregularities. A second kind is the result of wind stress on the surface.

As pointed out by Swallow (1976), the variability associated with ocean eddies had been observed for many years before their full importance was discovered but they were thought to be transient features associated with the wind or the internal tides. Swallow and Hamon (1960) first observed the currents in the deep open ocean with neutrally buoyant floats and discovered that the currents varied by 0.5 m s^{-1} rather than the expected 0.01 m s^{-1}. Further work revealed that, in total, the eddies in the ocean actually contain more kinetic energy than the mean flows.

(d) Gyres

The word 'gyre' is usually used to describe a circular current that is confined by or associated with bathymetric features. The term covers a wide range of spatial scales. Thus, the currents of the Gulf Stream, the North Atlantic current, the Canary current, and the North Equatorial current make up the North Atlantic subtropical gyre at the ocean basin scale. Over shallow banks such as Georges Bank in the Gulf of Maine there is a continuous current flowing around the bank located over the edge. This circulation is the Georges Bank gyre.

8.2.4 Rossby waves

Rossby waves are large, slow, horizontal oscillations in the open ocean (Fig. 8.16). Some of these have recently been associated with the low-frequency variations in the equatorial

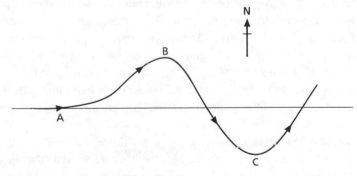

Fig. 8.16 The Rossby wave; displacements to the north or south in an eastward flow are turned back toward the original latitude by the generation of relative vorticity.

Box 8.01 Conservation of vorticity in Rossby waves

The conservation of potential vorticity is expressed by Eq. 8.05. If we assume that the thickness H stays constant, the ratio in Eq. 8.05 must remain constant for a water parcel as it moves. Thus we can write

$$f_A + \zeta_A = f_B + \zeta_B \tag{8.11}$$

where the subscripts A and B refer to different latitudes. At the beginning of the path shown in Fig. 8.16 we assume that ζ_A is zero. Thus at the latitude B the relative vorticity must be equal to the difference in the planetary vorticities of the two positions:

$$\zeta_B = f_A - f_B \tag{8.12}$$

Since in our example $f_B > f_A$ the relative vorticity will be negative or clockwise. Its magnitude can be easily calculated. Assume, for example, that the latitudes at A and B are 30° N and 40° N, respectively, where the values of f are 0.73×10^{-4} s^{-1} and 0.93×10^{-4} s^{-1}. The difference is the value of ζ_B and is -0.2×10^{-4} s^{-1}.

The simplest form of the Rossby wave is the one that is barotropic or constant throughout the depth. Following von Schwind (1980) these waves have a phase speed of

$$C = -\beta \lambda^2 / 4\pi^2 \tag{8.13}$$

where λ is the wavelength and β is the north–south gradient in the planetary vorticity given by Eq. 8.08. For $\beta = 2 \times 10^{-11}$ and $\lambda = 1000$ km the phase speed is 0.5 m s^{-1} toward the west. In fact, the phase speed of all Rossby waves is toward the west. The Rossby waves existing in the upper layer of a two-layer ocean and those existing on currents travelling in directions other than east are more complicated than the simple ones discussed here and are explored in Apel (1987) and Gill (1982).

Pacific that lead to El Niño (Chapter 9). The waves can have amplitudes of hundreds of kilometers, wavelengths of hundreds to thousands of kilometers, and periods of tens of days. Unlike the surface wind-driven waves and the internal waves that arise because of the force of gravity, Rossby waves exist because of the north–south variation in the vertical component of the earth's rotation. This is the β effect which was shown in section 8.2.1 to be responsible for the westward intensification of the wind-driven circulation.

To help explain how this factor can generate a wave motion, consider the eastward flowing current shown in Fig. 8.16. The water at A, besides moving east, is rotating around to the left with the rotation of the earth at a rate of $\Omega \sin \phi_A$ (as was pointed out in the discussion of the Coriolis effect in section 4.2 and Fig. 4.01). Let us assume that the water is deflected northward toward B. The water carries with it the rotation $\Omega \sin \phi_A$ but it is moving north where the vertical component of the rotation is $\Omega \sin \phi_B$, which is greater than $\Omega \sin \phi_A$ because $\phi_B > \phi_A$. So the water in the current at B is surrounded by water that is turning faster to the left. At B the reference frame is the earth, and the water imported from the south is rotating, relative to the local conditions, to the right or back toward the latitude where it began.

If the water now heads south, passes the original latitude, and heads for C, it will arrive in a region where the rate of rotation $\Omega \sin \phi_C$ is less than the water in the current. The water in the current then appears to turn to the left relative to the local vertical and again head back toward the original latitude. The wave motion is created because the poleward increase in the earth's rate of rotation provides a restoring force to north–south displacements of eastward currents. The same argument presented in terms of potential vorticity is given in Box 8.01.

That Rossby waves cannot be formed on a westward-flowing current can be easily seen by thinking of the current in Fig. 8.16 going in the opposite direction. A northward displacement generates clockwise rotation as in the eastward current, but in this case it tends to push the water further north rather than back toward the original latitude. Similarly, a southward displacement of a westward-flowing current generates counter-clockwise rotation that also tends to push the water further south. Thus, for a westward-flowing current displacements to the north or south do not generate a restoring force to push the water back to the original position and Rossby waves on such a current are not possible.

Rossby waves can also be formed when the potential vorticity is changed because of a change in depth. These topographic Rossby waves operate the same way as the ones that depend on the β effect except that we assume f to be constant and H to vary in Eq. 8.05. If we assume in Fig. 8.16 that the depth of the bottom is less in the northern half of the diagram than in the southern half, then the topographic effect will be capable of generating a wave. Consider the flow of water toward the east as before. A northward displacement causes H to decrease. For the ratio of Eq. 8.05 to remain constant, the numerator must also decrease, which means a decrease in ζ since f is assumed to be constant. The decrease in ζ is a clockwise rotation tending to restore the water to the original position. A displacement to the right leads to an increase in H and an increase in ζ. The counter-clockwise rotation pushes the water back toward the original position and a wave is formed. Like the β effect waves, the topographic Rossby waves can exist only for certain directions of flows. A flow along an isobath with shallow water on the right rather than the left cannot support a barotropic Rossby wave. In some cases both the topographic and the β effects are present, such as in the eastward current over a bottom sloping up to the north. The behavior of the Rossby wave will depend on the magnitude of the two effects.

8.3 DISTRIBUTION OF BIOLOGICAL PRODUCTION IN OCEAN BASINS

We have examined the consequences of differential solar heating at low and high latitudes that, in conjunction with the rotation of the earth, leads to westerly and trade winds. These winds, in turn, drive major anticyclonic subtropical gyres in each of the major ocean basins. Associated with them are powerful western boundary currents that transport large quantities of heat away from equatorial regions. These currents meander and cut off large gyrating bodies of water, the warm- and cold-core rings, that lead independent existences for several months.

We now examine the ways in which organisms have adapted to the various physical regimes. Some use the boundary currents for long-range transport between breeding grounds and feeding areas. Others make several circuits of an ocean gyre while growing to maturity. We shall see that the standard pattern of primary production for open-ocean situations (Chapter 3) is much modified by the presence of rings and gyres, with the result that primary and secondary production in the central basins is considerably higher than we thought a decade or so ago.

The biological phenomena associated with the major eastern boundary currents were dealt with in Chapter 5. These are regions of major coastal upwelling that are home for major stocks of sardines and anchovies. When we turn our attention to the western boundary currents and to the whole gyral pattern of circulation we find that well-known species like squid, eels, and salmon make migrations of thousands of kilometers by travelling with these currents. We also find that the larvae of some kinds of coastal invertebrates live long enough in the plankton to make, for example, the crossing of the North Atlantic before settling on the opposite side from where they were hatched. This movement leads to changes in our traditional ideas about gene flow between coastal populations on opposite sides of ocean basins.

8.3.1 Squid and the western boundary currents

Illex illecebrosus is caught in great quantities off the east coast of North America from Georgia to Newfoundland (Coelho 1985). From January to early March larvae occur in slope water along the northern edge of the Gulf Stream, roughly from Cape Hatteras to the latitude of Cape Cod. In May juveniles are found in warm waters along the southern slope of the Grand Banks, and from there they migrate to Canadian inshore waters in late June and July. They grow about 1.5 mm d^{-1} during the summer, and by November many of the males are mature, but the females are less advanced. There appears to be only one year-class and it is thought that the adults return south to spawn on the northern edge of the Gulf Stream, thus completing their life cycle in 1 year. The recapture of tagged squid confirms that this migration can occur.

Since squid are also caught inshore from Cape Hatteras to Cape Cod, it is inferred that part of the stock leaves the Gulf Stream at the latitude of Cape Cod and moves into the southward-flowing coastal currents. Warm-core rings (see section 8.4) may play an important part in this shoreward movement. The data supporting the view that the squid use the Gulf Stream to facilitate their migration are reviewed by Rowell and Trites (1985) (see also Trites 1983 and Coelho 1985). The center of the spawning area appears to be the Blake Plateau, off Cape Canaveral, Florida. The larvae and juveniles are thought to enter the Gulf Stream and be transported northward up to 1000 km per week, but many are probably caught up in frontal eddies that delay their northward progress and have the effect of mixing larvae and juveniles of different ages. Dawe and Warren (1992) found that the time series of catches in the United States, on the Scotian Shelf, and off Newfoundland all varied in the same pattern, albeit with more marked fluctuations in the more northern regions. It appeared that when the stocks were abundant in the south they migrated north in greater numbers. Analysis of environmental data showed that Newfoundland catches were inversely correlated with winter iceberg counts on the Grand Banks and positively correlated with high January sea-surface temperatures off the Scotian Shelf.

There is an analogous association between the Japanese squid *Todarodes pacificus* and the Kuroshio current (see map, Fig. 8.03). In this case there are three stocks breeding at different seasons: winter, summer, and autumn. The winter-spawning group is associated with the Kuroshio current. After spawning in the period January–April in the East China Sea, the larvae and juveniles travel north with the Kuroshio current, then turn inshore and are caught between the islands of Honshu and Hokkaido in summer.

The summer spawning is in another part of the East China Sea from which the larvae are entrained into the Tsushima current that flows north between the islands of Japan and the mainland. This current meets a southward-flowing cold coastal current, the Liman current, and the summer-spawned squid are fished along the boundary between the two. This example clearly illustrates the use of these western boundary jet currents as 'rapid transport,' enabling the eggs and larvae to develop in winter in warm water while the adults travel north with minimum energy expenditure to exploit the rich feeding grounds further north.

8.3.2 Eels and the North Atlantic gyre

The classic work of Schmidt (1922) on the life history and migrations of eels is well known. As a result of extensive collection of the translucent, leaf-like larvae (known as leptocephali) he concluded that there were two species of *Anguilla* in the North Atlantic: the European eel, *Anguilla anguilla*, and the American eel, *Anguilla rostrata*. (There are also at least 25 other genera.) The smallest larvae of both species were found in the region of the Sargasso Sea, with clear evidence of increasing mean size of European eel larvae as one moved closer to Europe, and increasing size of American eel larvae as one moved north from the Sargasso Sea toward Newfoundland. Schmidt therefore proposed the startling theory that all European eels are derived from a stock that breeds in the Sargasso Sea in winter, and that the leptocephali take over two years to reach the coasts of Europe, where they metamorphose into the adult form and migrate up the rivers and lakes to places far inland. No less startling was the view that this stock is maintained by eels that make the incredibly long breeding migration down the rivers of Europe and thousands of kilometers westward across the Atlantic Ocean to breed in the Sargasso Sea. One wonders whether eels have used the same breeding ground since the two sides of the Atlantic were much closer together, and have continued to colonize all the rivers available in Europe and North America, even though the continents have drifted apart.

In general, the case for the existence of two distinct species has been substantiated. Counts of muscle blocks (myomeres) in the larvae yield a mean of about 114 for European eels and about 106 for American eels. Counts of vertebrae in the adults yield similar differences in the mean values. A study of parasite distribution in the two species supports the view that an ancestral population inhabited both sides of the North Atlantic as well as Greenland, Iceland, and Scandinavia, but that the population split into European and American components when it was forced southward by the Pleistocene glaciation (Marcogliese and Cone 1993). A restriction-enzyme survey of the eels' mtDNA supports the existence of two distinct species (Avise *et al.* 1990). The present-day Iceland population appears to contain a small proportion of individuals resulting from hybridization between American and European eels, however.

Kleckner and McCleave (1985, 1988) made an intensive study of the distribution of Atlantic eels in the Sargasso Sea. They found that the smallest stages were confined to the southern Sargasso Sea, south of a front separating the warm, saline, permanently stratified surface water from more variable subtropical convergence zone water to the north.

After examining both historical data and new samples McCleave *et al.* (1987) concluded that the American eel is genetically distinct from the European eel, but that their breeding grounds have a considerable area of overlap. They suggested the existence of some reproductive isolating mechanism, since larval forms with intermediate characteristics have not been found. Small, newly hatched larvae of the American eels were found from mid-February to April and small European larvae from late February to mid-July. Both species enter the Gulf Stream. American eels grew at a rate of about 0.24 mm d^{-1} from February to October. A large fraction of the population was probably transported northwest in the Antilles current and entered the Gulf Stream north of the Straits of Florida by April. By May American eel leptocephali were abundant in the Gulf Stream opposite Cape Hatteras and beyond, and during July and August were common between Cape Hatteras and the southeast Newfoundland rise. From August to November the larvae were common in collections from slope water, inshore of the Gulf Stream, south to Cape Hatteras. It seems likely that the larvae migrate actively westward out of the Gulf Stream, after which they would be carried passively southward in the slope water.

European eels continue eastward in the North Atlantic drift, but the time it takes for them to reach the European estuaries has not been definitively determined. Schmidt (1922) estimated this period to be about 2.5 years, but Boëtius and Harding (1985) concluded that it may be possible for European eel larvae to grow to a size appropriate for metamorphosis in a little more than a year. Schmidt's estimate was influenced by larvae of intermediate size that can be found in the mid-Atlantic at any time of year. It seems probable that a proportion of the larvae travelling in the Gulf Stream and North Atlantic drift find themselves in return flows directed toward the mid-Atlantic (see Fig. 8.13). Here the feeding conditions are poor and their growth is retarded. In fact, they may never reach the shores of Europe. The successful colonists of European rivers may be those eels that travelled in the main stream and completed the journey in a little more than a year.

To complete our mental picture of the way in which eels make use of the major currents of the North Atlantic, we may suppose that the returning adults would travel west by a subtropical route, making maximum use of the Canary and north equatorial currents. In this way, their life history would be elegantly adapted to the North Atlantic gyre, with the successful group of larvae riding the western boundary and North Atlantic currents, while the breeding migrations of the adults follow the return flows. This story is a clear indication of how recent improvements of our understanding of the physical oceanography of the North Atlantic have thrown new light on the biology of these remarkable fish. It is now known that in an analogous manner the Japanese eel *Anguilla japonica* breeds at a site in the north equatorial current west of the Mariana Islands. From there the eels travel with the Kuroshio current, which returns them to their juvenile habitats in Japan (Tsukamoto 1992).

8.3.3 Salmon and the Alaskan gyre

The life history of eels involves breeding at sea and ascending the rivers to feed and grow. The migrations of salmon are in the reverse sense: they breed in the rivers but migrate out to sea to feed and grow. For many years the whereabouts of Pacific salmon, when they went to sea, were very poorly known, but in the decade of the 1960s an effort was made to obtain the missing information. A great deal of research was summarized by Royce *et al.* (1968). There are five kinds of salmon breeding along the Pacific coast of

North America. All belong to the genus *Onchorhynchus*: pink salmon, *O. gorbuscha*; coho salmon, *O. kisutch*; chinook salmon, *O. tsawytscha*; chum salmon, *O. keta*; and sockeye salmon, *O. nerka*. Of these, the migration of the pink salmon is the shortest and easiest to understand. Spawning takes place in the rivers from mid-July to mid-October. Fry hatch and emerge from the gravel in the following spring and immediately migrate down to the estuaries and coastal waters. They migrate out into the ocean proper between July and September at about 1 year of age. It is estimated that during the next year they swim between 5500 and 7500 km. They do so by swimming actively downstream with the Alaskan gyre. During the summer they move rapidly north and west, travelling parallel with the coast and feeding voraciously. In winter they move about 10° south into the west wind drift and in the following spring they swim rapidly toward their parent river, to spawn at 2 years of age. It is estimated that their average rate of travel is about 18.5 km d^{-1}, but they are substantially helped in this endeavor by the Alaskan gyre (see map, Fig. 8.03). During their travels they have exploited the marine plankton, which have grown to breeding size and have matured, large gonads.

Further to the west, the pink salmon stocks originating on the coast of Asia in East Kamchatka are believed to follow the East Kamchatka current south until they join the eastward-flowing Kuroshio extension and North Pacific current with which they travel through the winter. In the following spring they migrate back to their rivers of origin. In both these cases, the fish travel great distances with the help of the prevailing currents.

Sockeye salmon take longer to complete their life history, spending one or two winters in freshwater before running down to the sea. Those from southern British Columbia are thought to spend 2 years at sea, during which they make approximately two circuits of the Alaskan gyre. Sockeye salmon from rivers on the shores of the Bering Sea, on the other hand, stay at sea for 3 years and appear to make two or three circuits of an extended loop that takes in both the Alaskan gyre and the Bering Sea gyre.

8.3.4 Transport of invertebrate larvae across ocean basins

Before we leave the question of the biological importance of the major ocean currents, it is interesting to note that Scheltema (1966) obtained the first good evidence for the transatlantic transport of a larva of a benthic invertebrate. A few years before, the eminent marine benthic ecologist Thorson had concluded that very few invertebrate larvae had even the slightest chance to cross the larger ocean basins. The implication was that invertebrate populations in shallow water on the east and west coasts of the Atlantic were genetically quite distinct.

Scheltema (1966) found veliger larvae, probably of the gastropod *Cymatium parthenopeum*, throughout the Gulf Stream and North Atlantic drift. He calculated from drift-bottle experiments that the average rate of transport of water from the Bahamas to the Azores is 0.4 knots, and he succeeded in keeping the larvae alive in the laboratory for a time longer than that necessary for the transatlantic crossing. He pointed out that *C. parthenopeum* is found on the eastern side of the Atlantic from the Azores to South Africa and on the western side from Bermuda to Brazil. There is no need to postulate an introduction by humans: the survival of larvae throughout a transatlantic crossing may not be so very unusual.

Fig. 8.17 Schematic diagram of a Gulf Stream frontal eddy on the Georgia shelf. Reproduced with permission from Lee *et al.* (1981), Pergamon Press.

8.4 BIOLOGY OF EDDIES AND RINGS ASSOCIATED WITH MAJOR CURRENTS

As we have seen, the western boundary currents in both northern and southern hemispheres tend to be fast, deep, and relatively narrow, so that a great deal of energy is concentrated in a relatively small cross-section. These currents have a tendency to meander, and in doing so they form eddies and gyres of various kinds. One kind remains attached to the boundary of the current, and is known as a frontal eddy. Another kind breaks off as a distinct ring with an independent existence. For example, the Gulf Stream rings discussed in section 8.2.3 may persist as distinct entities for many months, and are prime targets for investigation of their physical and biological properties. Studies have been made of rings associated with the Gulf Stream in the western North Atlantic, the Kuroshio current in the western North Pacific, and the east Australia current in the western South Pacific.

8.4.1 Gulf Stream frontal eddies

The Gulf Stream (see Figs. 8.03 and 8.13), as it flows in a northerly direction offshore of Miami, Florida, is about 30 km wide and 300 m deep and has a flow in excess of 25 Sv. It flows close to the edge of the continental shelf until it reaches Cape Hatteras, by which time it has approximately doubled its flow. At that point it is approximately 50 km wide and about 1000 m deep. Surface velocity is in the range 2–5 knots and the temperature 25–28° C. It is bounded on the west side by waters of the continental slope and continental shelf that are much cooler than the Gulf Stream, and on the east side by slightly cooler waters of the Sargasso Sea with a temperature range of 20–24° C. As discussed in section 6.2.1 there is a strong front on the westward edge, sometimes known as the 'cold wall' of the Gulf Stream (Fig. 8.12).

A consequence of the meandering of the Gulf Stream is that its distance from a particular point on the continental shelf is constantly changing. At a time when this distance is greatest, it is common to find a frontal eddy between the Gulf Stream and the coast (Yoder *et al.* 1981). This eddy often takes the form of a finger-like extension of the Gulf Stream protruding into the shelf water and folding back to enclose a core of cold water (Fig. 8.17). This cold core differs from that contained in cold-core rings formed further north (see below) in being formed by upwelling of North Atlantic central water from deep in the Gulf Stream. A simplified explanation is that wherever the Gulf Stream in its meandering moves away from the coast, water from deep in the Gulf Stream upwells in the space created. Formation of these structures is a common event south of Cape Hatteras, a new one being formed on average once every fortnight. The upwelled water is rich in nitrate and Lee *et al.* (1981) estimated that this mechanism introduced about 55,000 tons of nitrogen annually to the outer shelf.

Yoder *et al.* (1981) made a study in April 1979 that showed clearly that a vigorous diatom bloom covering an area of more than 1000 km^2 occupied the cold core of upwelled water, in the region of the 200 m isobath. They speculated that this production helped explain why the shelf break south of Cape Hatteras is a breeding center for Atlantic menhaden (*Brevoortia tyrannus*) and for bluefish (*Pomatomus saltatrix*). Atkinson and Targett (1983), using acoustic methods of fish biomass assessment, found that larger fish biomasses were associated with areas of upwelling. The distribution of sea birds is also influenced by these frontal eddies. Haney (1986) found that the shearwaters *Puffinus gravis* and *Calonectris diomedea* and the storm-petrels *Oceanites oceanicus* and *Oceanodroma castro* were most abundant in the upwelled cold core of the eddy, while the black-capped petrel *Pterodroma hasiata* and the bridled tern *Sterna anaethetus* were most abundant in the oligotrophic warm filament inshore of the cold core.

More detailed follow-up studies (Yoder *et al.* 1983) showed that in winter and early spring, when the waters were not stratified, the upwelling events led to primary production of about 2 g C m^{-2} d^{-1} (for a total of about 175 g C m^{-2} during the 6 months of winter and spring). In summer and autumn, when the shelf water was stratified, upwelled water penetrated onto the shelf as a subsurface intrusion. Phytoplankton in this intrusion used the nitrate in about 10 days, and likewise fixed carbon at the rate of about 2 g C m^{-2} d^{-1}. Strong southwesterly winds were found to cause Ekman offshore transport that favored both the formation of an offshore meander of the Gulf Stream and the inshore movement of upwelled water at intermediate depths, to compensate for the offshore transport of the surface layer (Atkinson *et al.* 1984). It is now thought that this upwelling caused by the meandering offshore of the Gulf Stream is the primary mechanism for providing nutrients and stimulating production on the outer continental shelf of the southeastern United States. This effect operates south of Cape Hatteras, where the Gulf Stream runs relatively close to the coast. North of Cape Hatteras the Gulf Stream moves away from the coast and instead of cold-core frontal eddies we find warm-core ring formation on the shoreward side of the current (see section 8.2.3).

Tranter *et al.* (1986) showed that slope water intrusions occurred near Sydney, Australia, and that they seemed to occur when eddies or meanders of the East Australian current came within 90 km of the shelf break. They made comparisons with the upwelling events associated with the western boundary of the Gulf Stream, but pointed out that the mechanism producing the intrusions must be different, because the continental shelf of that

Fig. 8.18 Diagram of a vertical section through warm-core ring Bob to show the elevation of the 5, 10, and 15° C isotherms. After the Ring Group (1981).

part of Australia is much narrower and does not permit the formation of cold-core frontal eddies of the type found off Florida and Georgia.

8.4.2 Formation of Gulf Stream rings

North of Cape Hatteras the Gulf Stream leaves the edge of the continental shelf and begins to flow in a more or less easterly direction. Every month or two a meander cuts off distinct rings, as in Fig. 8.14. As we saw in section 8.2.3, cold-core rings form on the south side of the Gulf Stream, and warm-core rings on the north. Once free of the Gulf Stream they tend to move in a southwesterly direction, more or less parallel with the Gulf Stream, at a speed of about 3–5 km d^{-1}. Eventually, most make contact with the Gulf Stream and are reabsorbed into it, although some lose their identity by diffusion into the surrounding water.

8.4.3 Ecology of cold-core rings

From first principles it would seem that cold-core rings have an interesting potential to stimulate biological production. The water of the core is derived from the slope of the continental shelf and typically has a much larger concentration of plants, animals, and nutrients than the Sargasso Sea, into which it is propelled. In 1976–1977 a number of scientists who called themselves the Ring Group conducted four multidisciplinary research cruises to study a total of six rings, but the most intensively studied was ring Bob, which was formed in February and March 1977, interacted with the Gulf Stream in April and May, and finally coalesced with it off Cape Hatteras in September. Their results to 1981 were summarized in an article in *Science* (Ring Group 1981).

A cross-section of the ring soon after formation (Fig. 8.18) showed bell-shaped isotherms, with 10–16° C isotherms located about 600 m higher in the water column than their normal Sargasso Sea depth. One important effect was to bring nutrient-enriched water into the euphotic zone.

The biological characteristics of the ring changed more rapidly than the physical characteristics. In April the surface concentration of chlorophyll at the center was about 4 μg l^{-1} compared with less than 0.1 μg l^{-1} in the Sargasso Sea water, but the concentration declined by a factor of eight between April and August. As the surface waters were heated by the sun there was a shift to smaller species and a greater species diversity, so that the phytoplankton community came to resemble more closely that of the Sargasso Sea. By August the chlorophyll maximum had moved from the upper 20 m to a mean depth of 80 m and its concentration was less than 0.6 μg l^{-1}. Overall, it was estimated that primary production in the rings was about 50% higher than that in the Sargasso Sea and that these rings occupy about 10% of the area of the Sargasso Sea at any one time.

In general, cold-core rings were found to have a zooplankton biomass 1.3 to 1.8 times higher than the surrounding Sargasso Sea. As might be expected, the species composition (at least in young rings) was characteristic of slope waters. The standing stock of foraminiferans in cold-core rings can be as much as 18 times the background level in Sargasso Sea water. Species that become more abundant as a cold-core ring ages include the pteropod *Limacina inflata* and various myctophid fishes. They may join a ring after it has entered the Sargasso Sea, however. Zooplankters that are carried from the cool water of the slope to the Sargasso Sea in cold-core rings often do not thrive. Individuals of *Nematoscelis megalopsis,* for example, were driven deeper as the surface waters warmed, and failed to grow and reproduce. In a 17-month-old ring, none could be found. Backus and Craddock (1982) studied mesopelagic fishes along a transect from slope water, through a cold-core ring and into Sargasso sea water. They found that the slope water species tended to become more scarce as the ring aged, while several Sargasso Sea species invaded the ring from the top as the water warmed.

In general it seems that cold-core rings have the effect of boosting the primary production of the Sargasso Sea, to the advantage of a number of Sargasso Sea species that invade the ring from above. Species introduced from the slope water thrive for only a limited period.

8.4.4 Ecology of warm-core rings

The internal structure of a warm-core ring is in many respects the opposite of that of a cold-core ring (Fig. 8.19). Cold, nutrient-depleted waters are found at greater depth in the core than elsewhere, so isotherms and nutriclines are depressed at the center. Yentsch and Phinney (1985) discussed this structure in an interesting way (Fig. 8.20), pointing out that the outer part of the ring, the high-velocity region, is formed from the 'cold wall' of the Gulf Stream, which has the usual frontal structure of sloping isopycnals (see Chapter 6). It is to be expected that there will be transport of water along lines of equal density (isopycnal mixing; Woods 1977) and that this process will lead to the upwelling of nutrients in the high velocity peripheral region.

By analogy with what is known about cold-core rings, it might be supposed that warm-core rings, which have at their center nutrient-depleted Sargasso Sea water, would at first be biologically unproductive and would slowly evolve to take on the character of slope water in which they are immersed. One would not suppose that warm-core rings

Fig. 8.19 Cross-section of a warm-core ring lying between the Gulf Stream and the continental slope, showing isotherms (°C). After Csanady (1979).

Fig. 8.20 Diagram showing how the Gulf Stream, with its isopycnals sloping upward to the left, gives rise to a warm-core ring with its isopycnals sloping upward toward the perimeter. Arrows show flow of the Gulf Stream, isopycnal flow, and rotation of the warm-core ring. After Yentsch and Phinney (1985).

would appear in satellite images as regions of higher chlorophyll-a biomass than the surrounding water. Yet this is what happens from time to time (McCarthy and Nevins 1986; Tranter *et al.* 1980).

Yentsch and Phinney (1985), synthesizing the information available, have put forward the hypothesis that two distinct physical mechanisms regulate primary production in warm-core rings. The first has already been mentioned: isopycnal mixing brings deep, nutrient-rich water to the surface in the periphery of the ring. The authors show satellite images in which one can distinguish the low-chlorophyll core of Sargasso Sea water and the outer ring of Gulf Stream water with a higher chlorophyll content.

In a modelling study, Franks *et al.* (1986) invoked isopycnal flow in a rather different way. They proposed that as the ring ages, energy is lost due to friction and the permanent pycnocline, much depressed in the newly formed ring, gradually rises toward the level of the permanent pycnocline in the surrounding waters. As this movement occurs, the volume of the ring decreases and water is forced out along isopycnals. When this physical process was modelled along with the biological dynamics, a lens of enhanced phytoplankton biomass was formed at a depth of about 20 m in the ring center. This lens was attributed to upward movement of nutrient-rich water at the ring center.

The second mechanism proposed by Yentsch and Phinney (1985) as influencing primary production in warm-core rings is convectional mixing followed by stabilization. Since the warm-core rings, during their lifetime of several months, are lying north of the Gulf Stream it is inevitable that surface waters will become cooler than those at depth, which are derived from the Sargasso Sea, stimulating convective mixing and formation of a deeper well-mixed layer. This layer, having discrete boundaries and a uniform temperature, is referred to as the thermostad. Since isotherms are depressed in the center, convective mixing in the core may carry phytoplankton to greater depths than in the outer ring, reducing photosynthetic production. Conversely, convective mixing in the outer ring may bring to the surface water that has been enriched by the isopycnal flow of nutrient-rich water, thus stimulating phytoplankton production.

In addition to the changes resulting from the northward movement of Sargasso Sea water in a warm-core ring, seasonal changes have been observed in the rings related to the events typical of slope and shelf waters. Because of the distinctive features of the warm-core ring, the Gran effect (the formation in spring of a shallow mixed layer and its breakup in autumn) occurs out of phase with the surrounding water. As a result, there may be times when a warm-core ring is undergoing its 'spring bloom' while the surrounding shelf waters are not. Events are complex and incompletely understood, but empirical studies show that warm-core rings may undergo bursts of relatively high production associated with the vertical transport of considerable quantities of nitrate (Tranter *et al.* 1980; McCarthy and Nevins 1986; Hitchcock *et al.* 1987). Toward the end of its life, the thermostad of a warm-core ring may be completely covered by waters flooding in from the surrounding waters, thus making the mixed layer indistinguishable from surrounding slope waters. The process was modelled by Chapman and Nof (1988). Over the lifespan of a warm-core ring, from its formation from Gulf Stream meander to its eventual rejoining with the Gulf Stream further south and west many months later, its primary productivity seems to be not very different from that of the slope water surrounding it (Hitchcock *et al.* 1985). This conclusion is quite different from earlier suggestions that warm-core rings, being composed of nutrient-depleted Sargasso Sea water, are islands of low productivity surrounded by much more productive slope water.

Zooplankton and fish in warm-core rings
Species identification reveals that many zooplankton are advected into warm-core rings from the surrounding slope water (Roman *et al.* 1985), and physical oceanographic studies using salinity as a conservative property confirm that significant lateral exchange of water masses occurs (Olson 1986). In general, a newly formed ring is found to have a low zooplankton biomass in the core, consistent with its origin from the Sargasso Sea, but within a few months there is often a biomass of zooplankton as high as or higher than in the slope water. In general this increase is thought to be the result of *in situ* growth, resulting from the enhanced production of phytoplankton discussed earlier, and its consumption by

zooplankton. A concrete example of such production was obtained by Cowles *et al.* (1987). After gale-force winds had deepened the mixed layer and introduced nutrients to the surface waters of a warm-core ring in 1981, microzooplankton, mainly young cyclopoid copepods, doubled in biomass. Adult copepods showed high grazing rates and strong egg production. The response could be detected within 10 days of the storm event.

In contrast to all these changes, mesopelagic fish and siphonophores that characteristically form a nonmigratory acoustic deep scattering layer in the Sargasso Sea were found to be captured in the thermostad of a warm-core ring and could be traced acoustically throughout the life of that ring (Conte *et al.* 1986). There was no evidence of a significant change in abundance and their vertical distribution remained approximately constant for several months in spite of changes in temperature and salinity. On the other hand, acoustic studies in a warm-core ring of the Kuroshio system showed intense sound scattering from zooplankton and fish populations within the ring and very sparse acoustic signals outside it (Aoki and Inagaki 1992).

In the next section we shall consider the biological consequences of a warm-core ring coming into contact with the edge of the continental shelf.

Interaction of warm-core rings with the continental shelf

Satellite imagery shows that it is not uncommon for warm-core rings to make contact with the continental shelf and for large volumes of cold water to be entrained and dragged off the shelf by the rotational movement. It also seems likely that on the western or southern side of the rings considerable quantities of warm water are injected onto the shelf. It was natural, therefore, that when Markle *et al.* (1980) discovered fish larvae of tropical and subtropical origin on the Scotian Shelf they should attribute their presence to the action of warm-core rings. Similar observations have been made off southern New England.

Friedlander and Smith (1983) identified water advected offshore by a warm-core ring. Contained within the ring were abundant sand lance larvae (*Ammodytes* sp.) that had been carried off the shelf along with the copepods that form their food. The idea then developed that a strong advection of a shelf water mass at a time when the larvae of commercially important species of fish are present could have harmful effects on the year-class strength of those species.

Wroblewski and Cheney (1984) explored a warm-core ring off the coast of Nova Scotia and found that larval and juvenile white hake were being carried as much as 140 km seaward of the shelf break. They were in much poorer condition than the same species on the continental shelf and the authors thought that they represented a loss from their source populations. The ring being studied also contained many fish larvae of tropical and subtropical origin belonging to the same species reported as immigrants to the Scotian Shelf by Markle *et al.* (1980).

A modelling study (Flierl and Wroblewski 1985) showed that at a time when commercially important fish larvae are being carried by the residual current southwest along the Scotian Shelf, the advection caused by a stationary warm-core ring could cause a 35–50% drop in abundance, depending on the ring size, and a slowly moving ring could, in certain circumstances, remove most of a year-class. This initiative was followed up by Myers and Drinkwater (1989), who analyzed weekly satellite images for 1973–1986 to determine positions and numbers of warm-core rings, and location of the Scotian shelf-slope front during those years. These data were combined with estimates of the timing and duration of spawning and the duration of larval stages, to provide 'susceptibility functions' for 17 groundfish stocks. They found evidence that increased warm-core ring activity was associated with reduced recruitment in 15 of the 17 groundfish stocks. A similar analysis of seven pelagic stocks and one shellfish stock showed no consistent evidence of the same effect, but it was pointed out that recruitment data for the pelagic stocks are less reliable than for the groundfish stocks. As the authors said, it is possible that the true mechanism causing reduced recruitment may be not the warm-core ring itself but some hydrographic feature that is correlated with warm-core ring activity. Nevertheless, off-shelf transport due to entrainment in a warm-core ring is the best explanation available at present.

An interesting variant on the warm-core ring theme was described by Duncombe Rae *et al.* (1992). The Agulhas current, a western boundary current that flows southward along the east coast of South Africa, continues south or southwest beyond the tip of South Africa and turns eastward in a great loop called the Agulhas retroflection. Periodically, the retroflection pinches off rings of Indian Ocean water that then move northwest into the South Atlantic, spinning counter-clockwise. From time to time these rings interact with the Benguela upwelling system (see Chapter 5). Duncombe Rae *et al.* (1990) showed that in June–July 1989 a major cold filament from the Benguela upwelling front was entrained in an Agulhas ring so that a large volume of frontal water was advected to mid-ocean and prevented from returning to the shelf region.

Poor year-classes of anchovy were produced in the Benguela system in 1989 and 1990. By analysis of seabird stomachs it was shown that the most marked decrease in anchovy biomass occurred in June–July 1989. It was postulated that removal of larvae and young anchovy from the shelf by the Agulhas ring was a major factor in the depression of the 1989 year-class.

A study of an Agulhas warm-core ring surrounded by sub-Antarctic water (Dower and Lucas 1993) showed that chlorophyll concentration and primary productivity were lower within the ring than outside it, and a higher proportion of the phytoplankton inside the ring was less than $20\,\mu$ in diameter. This finding is similar to the characteristics of a Gulf Stream warm-core ring soon after it has been formed.

To summarize to this point, it appears that cold-core rings are responsible for introducing slope water species into the Sargasso Sea, and that warm-core rings introduce tropical and subtropical species not only into slope waters but even onto the continental shelf. Cold-core rings, when newly formed, have a productivity much higher than the Sargasso Sea, but as their surface layer warms and their cold-water fauna are driven deeper and deeper, their productivity falls to something close to the productivity of the Sargasso Sea. Warm-core rings, by analogy, were expected to begin life with a low productivity characteristic of Sargasso water, and to slowly come to resemble the more productive slope water. In practice it seems that warm-core rings have mechanisms for generating upwelling of nutrient-rich water especially near the perimeter in the high velocity zone. Stratification as a result of spring warming often leads to levels of primary production in the rings greater than that in the surrounding slope water and to subsequent rapid growth of zooplankton populations. Over the life of a warm-core ring its productivity appears to be about the same as that of the slope water.

Perhaps the most interesting aspect of the whole story is the sequence of events that takes place when a warm-core ring comes in contact with the continental shelf. Cold water is advected off the shelf in large amounts, and at times when planktonic fish larvae are present in the water, this movement leads to reduced larval survival and subsequently to reduced recruitment in the parent stock. On the other side of the ring, warm water is probably being

Table 8.1 Approximate values for biological and related measurements in three kinds of ocean environments. From Blackburn (1981).

Measurement*	Subpolar gyres	Equatorial divergences	Subtropical gyres
Mixed layer depth, summer (m)	15–30	25–65	15–20
Mixed layer depth, winter (m)	>120	25–65	65–120
Euphotic layer depth (m)	20–50	45–85	75–150
NO_3–N, 0 m (μg-at 1^{-1})	5–25	5–10	0–1
NO_3–N, 100 m (μg-at 1^{-1})	10–25	10–25	0–5
Primary production ($g\ C\ m^{-2}\ d^{-1}$	0.1–0.5	0.1–0.5	≤0.1
Chlorophyll (mg m^{-2})	15–150	15–30	5–25
Herbivore mean weight (μg C)	40–400	4–40	4–40
Zooplankton, 0–200 m (g m^{-2})	30–50	7–25	<7
Zooplankton, 0–1000 m (g m^{-2})	150	8–13	9
Micronekton, 0–200 m (g m^{-2})	0.6	0.6–0.8	0.3
Micronekton, 0–1000 m (g m^{-2})	2	1–3	1
Benthos (g m^{-2})	0.1–1.0	0.1–1.0	<0.05

*Primary production and chlorophyll integrated for the euphotic layer. For zooplankton and micronekton, 1 mL displacement volume assumed to equal 1 g.

injected onto the shelf and is the means of introducing adults and larvae of tropical and sub-tropical origin into temperate shelf waters.

8.5 ECOLOGY OF THE CENTRAL GYRES

In section 8.3 we discussed the major currents associated with the periphery of the huge sub-tropical gyres. We shall now look at the relatively quiet water that occupies the central gyral regions. The anticyclonic gyre of the North Pacific has been the subject of particularly in-tensive study and is the focus of a major debate about the general level of biological produc-tivity of the subtropical gyres. There is also a long history of observations in the Sargasso Sea.

8.5.1 Primary production in the subtropical gyres

Table 8.1 from Blackburn (1981) summarizes the data from more than a decade of studies that indicated that subtropical gyres are the least productive parts of the ocean. Primary pro-duction, measured by ^{14}C uptake, was believed to be less than 0.1 g C $m^{-2}\ d^{-1}$, with the re-sult that biomass of both phytoplankton and zooplankton is low, the water is very clear with the euphotic zone extending to 75–150 m in depth, and nitrate levels at the surface are in the range of 0–1 μg-at 1^{-1}. This low productivity extends to the fish in the water column and the invertebrates on the bottom. Consideration of these facts led to the description of the subtropical gyres as 'biological deserts.'

In section 3.2.5 we considered the mechanisms controlling productivity in such a situa-tion, approaching them from the point of view of vertical structure of the water column. Strat-ification is a permanent feature, with the thermocline separating a lower, cooler layer having high concentrations of nutrients from an upper, warmer mixed layer having an extremely low level of nutrients. In the upper parts of the mixed layer phytoplankton production is ap-proximately balanced by zooplankton consumption, and the ammonia excreted by the zoo-plankton provides a nitrogen source for a continued low level of phytoplankton production.

Nitrogen and other nutrients are, therefore, in some approximate steady state of recycling and the primary production based on this cycle is referred to as regenerated production.

At the same time, a certain amount of nitrate is transported upward through the nutricline by turbulent diffusion. This process leads to more rapid growth of the phytoplankton population and the formation of a zone of maximum phytoplankton biomass, the 'chlorophyll maximum,' just above the nutricline. The upper boundary of this zone is set by the supply of nutrients from below, and the lower boundary is set by the availability of light from above. This primary production based on nitrate-nitrogen is referred to as 'new production' and the ratio of new production to total production is called the 'f' ratio. Measurements of the f ratio for the North Pacific subtropical gyre suggested that its value lay in the range of 5–10%.

This view of the gyre as being in a relatively steady state with a low level of primary production and low f ratio has been challenged by a number of different measurements summarized in Eppley (1980). Use of a particle-counter to measure changes in particle volume suggests that phytoplankton in oligotrophic subtropical gyres are completing a new generation about once every 3 hours, which would lead to an estimated production rate an order of magnitude greater than previous estimates.

Measurements by means of particle traps of the rate of sinking of particles out of the euphotic zone were interpreted as reflecting the rate of new production, since in steady state the amount sinking out must be balanced by new production, rather than by production that is being recycled. The traps collected an average of 68 mg C m^{-2} d^{-1}. If the f ratio is 5%, then total production is 1.360 g C m^{-2} d^{-1} and if 10%, total production is 0.680 g C m^{-2} d^{-1}. Comparing these figures with the value of < 0.1 g C m^{-2} d^{-1} cited by Blackburn (1981), Eppley (1980) concluded that there is an order of magnitude uncertainty about the level of primary production in the large ocean gyres.

Shulenberger and Reid (1981; Fig. 8.21) used a much larger-scale approach than anyone preceding them. They pointed out that, in the open mid-latitude Pacific in summer, a subsurface oxygen maximum is found in which there are values up to 120% of saturation. This maximum lies below or within the pycnocline, and certainly below the warm mixed layer, which they refer to as a density cap. It is usually located above the chlorophyll maximum but

Fig. 8.21 Percentage oxygen saturation at depth of maximum value in the North Pacific. From Shulenberger and Reid (1981).

below the primary production maximum (see Fig. 3.01). As Fig. 8.21 shows, values in excess of 100% saturation are found throughout the North Pacific and the area of greatest values is centered on 40° N 160° W.

The authors argued that this oxygen supersaturation accumulates in summer as a result of photosynthesis, and is prevented from equilibrating with the atmosphere by the density cap. Equilibration occurs in winter when the mixed layer deepens and the supersaturated water is circulated in the mixed layer. Allowing 120 days for the build-up of the summer oxygen excess, they calculated that, at a minimum, the rate of photosynthesis must be greater than the rate indicated by ^{14}C incubations by a factor of about four.

Platt (1984) challenged the Shulenberger and Reid conclusions on the grounds that the oxygen concentration at the beginning of the summer season was not known, the time scale of accumulation was not known, and the data used to compare an instantaneous carbon flux with a time-averaged oxygen flux were inadequate from a number of points of view. Reid and Shulenberger (1986) published a rebuttal and held to their position. Platt and Harrison (1986) then drew attention to the fact that changes in oxygen concentration measure new production, P_{new} (since regenerated production is approximately in balance with consumption, and net oxygen change resulting from these processes should be zero), whereas ^{14}C measures total production P_T. Scaling up the oxygen changes to estimate P_T would lead to impossibly high estimates of total production.

In a subsequent review paper Platt *et al.* (1989) conceded that the evidence from oxygen accumulation, sediment trap studies, and direct determinations of the rate of nitrate uptake by phytoplankton in oligotrophic waters shows that the value of P_{new} in these waters is much higher than previously believed. This conclusion is very important in relation to the problem of CO_2 accumulation in the atmosphere and the role of the oceans in helping to counteract it. In the long term, organic carbon sinking out of the mixed layer must be balanced by new production, which involves removal of CO_2 from the atmosphere. (This topic will be discussed more fully in Chapter 9.) The explanation of the higher values of P_{new} lies almost certainly in the occurrence of intermittent episodes of upwelling of nutrients. We have to abandon the view that the central gyres are uniform and quiescent and think of them as filled with eddies of varying energy levels (Kerr 1985). Cyclonic eddies will raise the pycnocline and are capable of bringing about upwelling of nutrients. The problem is to get measurements of these transient episodes of elevated production either directly or by using methods that integrate their effects over time. Satellite monitoring may provide the key to direct measurements while sediment traps and methods involving monitoring oxygen accumulation in the water have the potential to integrate results over time.

Meanwhile, a few people have made direct observations of transient episodes of enhanced primary productivity. McGowan and Hayward (1978) examined biological and physical data from the central Pacific gyre over the period 1964–1974 (10 cruises). In the summer of 1969 the rate of primary production was double what it had been in the previous summer and during four subsequent summers. The zooplankton biomass was also significantly higher. The physical data showed a negative temperature anomaly at the surface and, at a depth of 150–250 m, high variance in temperature with frequent occurrence of temperature inversions. This variance was taken as evidence of internal wave activity. In all other years of observation, the internal wave activity appeared to have taken place at depths of 300–400 m. The occurrence of these internal waves nearer to the surface were thought to have induced upward mixing of cooler, nutrient-rich water into the euphotic zone.

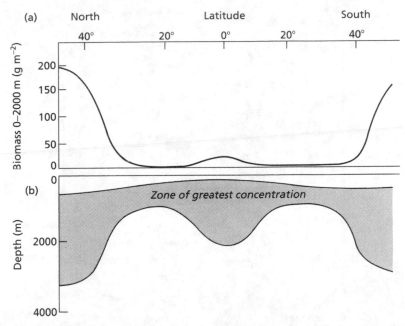

Fig. 8.22 (a) Idealized diagram of distribution of zooplankton biomass along a Pacific transect running north–south across the equator. (b) Vertical section along same transect showing zone of greatest concentration extending more deeply in regions of greater biomass. From Blackburn (1977), using data of Vinogradov.

More recently, Laws *et al.* (1987) measured primary production at a station about 550 km north of the Hawaiian islands. The mean gross primary production measured on three occasions between 21 August and 2 September 1985 was $450 \pm 37 \, mg \, C \, m^{-2} \, d^{-1}$, which is about twice the average productivity level recorded at that station over the years 1968–1980. The question of whether this occurrence was a short episode of increased productivity, or whether earlier measurements were in error, was left open.

8.5.2 Secondary production in the subtropical gyres

Table 8.01 includes data for the biomass of zooplankton and of micronekton. There are no routine methods for measuring secondary production, although some estimates will be made. However, a good indication of the low levels of secondary production can be seen from the fact that the biomass in the upper 1000 m of the water column is about 9 g wet weight m^{-2}, while in the subpolar gyres it is 150 g. Micronekton (animals about 1–10 cm long) and benthos also have biomasses much lower than in the subpolar gyres. Figure 8.22 shows a generalized relationship of zooplankton biomass to latitude in the Pacific. There are very high values at latitudes greater than 40°, extremely low levels in the subtropical gyres, and some increase in the area of the equatorial divergences. The zone of greatest concentration reflects the level of primary productivity. Where this production is low, the zooplankton are concentrated high in the water column, because most food is used before it reaches the deeper levels. By contrast, there is enough food in the subpolar gyres to support a dense population of zooplankton to depths of 3000 m.

The structure and dynamics of the fish community in subtropical gyres have been reviewed by Mann (1984). In the euphotic zone (0–100 m) are found fast-swimming

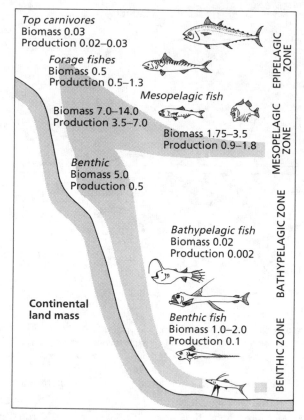

Fig. 8.23 Diagram summarizing mean values of fish biomass (kcal m^{-2}) and production (kcal m^{-2} y^{-1}) in the open ocean. Shading symbolizes decreasing biomass as one moves from the shelf break toward the center of the gyre. From Mann (1984).

predators such as the tunas, bill-fish, and swordfish. Most of them breed in tropical waters but make major excursions into the subtropical gyres. For example, bluefin tuna (*Thunnus thynnus*) breed in the Caribbean area, and medium-size fish migrate north in summer as far as the latitude of Cape Cod, but large, older fish may migrate as far as Newfoundland or cross the Atlantic and spend the summer in Norwegian waters. These large predators feed on a variety of smaller fishes, on squid, and on larger crustaceans such as euphausids.

The zone from 200 m to a depth of 1000 m is termed the 'mesopelagic zone.' From acoustic studies we know that there is a 'deep scattering layer' caused by small fish (mostly < 10 cm) and larger invertebrates. Many of these animals migrate to the euphotic zone to feed at night. This layer is more or less continuous all the way across the major ocean basins. Finally, below 1000 m is the bathypelagic zone occupied by fishes with dark coloring, small eyes, weak musculature, and large mouths. Some, like the angler-fishes, have elaborate lures to attract prey. They appear to be adapted to life in a food-poor environment, expending little energy on swimming but taking a large meal whenever something comes within range. Figure 8.23 summarizes the biomass and production data (in kcal m^{-2} and kcal m^{-2} y^{-1}, respectively) for the different depth zones of a subtropical gyre. Using the approximate conversion 1.0 g wet weight equals 1.0 kcal, we see that the greatest values are in the mesopelagic fishes. The top carnivores, which are the main commercially important fishes, produce on average only 0.02–0.03 g m^{-2} of new tissue per annum. In addition, the smaller fishes on which they feed are thought to produce 0.5–1.3 g m^{-2} y^{-1}. This growth may be compared with 5–10 g m^{-2} in many continental shelf areas and 10–20 g m^{-2} in particularly favorable areas such as Georges Bank.

It is not possible at this stage to use the information on fish productivity to throw light on the question of how much primary production occurs in the upper layers of the ocean. As we have seen, the major uncertainty is about the level of 'new' production. Fish that live and feed in the mixed layer can be regarded as part of the cycle of regenerated production, while fish that feed in the upper layer but migrate to depth during the day may contribute to the removal of nitrogen from the euphotic zone, and hence form part of the system that is fed by the upwelling of new nitrogen. More work is needed to unravel the complicated network of nutrient transfers.

8.6 SUB-ARCTIC GYRES

North of the main anticyclonic subtropical gyres in both the Atlantic and Pacific are smaller, cyclonic sub-Arctic gyres (Fig. 8.03). In or near to these gyres are ocean weather stations, some of which have been extensively used to obtain time series of oceanographic data. Parsons and Lalli (1988) analyzed data from ocean weather stations P (in the Pacific) and B, I, and J (in the Atlantic) (see Fig. 8.03), together with data from the Continuous Plankton Recorder Programme (Colebrook 1979) (see Chapter 9), to synthesize a comparative view of plankton ecology in northern parts of the two major ocean basins.

The dynamics of the spring bloom in temperate waters was dealt with in section 3.3.2 using mainly material from the Atlantic Ocean. We shall now see that there seem to be systematic differences in the dynamics of the bloom between Atlantic and Pacific basins. The primary production data indicate that at ocean weather station I (OWSI) in the Atlantic, the arrival of the spring bloom in April is marked by a sharp increase in primary productivity from about 20 to over 300 mg C m^{-2} d^{-1}, while at OWSP in the Pacific there is a slow increase in productivity between January and July, over about the same range of values. Patterns of biomass change are also different in the two basins. In the Atlantic phytoplankton biomass changes by an order of magnitude between winter and summer, while in the Pacific it only doubles during the same period.

Differences between the two basins have their origins in the different winter depths of the mixed layer. In the North Atlantic it is greater than 200 m, while in the North Pacific it is about 100 m. (These differences are, in turn, explicable in terms of the global hydrological cycle; see Chapter 10.) As a result, the phytoplankton in the Pacific are exposed to a more favorable light regime during winter and are able to maintain a moderate level of biomass and production. About 80% of this biomass is in the form of small (<20 μm) flagellate cells. These cells are grazed by protozoa, which in turn are preyed upon by two or three common species of copepod—*Neocalanus plumchrus*, *Neocalanus cristatus*, and *Calanus pacificus*—that are relatively large and have only one generation per year. Hence, modest levels of both phytoplankton and zooplankton biomass are maintained through the North Pacific winter. When the surface waters warm in spring and the mixed layer becomes shallower, the zooplankton grazing almost keeps pace with the increase in phytoplankton productivity, so that phytoplankton biomass increases only slowly.

In the North Atlantic, by contrast, the mixed layer depth is sufficiently great that phyto-plankton production virtually ceases during winter. The dominant copepod, *Calanus fin-marchicus*, is adapted to this period of famine by having overwintering larvae living in cold deep water greater than 300 m in depth. When the spring bloom of diatoms starts in surface waters there is a lag in the start of zooplankton grazing. Phytoplankton biomass increases to high levels and much of it may sink without being grazed in the plankton. Then, between April and June, the *Calanus finmarchicus* larvae rise to the surface layer, complete their de-velopment, and begin breeding. In most places these calanoids complete several generations in quick succession. Zooplankton production is much more narrowly pulsed in the Atlantic than it is in the Pacific.

In spite of these ecological differences, the two areas seem to have about the same to-tal levels of phytoplankton and macrozooplankton production, and about the same popula-tions of predators (e.g., euphausids, jellyfish, chaetognaths, myctophids, squid). It is thought that the more constant biomass of plankton in the Pacific system supports a greater amount of pelagic fish production (salmon, for instance) while the sinking of the diatom bloom in the Atlantic may support a larger benthic fish production.

8.7 CONCLUSIONS

We have seen in this chapter that ocean circulation is closely linked to atmospheric circu-lation. The difference in solar heating between low and high latitudes, in conjunction with the rotation of the earth, leads to westerly and trade winds, which drive major anticyclonic subtropical gyres in each of the ocean basins. Associated with these gyres are powerful western boundary currents that transport large quantities of heat away from equatorial re-gions. These currents meander and cut off large gyrating bodies of water, the warm-core and cold-core rings, that lead independent existences for several months. Organisms have adapted to exploit the properties of the currents and the rings. For example, squid breed in warm subtropical waters, then 'ride' the western boundary currents to exploit the higher levels of productivity in temperate latitudes. Pacific salmon swim thousands of kilometers with the currents of the sub-Arctic gyres and exploit the abundant food supply during their period of rapid growth. Eel larvae rely on the Gulf Stream to carry them from their sub-tropical breeding grounds to the coasts of Europe.

The anticyclonic gyres are permanently stratified (though with seasonal changes in the thickness of the mixed layer) and surface waters have constant low levels of nutrients. It was once thought that primary production was uniformly low, relying mostly on the recycling of nutrients between the grazers and the phytoplankton. It is now thought that various mecha-nisms, at present little understood, lead to transient episodes of upwelling of nutrient-rich water, so that even in these gyres the level of 'new' production is much higher than previ-ously supposed.

The sub-Arctic gyres have alternating periods of summer stratification and winter mix-ing so that there is a seasonal injection of nutrients into surface waters. In the North Atlantic it appears that a bloom of phytoplankton develops when the zooplankton populations in sur-face waters are at a low level, so that much of the plankton biomass sinks without being grazed. In the Pacific, on the other hand, the depth of the winter mixing is more restricted, and phytoplankton are able to maintain a modest level of primary production throughout the winter. This condition supports a population of zooplankton that, when the spring bloom begins, is able to keep pace with the growth of phytoplankton and prevent the accumulation of biomass. Hence, biological production in the North Pacific appears to be less strongly pulsed than in the Atlantic.

While the major ocean gyres and their associated ring formations are well recognized and have been extensively studied, it is now believed that eddies permeate all parts of the oceans. Their biological properties have been little studied, but it is thought that some of them may be responsible for the transient upwelling events that stimulate the primary productivity of surface waters far above their long-term mean values. One of the challenges for oceanography in the coming decade is to develop methods of locating and studying these transient events.

9

Variability in Ocean Circulation: Its Biological Consequences

9.1 INTRODUCTION

During the present century our technological advances in fishing have brought us to the point where it is possible to overfish a stock so seriously that it is no longer biologically viable. Fisheries scientists began to try to define the maximum catch that could be taken on a yearly basis while still permitting the stock to remain vigorous; they called it the maximum sustainable yield. At first these calculations were done with the environment of the fish (e.g., physical and chemical conditions, food supply, predator pressure) held constant. Gradually, fisheries scientists became aware that these assumptions were unrealistic. Large fluctuations in abundance occurred that appeared to be unrelated to pressures of the fishery. In the English Channel between 1925 and 1936, the herring stock went into decline and was replaced by pilchards. In various upwelling systems around the world sardines were replaced by anchovies, and vice versa, and cores taken from anaerobic sediment showed that these changes had also occurred before human fishing pressure had become a significant factor.

Perhaps the best-known fluctuations of all are those associated with the El Niño phenomenon off the coast of Peru. Changes in oceanographic conditions (see Chapter 5) led on many occasions to drastic declines in numbers of anchoveta, with spectacular consequences in terms of deaths of dependent sea birds and mammals.

Not all changes have been negative. During the period 1940–1970 the cod stocks greatly expanded their range and abundance on the west coast of Greenland, and in 1960–1970 there was a great expansion of the stocks of cod and related species in the North Sea.

Contemporary fisheries management policy, therefore, seeks to control fishing effort against a background of natural fluctuations in stock sizes. For many years the fluctuations seemed totally unpredictable, but as people began to study oceanographic processes on larger and larger scales, it gradually became apparent that many local changes in fish stocks were related to large-scale processes in the ocean. In this chapter we shall examine two of these processes. In the Pacific Ocean we shall see that coupling between the ocean and the atmosphere causes an oceanographic oscillation between Peru and Indonesia—the southern oscillation—that affects the fish stocks of Peru and many parts of the North Pacific. In the Atlantic Ocean we shall see that changes in the pattern of the westerly winds running along the boundary between the Arctic air mass and the subtropical air mass give rise to important shifts in weather patterns. From about 1890 to 1940 conditions in the northeast Atlantic became progressively warmer; southern species of invertebrates and fish extended their range northward, several stocks expanded dramatically, and the balance of species present in the plankton of the English Channel changed radically. Between about 1940 and 1980 there was in the same area an increase in the incidence of cold northerly winds, and the biomass of phytoplankton and zooplankton, though variable, showed a steady downward trend. A number of important fish stocks declined (although at least one showed dramatic improvement) and then toward the end of this period the plankton and fish community of the English Channel reverted to its pre-1930 condition.

To try to understand these changes, it is necessary to study the major features of the dynamics of the earth's atmosphere, and the way in which it interacts with the ocean currents.

9.2 PHYSICAL VARIABILITY IN THE PACIFIC AND ATLANTIC OCEANS

9.2.1 El Niño–southern oscillation (ENSO)

The anchovy fishery off Peru and Ecuador used to yield, in a good year, about 10^7 tons of fish. The fishery exists because equatorward winds along the coast cause nutrients to be upwelled from below the shallow thermocline as described in section 5.2.2. The catch rate is not constant throughout the year, because in November and December warm water from the equatorial region moves south and disrupts the upward flow of nutrients. This change in the oceanographic conditions, because it occurs around Christmas, is called El Niño after the Christ child. In normal years the disruptive phenomenon lasts only a few months and the high nutrient upwelling returns, but every 3–7 years the high nutrient flow is cut off for up to a year. This condition has a devastating effect on the anchovy fishery, which can decline to one-fifth of the peak catches. The term 'El Niño' is now used exclusively for this catastrophic phenomenon.

Until about 30 years ago El Niño was believed to be caused by a change in the upwelling-favorable winds and thus an event local to the coasts of Peru and Ecuador. It is now known that El Niño is a small part of an ocean-wide oscillation in the atmosphere called the southern oscillation (SO). In this oscillation the usual pressure gradient from the region of high atmospheric pressure in the southeast Pacific Ocean to the region of low pressure in the area of Indonesia becomes higher or lower than average. A lower-than-average pressure gradient favors El Niño conditions, while the higher-than-average situation has been called La Nina. The combination of these phenomena with the southern oscillation has led to the acronym ENSO.

The southern oscillation is associated with large changes in the climate of the equatorial region and to a lesser extent the subtropical regions. It has been studied intensively over

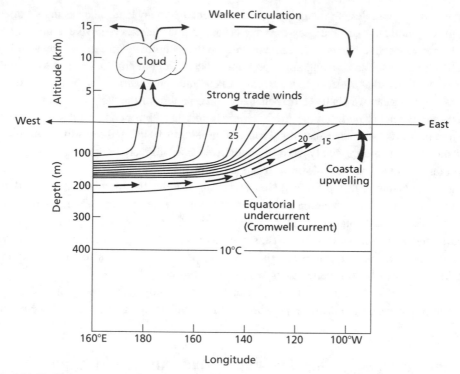

Fig. 9.01 Vertical cross-section of the atmosphere and the ocean viewed from the south showing zonal Walker circulation in the atmosphere and the thermocline that shallows in the east in response to wind stress at the surface. Also noted are the coastal upwelling off the coast of South America and the Equatorial undercurrent flowing east at the base of the thermocline.

the past 20 years by both sea-going oceanographers and theoreticians, and there have been tremendous increases in our knowledge and understanding of the processes involved. These processes are summarized in the reviews by Enfield (1989), Canby (1984), and Kessler and McPhaden (1995), and in a special volume of *Oceanus* (Vol. 27, No. 2, 1984). Our description of ENSO events draws from these articles and begins with a review of the circulation, near the equator, in both the atmosphere and the ocean.

As we saw in section 3.2.6 and Fig. 3.06, the trade winds on either side of the equator in the middle of the Pacific Ocean converge at the intertropical convergence zone (ITCZ), which lies at about 5° N in February and 10° N in August. The South Equatorial current lies from about 20° S to ≃5° N. Between its northern limit and ≃10° N is the eastward Equatorial counter-current, and the North Equatorial current flows to the west between 10° N and 20° N (see Fig. 3.06).

The circulation in both the ocean and the atmosphere along the equatorial Pacific is shown in a vertical section viewed from the south in Fig. 9.01. In the atmosphere the trade winds blow westward over the sea surface toward a low-pressure area in the western Pacific located at about 180° W. In the low-pressure region the air rises because of the heat gained from the warm ocean. The rising air loses its moisture through rain and eventually circulates back toward the east at upper levels and descends over the eastern Pacific high-pressure cell to complete the circuit. This vertical east–west circulation cell is one of a series that circle the equator known as the Walker circulation (Webster 1983).

In the ocean the trade winds push the water toward the west (Coriolis effect is zero at the equator) in the South Equatorial current. The wind stress toward the west is balanced by a pressure gradient in the opposite direction in the form of a slight rise in sea level toward the west. The pressure gradient is not transmitted into the deep ocean because the density structure in the upper layer of the ocean adjusts to cancel the pressure difference. This change is accomplished by an increase in the thickness of the low-density mixed layer from 30 to 50 m in the east to ≃150 m at the middle of the ocean. In the western half of the ocean the pressure gradient is mainly balanced by the temperature increase toward the west in the mixed layer, which stays roughly constant in thickness (see Fig. 9.01).

The large increase in the sea surface temperature from ≃20 °C in the east to ≃30 °C in the west is due primarily to upwelling along the equator. The upwelling is created by horizontal Ekman transports away from the equator (Fig. 3.06), created by the easterly trade winds. The upwelling is especially large near the equator because the Coriolis force becomes very small. For example, the equation for the Ekman transport (Eq. 5.06)

$$M = -\tau/f \tag{9.01}$$

indicates that for a given wind stress τ the transport M increases as f, the Coriolis parameter, decreases. When f is zero the equation predicts an infinite transport that is, of course, not realistic but high Ekman transports close to the equator are observed. In the eastern Pacific the thermocline lies at ≃50 m and upwelling brings water from below the thermocline up into the mixed layer. Toward the west, where the thermocline gets deeper, the upwelling water comes more and more from the warm mixed layer rather than the cool subthermocline waters. This change in the temperature of the upwelled water is what causes the east–west rise in the temperature of the mixed layer and the sea surface.

In the El Niño, or warm phase of the ENSO cycle, one of the first indications that the normal situation is changing is a decrease in the trade winds in the central and western Pacific. The accompanying decrease in the east–west pressure gradient along the equator allows the thermocline to relax from its normal tilt by rising in the west and descending in the east. This change propagates from the west to the eastern side of the ocean in 2–3 months in the form of equatorially trapped baroclinic Kelvin waves. These are the same waves we examined in section 7.2.3 but in this case they are trapped at the equator (Gill 1982) by the vanishing of the Coriolis force rather than by a coast. Also, they move much more slowly because they are baroclinic (only in the upper layer) rather than barotropic. At the eastern side of the ocean these waves split and continue poleward along the west coasts of North and South America.

As the east–west tilt in the thermocline decreases, the mixed layer in the east becomes deeper and the water being upwelled along the equator is drawn from warmer sources. This movement causes the surface temperature along the equator to increase. Figure 9.02 shows the extent and degree of this warming at the 'mature phase' of El Niño as calculated by Rasmussen and Carpenter (1982), who averaged temperature anomalies from the six El Niño episodes between 1951 and 1973. The figure indicates that a significant change in surface temperature occurred over an area roughly 10,000 km long and 4000 km wide. The maximum of the averaged anomaly lies on the equator at ≃130° W longitude and is ≃1.6 °C. In the severe El Niño of 1982–1983 the maximum observed anomaly was close to 6 °C.

As the sea surface temperature increases, the region of ascending moist warm air over the western Pacific (Fig. 9.01) expands toward the east. The volume of rising moist air

Fig. 9.02 Temperature anomalies in the equatorial Pacific Ocean, averaged from the 'mature phase' of six El Niño episodes between 1951 and 1973. Adapted from Rasmussen and Carpenter (1982).

consequently increases. As the air rises it cools and the moisture condenses to form clouds in the upper atmosphere. Consequently, the heat that caused the water to evaporate in the first place (the latent heat) is released. This creates a large additional source of energy and the convection grows in intensity, which in turn increases the already anomalous circulation in the atmosphere. Thus, the chain of events that started with an anomaly in the wind speed ends up, through positive feedback, increasing the anomaly in the wind speed.

The coastal upwelling off Peru and Ecuador shown in Fig. 9.01 at the far eastern end of the section continues as the trade winds diminish. With the presence of the deeper mixed layer, the upwelling water at the coast is the warm nutrient-poor water from the mixed layer rather than the cool nutrient-rich water from beneath the thermocline. Thus the warm water and the disappearance of the anchovy fishery called El Niño is seen to be a relatively small part of the ocean-wide changes in both the ocean and the atmosphere, rather than a local phenomenon as originally believed.

The end of El Niño, like the beginning, appears to start in the western Pacific. A rise in the thermocline, thought to originate north and south of the equator, emerges in the west and propagates toward the east along the equator. This process raises the cooler subthermocline water closer to the surface. When combined with the equatorial upwelling, this leads to a decrease in the sea surface temperature. The lower temperatures slow down the strong convection in the atmosphere and lead subsequently to an increase in the trade winds back to the original situation.

The rise in the thermocline at the western end of the equatorial Pacific, triggering the end of El Niño, is generated by the warm phase of El Niño. The change in the wind pattern along the equator, including the decrease in the easterly trade winds and the eastward movement of the convection region, creates regions north and south of the equator where the upper layer thickness is less than normal. These shallower thermoclines move westward as slow-moving Rossby waves until they hit the western boundary, where they form baroclinic Kelvin waves. These move the shallow thermocline toward the equator and then along the equator from the west to the east. In this scenario, the return of the cold phase of El Niño is triggered by events in the warm phase. This leads to the idea that the whole phenomenon is an oscillator.

The contention that ENSO may be a natural oscillator—that is, a self-contained closed loop of variations and interactions between the ocean and the atmosphere—is made by

Graham and White (1988) and Enfield (1989). They point out that an oscillator requires first a mechanism that makes small instabilities grow (positive feedback), and mechanisms that eventually damp out the instability (negative feedback). The timing of the oscillation is controlled by the timing of the feedback mechanisms. The long period, 3–7 years, of the ENSO oscillator is thought to be controlled by the slow rate that the shallow thermocline anomalies, generated north and south of the equator during the warm phase, move westward. The changes in the thickness of the upper layer move east along the equator much more rapidly in baroclinic Kelvin waves. McCreary and Anderson (1991) present an overview of numerical studies of these processes.

To gain further understanding of the physical processes of ENSO, a large-scale observing program known as Tropical Ocean-Global Atmosphere (TOGA) was begun in 1985 (McPhaden 1993). One of the main components of this program is the Tropical Atmosphere Ocean (TAO) Array, which consists of 65 moorings spread across the equatorial Pacific between 8° S and 8° N. Temperatures and velocities of air and water from these moorings are telemetered to shore in real time, where they are used to map the state of the ocean and used as input into climate prediction models.

A good example of the power of these data to reveal the important processes in unprecedented detail is given by Kessler and McPhaden (1995). They show in detail how warmer conditions in the western Pacific in 1991–1993 were generated not by a Rossby wave reflecting off the western boundary but by westerly winds that are normally found further west over Indonesia. These winds also generated Kelvin waves that caused the warmer conditions and the propagation of the deeper thermocline eastward. The anomalous conditions, however, did generate a Rossby wave that, when reflected off the western boundary, caused the warm event to end by raising the thermocline along the equator.

McPhaden (1993) says these data show there are common features in all of the warm events. Important differences still exist between the events, however, that make it difficult to develop a comprehensive theory to explain all the events.

9.2.2 Teleconnections, patterns, and oscillations

The substantial variations in the atmosphere and ocean that are part of the El Niño–southern oscillation appear simultaneously with abnormal meteorological and/or oceanographic conditions in regions far removed from the equatorial Pacific. These remote associations are termed 'teleconnections' but the processes linking the remote anomalies to the equatorial Pacific are still under study. The terms 'pattern' and 'oscillation' have also been commonly used to specify large long-term changes in the atmosphere that may or may not be part of the recently discovered ENSO teleconnections.

In the ocean, the best-known remote connections with El Niño are changes in the conditions off the west coasts of both North and South America. In the Gulf of California, for example, interannual variations in the sea level are correlated with the occurrences of El Niño and the California current is observed to slow or reverse direction during strong El Niños (Mysak 1986; Enfield 1989). These changes are transmitted along the coast by the baroclinic Kelvin waves travelling north along the west coast from the equator, where they are generated when the thermocline along the equator is either increasing or decreasing its downward tilt toward the west.

The unusually large El Niño in 1982–1983 led to unusually warm conditions along the west coast of North America as far north as Alaska. Mysak (1986), however,

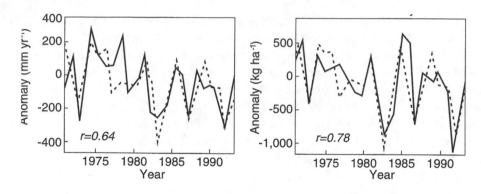

Fig. 9.03 Solid lines: Annual variation in the Zimbabwe rainfall and maize yield of 1970–1993. Dashed lines: Sea surface temperature anomalies of the eastern equatorial Pacific. r is the correlation coefficient of the pairs of the curves. After Cane *et al.* (1994).

presents two reasons why baroclinic Kelvin waves could not propagate as far north as Alaska. The thermocline, on which they travel, is too deep near the equator and farther north the thermocline rises to the surface, which stops such waves. Mysak argues that the warm conditions in the northeast Pacific were caused by the intensification of the Aleutian low. This strengthening led to stronger-than-normal winds from the southwest along the west coast of North America, bringing an unusually long period of down-welling, high coastal sea levels, and a strong northward transport of warm water along the coast from California to Alaska. Mysak (1986) also makes the point that although in-tensification of the Aleutian low often follows an ENSO event, the historical records show that weak ENSO events have been followed by strong anomalies in the Aleutian low, and that ENSO events have occurred without being followed by intensification of the Aleutian low. Hence, the correlation between the two is strong but not by any means complete.

In the atmosphere there are strong correlations between the warm period of ENSO and weather changes along the equator. Droughts in Australia, Brazil, and Africa occur at this time along with unusually high amounts of rain along the equator east of 160° E. Higher rainfall is also common in Ecuador and northern Peru. Descriptions of some of the disas-ters inflicted by the altered weather caused by the 1982–1983 El Niño are given by Glantz (1984) and Canby (1984). The abnormal weather is thought to be primarily due to the changes in the position and intensity of the atmospheric convection, which moves eastward during the warm phase of ENSO from Indonesia toward the central Pacific. This adjustment affects the adjacent Walker cells, resulting in changes in the equatorial weather remote from the central Pacific.

Knowledge of these teleconnections has led to attempts to forecast the effects of El Niño on crop yields in various parts of the world. Cane *et al.* (1994) quantified the relationship between February–March sea surface temperature anomalies in the eastern equatorial Pa-cific and the yield of maize in Zimbabwe (Fig. 9.03). More than 60% of the variance in yield is accounted for by this El Niño index. The authors then used the Zebiak and Cane (1987) predictive model of ENSO to forecast the maize yield in Zimbabwe a full year ahead. They were able to obtain a correlation coefficient of $r = 0.57$, significant at the 99.5% level, be-tween the sea surface temperature anomalies and the maize yield one year later. Forecasts

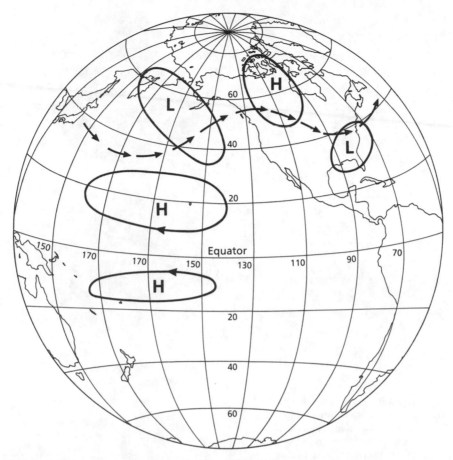

Fig. 9.04 Pattern of anomalous pressure distribution in winter during the warm phase of El Niño. Arrows indicate mean path of westerlies. After Horel and Wallace (1981).

made from sea surface temperatures 6 months ahead of the harvest give an even better correlation, $r = 0.71$.

In the temperate zones, teleconnections with ENSO also exist but the correlations are usually not as strong as they are with the abnormal events near the equator. The pattern of anomalous atmospheric pressure related to the warm phase of El Niño (suggested by Horel and Wallace 1981) is given in Fig. 9.04. These authors first constructed indices for various features of the southern oscillation over the 28 years between 1951 and 1978 and then calculated the correlations between these indices and various atmospheric parameters throughout the northern hemisphere during winter. During the warm phase of El Niño, regions of higher than normal pressure are found in the western Pacific on either side of the equatorial region of high rainfall. Toward higher latitudes a sequence of alternating low- and high-pressure anomalies form along a curving eastward path. These anomalies, which are strongest in the winter, cause the path of the mid-latitude westerlies to be distorted as indicated by the line of arrows in Fig. 9.04. The distortion brings westerlies further south in mid-ocean, causing a decrease in the rainfall at Hawaii; at the west coast of North America the westerlies are further north than usual and bring warmer air to Alaska and the Canadian prairies. On the eastern side of North

Fig. 9.05 Atmospheric pressure distributions over the northern hemisphere during winter (December, January, and February) for (a) the positive phase of the North Atlantic oscillation, (b) the normal distribution, and (c) the negative phase of the North Atlantic oscillation. Air flow is clockwise around the highs and counter-clockwise around the lows. Adapted from Fang and Wallace (1994).

America the air flow is more northerly than usual and brings more of the cold Arctic air to the eastern United States.

The pattern in Fig. 9.04 is an average over many realizations and does not accurately describe the teleconnection pattern for each unique El Niño episode. For example, the 1976–1977 El Niño was associated with severe drought in the western United States and cold in the southeast but during the 1982–1983 episode the west coast was hit by an unusual number of storms and the east was unusually mild (Namias and Cayan 1984).

The alternating series of anomalous highs and lows shown in Fig. 9.04 constitutes one of the better-known long-term weather patterns. It is called the Pacific/North American pattern and has been used successfully for many years as an aid in long range forecasting. Another pattern, well known since the whaling days in Baffin Bay, is the North Atlantic oscillation described by van Loon and Rogers (1978).

The best-known effect of the North Atlantic oscillation is that warm winters in Europe coincide with cold winters in Labrador and West Greenland, and vice versa. In Fig. 9.05, the normal winter sea-level pressure distribution over the northern hemi-

Fig. 9.06 The North Atlantic oscillation winter index based on the average pressure differences between the Azores and Iceland during the winters of 1895–1994. The heavy line is the 5-year running mean. Adapted from Rogers (1984) and updated with data from Drinkwater (personal communication).

sphere is presented along with the distributions found at the two extremes of the oscillation. The average distribution (Fig. 9.05b) contains two low-pressure regions: the Icelandic low over Iceland and southern Greenland, and the Aleutian low in the North Pacific Ocean. The two main high-pressure regions are the Azores high over the central North Atlantic Ocean and the large Siberian high over Russia. Under these normal conditions, the western side of the counter-clockwise circulation around the Iceland low brings cold air out of the Arctic over eastern Canada, Baffin Bay, and the Labrador Sea. Under the eastern side of the low, warm moist air from the eastern North Atlantic flows over western Europe.

The pressure distribution associated with the positive phase of the North Atlantic oscillation is shown in Fig. 9.05(a). In this case the Icelandic low is deeper than normal, resulting in colder than normal winters under the western side of the low and warmer than normal winters under the eastern side of the low. The pressure distribution of the opposite phase of the oscillation in Fig. 9.05(c) shows the Siberian high extending across the Arctic Ocean over northern Canada and the Icelandic low further south than normal over the North Atlantic Ocean. Under these conditions, the air flow over the Labrador Sea and Davis Strait is easterly and much warmer than in normal winters. Over northern Europe the air flow is northerly, rather than southerly, and much colder than during the winters when the pressure distribution is normal or characteristic of the other phase of the oscillation.

The state of the North Atlantic oscillation is often indicated by the atmospheric pressure difference at sea level between the Azores and Iceland. Positive values of this difference are associated with the more intense Icelandic low shown in Fig. 9.05(a) and negatively correlated with the situation shown in Fig. 9.05(c). A time series of these differences for the years between 1895 and 1994, shown in Fig 9.06, indicates that either the anomalous high or the low pattern can exist for many years at a time. From the mid-1950s to the early 1970s, for example, the pressure difference was well below average, bringing cold air to Europe. Since the early 1970s, on the other hand, the winters with a positive index have dominated with warm European winters but some of the severest winters on record occurring off Labrador and Greenland.

9.2.3 Interdecadal variability in the North Pacific

The Aleutian low-pressure system begins to form over the northern Pacific Ocean during the autumn months, intensifies during the winter months, and breaks down by early summer. Beamish and Bouillon (1993) used the winter area that was less than 100.5 kPa as an annual index of the intensity of the Aleutian low, smoothing this index with a LOWESS smoother using a bandwidth of 0.20. They found that the area peaked about 1940, had minima in 1950 and 1972, and then increased until well into the 1980s. They postulated that when the Aleutian low was extensive and intense, the wind stress around the sub-Arctic gyre system would be stronger, leading to more intense upwelling in the center of the gyre and bringing nutrient-rich water to the surface that would be transported toward the periphery, thereby stimulating biological production. Both physical and biological evidence support this hypothesis (Trenberth 1990; see also section 9.3.2).

On the continental shelf of the northeast Pacific, intensification of the Aleutian low is accompanied by warming of surface waters. Royer (1989) showed that in the Gulf of Alaska, along an east–west line at 55° N, an interdecadal fluctuation occurred in sea surface temperature, with warm anomalies peaking in 1958, cool anomalies reaching their lowest in 1972, and warm anomalies peaking once again in 1985. While the winds associated with the intensified Aleutian low cool the mid-ocean water, they also drive warm water northward along the continental margin.

Indications have also been found of intensification of the North Pacific subtropical gyre after the mid-1970s. Tanimoto *et al.* (1993), in an analysis of North Pacific sea surface temperatures on a range of scales, found an interdecadal change in temperature of the subtropical gyre, with strong negative temperature anomalies in the late 1950s and again in the 1980s. They showed how these anomalies interacted with the 2–5 year time scale characteristic of the El Niño cycle. When north–south temperature profiles in mid-Pacific for 1966–1975 are compared with those for 1976–1985 (Watanabe and Mizuno 1994), there is clear evidence of intensification of the subtropical gyre during the latter period, with cooling of water near its center to a depth of 400 m, indicating stronger Ekman pumping.

Polovina *et al.* (1994) reported a significant change in the physical and biological properties of Pacific water northwest of Hawaii. The change, which began in the mid-1970s, peaked in the early 1980s and ended by the late 1980s. It involved an increase in average mixed layer depth, and an increase in frequency of deep mixing events. This process led to an increased input of nutrients to the euphotic zone, increased phytoplankton productivity, and, with appropriate lags, increasing productivity of reef fishes, squid, flying fishes, lobsters, sea birds, and seals.

9.3 BIOLOGICAL VARIABILITY IN THE PACIFIC OCEAN

The biology of the Peruvian upwelling system and its interannual variability were treated in Chapter 5. We saw that at the height of the 1982–1983 El Niño event the upwelling waters off Peru were at 29 °C instead of the usual 16–18 °C, had very low nutrient concentrations, and supported a primary productivity that was only about 5% of normal. It is well established that El Niño years are times of greatly reduced catches of anchoveta, although the exact cause is uncertain. A proportion of the stock migrates south and finds cooler waters less affected by the abnormal surface warming. Others remain in the affected area and die from thermal shock, a lack of suitable food, or perhaps a combination of the two. Since

the economy of Peru is tied quite strongly to the anchoveta industry, the social consequences of ENSO events are serious, and there is a great deal of interest in trying to predict
their occurrences. This situation is perhaps the clearest example in the world of the influence of physical oceanography on fish productivity. It underlines the need for a better understanding of physical–biological interactions in the sea. Meantime, evidence is
accumulating from other areas that year-to-year variability in ocean circulation contributes
to the variability in biological events.

As was mentioned in section 9.2.2, an ENSO event is normally accompanied by an intensification of the Aleutian low-pressure system in winter. Simpson (1992) confirmed
this coupling for the ENSO events of 1940–1941, 1982–1983, and 1991–1992. The effect
is to change the California current by (i) enhanced onshore transport of sub-Arctic water
from the northwest, (ii) increased transport of Equatorial Pacific water from the south, especially in the California undercurrent, and (iii) cessation of upwelling. The volume of
water flowing south along the California coast is reduced (Chelton *et al.* 1982). This decrease is accompanied by a rise in sea level, warmer water, and the appearance of tropical pelagic red crabs and southern species of fish such as tuna, marlin, barracuda, and
bonito (McGowan 1985; Anon. 1992). The elevated temperatures may cause the decomposition of giant kelps. In the absence of upwelling, plankton production is low, and sea
lions migrate northward.

The effect extends further north into Oregon and beyond and has been referred to as 'El
Niño North' (Wooster and Fluharty 1985). During the 1983 El Niño, a thick blanket of water up to 7 °C warmer than usual lay over the Oregon shelf to approximately 100 nautical
miles from the coast. Upwelling brought up only warm, nutrient-poor water and the biomass of zooplankton was reduced about 70%. Species characteristic of waters further south
appeared in the plankton, and the northern anchovy spawned early (Miller *et al.* 1985). Juvenile salmon were less abundant, and catches of adult coho salmon were the lowest since
recordkeeping began (Pearcy *et al.* 1985). The reproductive success of cormorants and
murres along the Oregon coast was sharply reduced (Graybill and Hodder 1985).

These effects are caused by a pushing northward of the boundary between the subtropical water of the Pacific gyre and the sub-Arctic water of the Alaska gyre (Fig. 9.07). Fulton
and Lebrasseur (1985) documented this transition from characteristic species of zooplankton. The Pacific gyre is characterized by many small species of *Calanus, Mesocalanus,* and
Acartia, and the mean spring biomass in the California current is 72 mg m^{-3}. The Alaskan
gyre contains large copepods such as *Neocalanus plumchrus* and *Neocalanus cristatus,* and
the spring biomass is about 147 mg m^{-3}. Between normal years and El Niño years, the
boundary can move about 15° in latitude from Cape Mendocino to the Queen Charlotte Islands. One effect of 'El Niño North' is to deprive juvenile pink salmon of the large zooplankton on which they thrive.

The effect is not confined to coastal waters. Karl *et al.* (1995) reported an increase in
stratification and a decrease in vertical mixing north of Hawaii in the El Niño years
1991–1992. Near-surface primary production decreased by two-thirds compared with the
previous two years, and abundance of the nitrogen-fixing *Trichodesmium* increased. This
released the phytoplankton from nitrogen limitation and permitted an increased productivity per unit of chlorophyll-a. We may also note that while production based on upwelled nitrate is referred to as new production, in contrast to regenerated production based on
ammonium (section 3.5), we need a third term for production based on nitrogen fixation.

Fig. 9.07 Diagram illustrating the interannual variation in the position of the boundary between the sub-Arctic gyre and the California current portion of the subtropical gyre, in the North Pacific. Adapted from Fulton and LeBrasseur (1985).

To summarize to this point, it seems that there is a North Pacific interannual variability that frequently appears to be related to ENSO. The Aleutian low-pressure system intensifies, and there are unusually strong winds from the southwest along the west coast of North America. This action piles up warm water inshore, elevates sea levels, and forces the boundary between the sub-Arctic water of the Alaskan gyre and the warmer waters of the California current section of the subtropical gyre to move northward, sometimes as much as 1500 km.

9.3.1 Effects on fish stocks

(a) The time scale of El Niño events

The effect on fish stocks of interannual fluctuations in the Aleutian low-pressure system at times of El Niño events is now well established. Robinson (1994) confirmed by field data and modelling that a strong Aleutian low in winter has the effect of delaying the onset of upwelling in southern British Columbia; this delay is the prime natural cause of fluctuations in stocks of Pacific hake and Pacific herring. Along the California coast, 'El Niño North' conditions have the effect of decreasing plankton production, which would be expected to have a negative effect on fish species feeding in the plankton. In addition, there is the possibility that reduced offshore Ekman transport is beneficial for some species. If, in normal years, large numbers of larvae are advected offshore and lost to the breeding population, re-

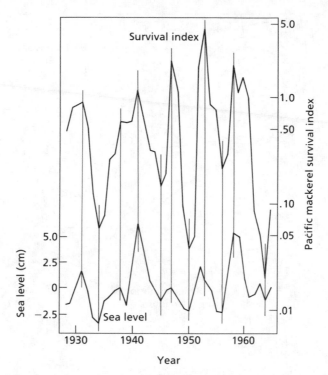

Fig. 9.08 Survival index for Pacific mackerel (upper line) and sea-level anomalies along the California coast. High sea levels indicate El Niño events. After Sinclair *et al.* (1985).

duction in the amount of upwelling may be beneficial. Pacific mackerel (*Scomber japonicus*) may be a case in point (Sinclair *et al.* 1985), as shown in Fig. 9.08. Survival of Pacific mackerel larvae shows a highly significant correlation with mean sea-level height, which is itself associated with reduced upwelling. Similar correlations have been demonstrated for Pacific hake (*Merluccius productus*) off California (Bailey 1981) and Pacific herring off Alaska (Pearcey 1983).

Using a physical oceanographic model of the Alaskan gyre, Thomson *et al.* (1992) showed that variation in the Aleutian low-pressure system had a significant effect on the latitude at which salmon approach the coast after a period of feeding in the open ocean. Fish approaching the Fraser River to spawn have two possible migration routes. One takes them south of Vancouver Island, through the Juan deFuca Strait, where they are equally available to U.S. and Canadian fishermen. The other takes them north of Vancouver Island and into the Fraser River by way of Canadian waters. Prediction of the percentage following each route is, therefore, of considerable economic interest to the two countries. Xie and Hsieh (1989) found that the percentage taking the northern route was greater in years having higher surface temperatures in March and high river run-off, both features being associated with a strong Aleutian low. Hamilton and Mysak (1986) pointed out that a strongly flowing coastal current (which changes with the strength of the Aleutian low) interacts with bottom topography to produce the Sitka eddy (Fig. 9.09). They noted this feature influences the migration route of salmon approaching the Nass and Skeena Rivers in northern British Columbia.

Interesting El Niño effects have been observed on the Asian coast as well (Yamanaka 1985). El Niño is associated with reduced sea surface temperatures, at times as much as

Fig. 9.09 Diagram showing position of the Sitka eddy and its postulated effect on the migration of salmon returning to the Nass and Skeena Rivers. (a) Weakly developed eddy, salmon follow the coast of Alaska; and (b) strongly developed eddy, salmon pass more directly to the coast of British Columbia. After Hamilton and Mysak (1986).

7 °C. Bluefin tuna (*Thunnus thynnus*) normally spawn southwest of Japan and a strong fishery is located off the coast of Japan. After spawning, the fish are believed to follow the Kuroshio current and the North Pacific current toward the mid-Pacific, where considerable quantities are caught by longlining. During El Niño years coastal catches decrease and mid-Pacific catches increase, suggesting an eastward shift in the population. At the same time, a number of cold-water species that normally stay well north of Japan appear off the Japanese coast.

If we consider the North Pacific as a whole, there seems to be a counter-clockwise shift in species distributions during El Niño years. Off North America, warm-water species extend their ranges northward, while off eastern Asia cold-water species extend their range southward.

(b) Interdecadal time scales

We saw in section 9.2.3 that some marked interdecadal changes have been identified in the oceanography of the North Pacific. The Aleutian low-pressure system and the associated sub-Arctic gyre were unusually vigorous around 1940 and again in the 1980s. The most recent long-term intensification of the low began in 1972, with a particularly rapid change

Fig. 9.10 Catches of sardines from three stocks: the Far Eastern, the Chilean, and the Californian, 1910–1988. From Kawasaki (1992).

during the winter of 1976–1977. Similarly, the subtropical gyre deepened and intensified between the decades of 1966–1975 and 1976–1985. Some remarkable changes in fish stocks appear to follow the same pattern. Kawasaki (1992) showed that three geographically separated sardine stocks—from the Far East, Chile, and California—followed the same pattern of abundance, being most plentiful in the 1940s and the 1980s, when the two North Pacific gyres were most vigorous (Fig. 9.10). Beamish and Bouillon (1993) showed that total landings of sockeye, chum, and pink salmon for the North Pacific also showed a similar pattern, with peak landings in the 1940s and the 1980s. More remarkable still was the fact that each species taken separately showed the same pattern (Fig. 9.11). When the landings of Canada, the United States, Japan, and the USSR were examined separately, they also showed the same general trends, albeit with more variability. The authors examined the hypotheses that fishing effort or management practices could have accounted for the similarities in patterns of landings in different countries and different species, but were forced to reject them. They concluded that environmental variables acting on the marine production systems provided the most plausible explanation.

Beamish (1993) also made an intensive study of the period 1976–1978, when the Aleutian low was changing most rapidly and a marked cooling of waters offshore was accompanied by a warming of coastal waters. He showed that strong year-classes and above-average survival occurred in both salmon stocks and in the majority of non-salmon species spawning from California to the Bering Sea. These fish had a wide range of life history types. Beamish concluded that the effect was attributable to improved ocean productivity caused by changes in the intensity of the Aleutian low. In a subsequent analysis of physical events in the same time period, it was shown that in the winter of 1976–1977 there was indeed an abrupt shift in the atmosphere–ocean system over the North Pacific (Miller *et al.* 1994). The Aleutian low deepened, causing storm tracks to shift southward, with increased storm

Fig. 9.11 Catch data for sockeye, chum, and pink salmon, and total for all salmon, for the entire North Pacific. Plots show annual catches and a line generated by a LOWESS smoother (bandwidth 0.20) fitted to the data. From Beamish (1993) and Beamish and Bouillon (1993).

intensity. Sea surface temperature cooled in the central Pacific and warmed off the coast of western North America. When 40 environmental variables were normalized and incorporated in a single time series, 1976–1977 emerged as a year of exceptional change. The climate system fluctuated around this perturbed state for the following 10 years.

In southern California during this same period (1976–1979), the southward-flowing California current was reduced, apparently by the strengthening of the Aleutian low with its attendant northward airflow along the coast, and there was a marked reduction in the amount of upwelling. Lange *et al.* (1990) reported that during this time the sedimentation of diatoms in the Santa Barbara basin decreased 10-fold compared with the average for 1954–1972.

9.3.2 Effects on coral reefs

The warming of coastal water associated with a strong El Niño has had serious consequences for the coral reefs of Costa Rica, Panama, Colombia, and the Galápagos Islands (Glynn 1985, 1990). In 1982–1983 surface waters were at 30–31 °C for 5–6 months. This warming caused the corals to lose their symbiotic zooxanthellae, giving them a bleached appearance, after which they died within a month. In the most dramatically affected regions, coral mortality was 90–95%. It is thought that the unusually high temperatures experienced from 1976 to 1983 may already have weakened the corals.

Interesting biological interactions were also observed with the coral reefs. Many species not resistant to attacks of the sea star *Acanthaster* grew in places where they were encircled by the branching coral *Pocillopora*, which has effective nematocysts and symbiotic crustaceans (*Trapezius* and *Alpheus*) capable of repelling *Acanthaster*. Cases were documented in which *Pocillopora* succumbed to the high temperatures, allowing the sea stars to attack species that were known to have grown uninterruptedly in the same place for 200 years.

9.4 SUMMARY FOR THE PACIFIC OCEAN

The cycle of physical events comprising an El Niño–southern oscillation is summarized in section 9.2.1. The most dramatic biological event is failure of the production system supporting anchoveta off Peru. The southern oscillation causes the thermocline to move much deeper, with the result that water upwelled in the coastal zone is warm, and nutrient-poor. Primary production is greatly reduced, and with it most kinds of secondary production. In extreme conditions, such as those experienced in the 1983 El Niño, tropical coral reefs may suffer severe damage from the elevated water temperature.

While the changes just described are first seen close to the equator, off the Ecuador–Peru coast, they gradually spread to higher latitudes. A rise in sea level and anomalously high sea-surface temperatures have been recorded along the North American coast from California to Alaska. They are loosely referred to as 'El Niño North,' but it seems that their occurrence is not always correlated with an El Niño. Intensification of the atmospheric low-pressure system over the Alaskan sub-Arctic gyre (the Aleutian low) causes an intensification of onshore winds from the southwest, a rise in sea level, and an increase in strength of the northward-flowing coastal current. These events usually, but not always, happen in El Niño years. As in Peru, coastal upwelling in the California current system during El Niño tends to bring to the surface warmer, nutrient-poor water and biological productivity decreases. Paradoxically, some species of fish, such as mackerel, hake, and herring, seem to have better survival under these conditions, producing bigger year-classes. It is suggested that in non–El Niño years they lose a large proportion of the newly hatched stock through offshore Ekman transport. During 'El Niño North' these species benefit from the reduced offshore transport. Similar arguments have been produced for the improved recruitment of crab stocks during El Niño years.

Changes in the strength of the northward-flowing coastal current are correlated with changes in the survival of cod and in the migration routes of salmon. Strong currents appear to carry a high proportion of the cod away from their nursery grounds, leading to poor year-classes. Factors associated with 'El Niño North,' such as warmer surface waters and higher rainfall along the coast (hence more run-off from the rivers), are thought to cause salmon to change their migration routes, and therefore change their vulnerability to fishing fleets along different parts of the coast.

We thus arrive at a concept of oscillations in the atmosphere–ocean system that benefit some species of fish in El Niño years and other species in non–El Niño years. This idea is important to fisheries managers. For a long time overfishing was most frequently blamed for declines in fish stocks. Later, biological interactions such as changing food availability or predation by one species on the young stages of another were implicated. It is only recently that it has become apparent that major physical changes on the scale of whole ocean basins occur routinely and have drastic effects on the year-class success of a wide variety of fish species.

A desirable economic aim of a fishing industry is to have a relatively constant annual yield from a stock, so that capital equipment and staff can be used to the best advantage year after year. Study of the large-scale physical processes and their coupling to biological events reveals that major year-to-year fluctuations in fish stocks are a natural phenomenon. Studies of the abundance of fish scales preserved in anaerobic basins in the California current system confirm this (Soutar and Isaacs 1974).

Evidence is accumulating for the existence of decadal-scale changes superimposed on the interannual changes associated with El Niño. In the atmosphere these changes are apparent as changes in the intensity of the Aleutian low-pressure system in winter. It is thought that, as the low-pressure system deepens, the winds driving the Alaskan gyre strengthen, so that the gyre becomes more energetic. This intensification, in turn, leads to more Ekman upwelling in the center, so that more nutrients are brought to the surface and ocean productivity increases. This process was most active in the 1940s and the 1980s with particularly rapid change in the system being recorded in 1976–1978. Many fish stocks showed good recruitment during years when the Aleutian low was deep and extensive, and poor recruitment when it was shallow and limited.

9.5 BIOLOGICAL VARIABILITY IN THE NORTH ATLANTIC

9.5.1 Evidence from long-term plankton records

In the years between the two world wars Alister Hardy developed a 'continuous plankton recorder' (CPR). It consisted of a towed body with a tunnel through it, along which sea water flowed. Across the tunnel was stretched a fine gauze mesh that unrolled continuously from a spool below and wound onto a propeller-driven spool above that was immersed in formalin. In this way, plankton was sampled and preserved continuously so long as the body was in forward motion. After World War II, a mechanism was put in place to have these recorders towed by merchant ships and ocean weather ships on regular routes (Glover 1967). Samples were collected at a standard depth of 10 m and an attempt was made to obtain records at monthly intervals from 12 different geographical areas (Colebrook 1986). For zooplankton, the record extends from 1948 to the present day, and recent analyses are based on 24 species in 247 series. For phytoplankton, a change in the method of enumeration was made in 1958, so analyses start from that time. Nevertheless, records are available for 24 species in 263 series.

This study showed that there are strong differences between the open ocean, the shelf, and the North Sea in the seasonal patterns of phytoplankton biomass. Stratification and bloom development begin very early in the North Sea, where the water column is shallower and the salinity is lower, a little later on the shelf, and latest of all in the open ocean (see Chapter 3). However, the open ocean is warmer in winter than is the North Sea, and once stratification is established, the phytoplankton grow faster. Thus the region of maximum

Fig. 9.12 Abundance data for *Pseudocalanus elongatus* from 12 different areas, combined. The data were normalized to zero mean and unit variance. The plot is of the first principal component, after principal components analysis. From Colebrook (1986).

phytoplankton abundance showed a regular seasonal shift from coastal waters, out onto the shelf and then into the open sea.

The seasonal patterns of zooplankton distribution were very different. In general, the highest densities appeared on the shelf at all times of the year, the least in the offshore waters. The densities of summer populations in each area appeared to be mainly a function of the densities of overwintering populations in those same areas, and not a function of the food supply or the dynamics of the water column. As Colebrook (1986) expressed it, phytoplankton achieve something approaching steady state in their response to seasonal changes in temperature and stratification but zooplankton do not. Instead, they appear to be limited by their population growth rates. As has been suggested elsewhere, zooplankton on the shelf tend to be unable to utilize fully the phytoplankton produced during a bloom. Instead, much of the phytoplankton biomass sinks to the bottom and nourishes the benthos. Davies and Payne (1984) documented this process for the North Sea.

Against this background information we may now look at long-term trends in the records. For the zooplankton, almost all species in all types of habitat, from shallow coastal to deep open ocean, showed a downward trend in numbers from 1948 to about 1980 (though an upward trend became apparent after 1980). Principal components analysis was used to obtain an expression of the common pattern for all areas using data standardized for zero mean and unit variance. The data for *Pseudocalanus elongatus* are shown in Fig. 9.12. The same technique was used to show the trends in total zooplankton, in each of the 12 areas sampled. In every case the 36-year trend was toward lower numbers, although there was an upward turn in the past few years that has since proved to be important. When the time series for the various taxa were compared on a seasonal basis it was found that the similarity was greatest in the winter data. From this Colebrook (1985) concluded that the factor responsible for the decline in zooplankton numbers probably operates primarily in winter.

Fig. 9.13 Time series of the abundance of zooplankton and phytoplankton for 1958–1984, derived from principal components analyses of the data for all species and all areas. From Colebrook (1986).

In spite of the quite different seasonal and geographic patterns displayed by phytoplankton and zooplankton in the CPR data, it turned out that total abundance of phytoplankton and zooplankton followed the same 20-year trend of decreasing abundance from about 1960 to 1980 (Fig. 9.13). It would be very surprising if the decrease in zooplankton were not a consequence of the decrease in availability of phytoplankton as food.

Referring to Fig. 9.06, we see that the winter index of the North Atlantic oscillation was negative from 1955 to 1970. This means that cold northerly winds should be more common over the eastern Atlantic. Dickson *et al.* (1988a) investigated the situation by calculating a slightly different index, the mean atmospheric pressure difference between 20° W and 10° E over the latitude interval 35–65° N, reporting monthly averages for the years 1873–1981. They found a regular seasonal pattern, with the pressure difference increasing during the first half of the year, as the Azores–Bermuda high became stronger. They then showed that during the decade of the 1970s the strengthening of that high started earlier and was more intense than it had been during the 1950s. The result was much stronger northerly winds over the coastal waters of western Europe.

This analysis was confirmed by the records of light vessels in the North Sea, many of which showed, between the 1950s and the 1970s, increases of 20–55% in the incidence of gale force winds during the springtime. The regions of maximum increase were close to the east coast of England, and were the same areas in which the plankton biomasses showed the most strong decline.

When the evidence from the winter index of the North Atlantic oscillation and the analysis of Dickson *et al.* (1988a) are considered together, it seems that there was a period from about 1950 to 1980 when the tendency for cold northerly winds to blow over the eastern Atlantic in winter and spring was progressively increasing. This finding leads to the conclusion that the most probable cause of decline in the phytoplankton stocks was strong wind mixing of the surface layers of the ocean in spring that, as we saw in Chapter 2, leads to a delay in the formation of the spring phytoplankton bloom because the phytoplankton are mixed to below the critical depth. Figure 9.14 shows contours of phytoplankton density by month and by year for an area where the spring gales

(a)   *West-Central North Sea*   (b)   *East-Central North Sea*

Fig. 9.14 Contour plots by month and year of the total biomass of 12 'spring species' of phytoplankton from continuous plankton recorder data. The west–central North Sea area was more subject to gale force northerly winds in winter and spring than the east–central North Sea area. Dark shading indicates maximum biomass. From Dickson *et al.* (1988a).

increased greatly, and for an area where the increases in northerly winds were less marked. The former shows both a decline in spring biomass and a 1-month delay in on-set of the bloom during the mid-1970s, while the latter shows a decline in biomass but not a delay. Both areas showed a partial recovery during the 1980s. These observations amount to powerful evidence in support of the view that climatic factors caused the de-cline in phytoplankton biomass between 1958 and 1980 by delaying or diminishing the extent of spring stratification and hence the ability of phytoplankton stocks to increase their biomass by growth and multiplication.

The close correlation between the trend of zooplankton biomass and that of the phyto-plankton makes it difficult to dismiss the view that zooplankton biomass is controlled by phytoplankton production. A delay in the spring bloom and a reduction in the phytoplank-ton biomass present throughout the growing season of the zooplankton would be likely to reduce the rate of growth and reproduction of the zooplankton.

Lamb (1969) developed an analysis of climatic trends in Great Britain using an index that might be regarded as roughly the complement of the index used by Dickson *et al.* (1988a). He called it the 'frequency of westerly weather.' Westerly weather in Great Britain is characterized by surface winds that are predominantly southwest to west and by rapidly changing weather produced by a series of depressions that travel across the country bring-ing frontal rain. In general the temperature is higher than when northerly winds predom-inate. Lamb (1969) found that the number of days per year on which this type of weather occurred increased from the year 1890 to 1925, decreased for 20 years, peaked again in 1950, and decreased again for more than 30 years. Thus, westerly weather was decreasing during the period that the incidence of northerly winds was increasing. In general the changes represent a period of climatic warming in the northeast Atlantic in the first 50 years of the 20th century, and a cooling trend in the period 1950–1980. Perhaps not coincidentally, the first 50 years of the 20th century showed an increase in mean global temperature, and the following 30 years a decrease (Fig. 10.01).

Lamb's index for frequency of westerly weather was used by Aebischer *et al.* (1990) to show that there was an interesting parallel between this weather pattern, the changes in phytoplankton and zooplankton in the North Atlantic and North Sea (see Fig. 9.10), changes in the recruitment of herring (*Clupea herengus*) in the northern North Sea, and various indices of breeding success of kittiwake gulls (*Rissa tridactyla*) nesting on the east coast of northern England. All showed the same trend of decreasing values from 1955 to 1980 and an increase during the 1980s. Spectral and cross-spectral analyses showed strong coherences at low frequencies (wavelengths greater than the length of the series) and at a periodicity of roughly 3–4 years. The relation between the plankton and climatic variables has already been discussed. The relationship between environment and herring is complicated by overfishing, which occurred during the period, but a statistical analysis to remove this effect still showed a significant correlation between herring recruitment and weather. For the birds, there appears to be a direct effect of weather on breeding success, but the availability of herring for food was also shown to be a factor.

9.5.2 Evidence from fish stocks

Fish stocks are noted for their variability, but some of it is attributable to overfishing and some to man's destruction of their habitat (for example, salmon spawning beds). However, the view that a very large proportion of the variability of the stocks is related to natural variability of the marine environment is gaining acceptance. Many examples of correlations between changes in fish stocks and physical changes in the North Pacific were given in section 9.3.1, and historical evidence that such changes have occurred in the past, independently of man's fishing efforts, was obtained by counting fish scales from the sediments of anoxic basins off the California coast (Soutar and Isaacs 1974). For the North Atlantic, the leader in the search for evidence of physical factors influencing fish stocks has been Dr. David Cushing (Cushing and Dickson 1976; Cushing 1979; Cushing 1982).

North Sea gadoid stocks

Between 1960 and 1970, a five-fold increase occurred in the stocks of gadoids (cod, haddock, whiting, coalfish, and Norway pout) as a result of good larval survival leading to strong year-classes. Cushing (1982) invoked his match/mismatch hypothesis (Cushing 1975) to explain the good year-classes. His hypothesis states that a year-class of fish is strong if the timing of the zooplankton maximum matches the timing of the appearance of the larvae requiring the zooplankton as food, and vice versa. He produced evidence to show that a delay in the zooplankton peak of biomass is favorable to the survival of cod larvae, and suggested that the strong northerly winds of the 1960s had the effect of delaying the development of plankton in spring, to the advantage of the cod.

Portuguese sardine stocks

An interesting link with the story about increasing strength and frequency of northerly winds is found in the Portuguese sardine fishery. It was mentioned briefly in section 5.7.2 that northerly winds produce upwelling off the coast of Spain, near Cape Finisterre. Similar upwelling also occurs further south, off the coast of Portugal, and Dickson *et al.* (1988a) found that an April–September upwelling index at Porto showed an increasing trend from 1950 to 1980. They attributed it to the same increase in northerly winds that is believed to have caused the decrease in plankton in the northwest Atlantic. One might have thought that increased upwelling meant increased plankton production and increased sardine production, but the story is not that simple. The joint sardine catch of Spain and Portugal (for which statistics are available) shows a *negative* correlation with the upwelling index. Moreover, the negative correlation is strongest with the April upwelling index over the previous three years. Since April is one of the spawning seasons of the sardine, it appears that weak upwelling may permit the formation of a planktonic population suitable in quality and distribution as food for the newly hatched sardines. In other areas there are examples of *positive* correlations between upwelling index and sardine catches, and this relationship may depend on the range of strength of upwelling that occurs. Whatever the explanation, it is clear that changing weather patterns have exerted a marked effect on the sardine stocks of Portugal.

Deep-water stocks and the great salinity anomaly

The trend toward cold northerly winds in winter over the coasts of western Europe turns out to have been part of a larger pattern of weather anomaly in the North Atlantic during the 1950s and 1960s (Dickson *et al.* 1975). When the average atmospheric pressure over Greenland during the period 1956–1965 was compared with the average for 1900–1939, it was found that there was a positive pressure anomaly over Greenland during the later period. It was particularly pronounced during winter months, when it amounted to an average difference of 7 mbar. The pressure ridge first formed in the early 1950s and persisted and intensified through the late 1960s. It was accompanied by abnormally strong and cold northerly winds, especially in winter. Sea ice in the Greenland Sea expanded its range each year and reached a maximum in 1968.

One result of these changes was that abnormal amounts of polar water were brought south to join the East Greenland and East Icelandic currents so that they became cooler and fresher (Dickson *et al.* 1988b; Cushing 1988). The East Islandic current, normally ice-free, began to transport and preserve drift ice and the East Greenland current increased greatly in volume. Furthermore, the surface salinities north of Iceland decreased below a critical value so that further cooling made them lighter and they did not mix with underlying layers. Below the surface ice there formed a layer of cold, fresher water about 200–300 m deep that was isolated from underlying layers. Although the effect was first seen in 1962, this water mass reached its minimum temperature in 1967 and minimum salinity in 1968. It retained its identifiable characteristics for nearly two decades, while making an enormous journey around the sub-Arctic gyre (Fig. 9.15). Christened by Cushing 'the great slug,' it travelled around the southern tip of Greenland (1969–1970), south along the Labrador coast to the Grand Banks (1971–1972), turned eastward along the sub-Arctic front to pass the weather ship *Charlie* in the mid-Atlantic in about 1974, and on to the northwest of the British Isles in 1975 and 1976. Minor portions of the slug are thought to have penetrated

Fig. 9.15 Route followed by the great salinity anomaly, from its formation north of Iceland in 1961–1962 to its return to the east coast of Greenland in 1981. Modified from Dickson *et al.* (1988b).

many parts of European coastal waters, but the signal rapidly became confused with local salinity variations. By 1977 the main anomaly reached the ocean weather ship *Metro* off the Norwegian coast and in 1978 and 1979 it was off the North Cape and to the west of Spitzbergen. Finally, although the evidence is incomplete, it appears that part of the same slug of water of anomalously low temperature and salinity entered the southward-flowing East Greenland current in 1981–1982. At about the same time these changes also appear (Lazier 1988) to have found their way into the dense water that flows over Denmark Strait to form the bottom water in the Labrador Sea.

Naturally, the slug did not retain absolute integrity, as its signal was reduced by mixing with adjacent waters as it travelled. Nevertheless, from an original water mass just over 2 °C cooler and 0.55% less saline when it left the north of Iceland in 1968, it was still 0.7 °C cooler and 0.11% less saline off the North Cape a decade later. It is possible to calculate that there was a 'salt deficit' of approximately 72×10^9 tons as it passed south along the Labrador coast, and a deficit of 47×10^9 tons as it passed through the Faroe-Shetland channel to the north of Great Britain (Dickson *et al.* 1988b). Since the slug originally approached Iceland from the Arctic Ocean, it is assumed that high-latitude regions north of the

Greenland Sea became correspondingly saltier. The total distance travelled between 1968 and 1979 is of the order of 10,500 km, which indicates a mean speed of about 3 cm s^{-1}, which is consistent with other estimates of the mean circulation speed of the subpolar gyre. What are the biological consequences of this major perturbation of the North Atlantic?

There were marked changes in the plankton community north of Iceland during the formation of the great salinity anomaly. Astthorsson *et al.* (1983) showed that on various sections off the north coast of Iceland the primary production measured at 10 m depth in spring averaged 2.6 mg C m^{-3} h^{-1} during 1958–1964, but between 1965 and 1971 it was only 0.7 mg C m^{-3} h^{-1}. During the same periods the volume of zooplankton obtained from a standard net haul filtering 21 m^3 changed from 10–30 mL per haul to less than 5 mL. They speculated that the presence of the cold, low-salinity water caused a year-round stratification that prevented the renewal of nutrients in the surface layers prior to the spring bloom. This restraint in turn led to greatly reduced primary production and consequently a greatly reduced build-up of zooplankton populations.

On the Grand Banks during the passage of the 'great slug' in 1973, the continuous plankton recorder network indicated that the biomass of phytoplankton was reduced to less than one-third of the average from the previous 13 years, and that similar reductions occurred in the biomasses of copepods and euphausids.

Cushing (1988) examined the spawning success of 15 stocks of fish whose breeding grounds were thought to lay in the path of the salinity anomaly. He performed his analysis by noting the distribution of year-classes in the catches. In 11 of those stocks, the year-classes spawned during the low-salinity years were significantly below normal abundance. Icelandic summer herring had anomalously low abundance from 1965 to 1971. A Wilcoxon rank test of 36 year-classes showed that those associated with the salinity anomaly were significantly different from those not so associated. Similarly, Icelandic spring herring had poor year-classes over an even longer period. The effect was visible in cod stocks of East and West Greenland, and the southern Grand Banks, but not in the stocks off Labrador and the northern Grand Banks. Similarly, many stocks in the northeast Atlantic and northeast Arctic showed the effect, but some did not, and the reason is not clear.

Low salinity in itself is not thought to cause mortality of young stages, but lowering of temperatures would extend the period of larval development and increase the risk of mortality of predation and other sources. Cushing (1988) was inclined to look to reduced food supply as the chief cause of reduced breeding success. In seeking a connection between the physical characteristics of the slug and the reduced plankton production, he emphasized the possibility that the cooler water would lead to a delay in onset of summer stratification and hence a delay in the onset of the spring bloom. He suggested that cool surface waters would be more subject to intermittent downward mixing by strong surface winds during the spring. In other words, there would be an extended period of transient thermoclines, which would have the effect of delaying the spring bloom and reducing the biomass of both phytoplankton and zooplankton during the growing season.

Assembling the relevant data to relate physical with biological events on such a large scale is a difficult task, and it is not surprising that the story has many loose ends. Nevertheless, the case for the formation of the great salinity anomaly, its journey of 10,500 km around the sub-Arctic gyre, and its depressing effect on the recruitment of fish stocks in its path is a persuasive one.

Fig. 9.16 Chart showing major ocean currents around Iceland and Greenland and possible migration routes of cod. From Cushing (1979).

The rise and fall of the west Greenland cod fishery

Most of the biological evidence reviewed so far in this chapter deals with the period of the 1960s and 1970s when the North Atlantic winter pressure index (Fig. 9.06) was predominantly negative, so that cold northerly winds were prevalent in winter in the northeast Atlantic and plankton productivity was depressed. If we now look at earlier periods in the record, we find that the index was mostly positive in the 1920s and 1930s, and again in the 1940s and 1950s. This finding implies (Fig. 9.05a) that westerly weather rather than northerly winds predominated over the northeast Atlantic and the climate was warmer.

The history of the cod fishery in Greenland indicates that the stocks spread northward along the West Greenland coast between 1917 and 1936 (Cushing 1982). From Julianehaab, near the southern tip of Greenland, the fishable stocks spread north about 14° latitude, until by 1936 they were being caught far above the Arctic circle. The changing pattern of the North Atlantic weather from the 1950s onward seemed not to affect them at first. There were record catches in the early 1960s, but by the end of the decade catches were at a low level, and have not so far recovered.

The story is given added interest by the fact that fish were tagged in West Greenland from 1924 onward, and in the decade 1930–1940 significant numbers of tagged fish were found on the spawning grounds south of Iceland. When the West Greenland cod stock was at its most northerly distribution and the biomass of fish was approaching its maximum, large numbers of adults were returning to Iceland to breed. Although the tagging program continued in West Greenland in the 1950s and 1960s, very few tagged fish were recovered in Iceland after 1945. In 1963 developing cod eggs were traced in the Irminger current which flows from the southwest of Iceland across the Davis Strait to Greenland. This observation led to the hypothesis that under favorable circumstances the life history of cod in this area was as follows (Fig. 9.16).

Of the cod eggs produced on the major spawning ground south of Iceland, a portion were carried in the Irminger current to join the East Greenland current, which carried them to the southern tip of Greenland. Here the developing larvae and juveniles were carried north in the West Greenland current. In warm years these larvae and juveniles successfully grew to maturity as far north as Disko Bay and beyond, well into the Arctic circle. In cool years successful development was confined to the southern regions. At sexual maturity, the adults made a return journey in excess of 1500 km to the spawning grounds south of Iceland. This pattern appeared to reach its peak of effectiveness in the 1940s, but when the weather patterns changed during the 1950s, fewer and fewer fish seemed to return from Greenland to the Icelandic spawning ground. Yet as late as 1963, eggs were being carried to East Greenland by the Irminger current.

At this point we may recall that the 1960s and 1970s were the time of strong northerly winds to the east of Greenland. Cushing (1982) suggested that these winds may have disrupted the drift of eggs, larvae, and juvenile cod from Iceland to Greenland. This disturbance may have led to the final collapse of the West Greenland fishery in the late 1960s. Dickson and Brander (1993) showed that in the warming period of the first half of this century, there was a strengthening of easterly winds that would have assisted the drift of eggs and larvae from Iceland to Greenland, while in the cooling period after 1950 the reverse trend was apparent.

As in previous examples, one is struck by the large number of assumptions required to construct a coherent story. Nevertheless, the meteorological data and the fish catch data are well-established facts. They show clearly the nature of interannual variability in both physical and biological systems. It remains to establish the connections between the two. Until we have more detailed information on, for example, the connection between the incidence of strong northerly winds and levels of plankton productivity, we can invoke only such understanding as we have of the mechanism of the spring bloom and its relationship to the formation of a shallow mixed layer, as discussed in Chapter 3.

The Atlantic cod stocks of Labrador and Newfoundland

The stock of cod on the northern Grand Banks and the Labrador shelf was heavily exploited in the 1960s, suffered a precipitous decline in the 1970s, rallied briefly during the 1980s, then declined to such low levels that the Canadian government closed the fishery in 1992. Mann and Drinkwater (1994) suggested that three factors contributed to the collapse of the stock: (i) unfavorable environmental conditions, (ii) reduction in the range of age classes in the stocks, and (iii) overfishing. It was argued that although the stocks had survived equally adverse environmental conditions in earlier decades, reduction in stock biomass and in the range of year-classes had reduced the ability of the stock to survive adverse environmental conditions from about 1970 onward. The synergistic effect of environment and exploitation was deemed important, and a factor to be considered in the future management of this and other stocks.

A trend of increasing positive anomalies in the NAO index was noted from 1970 onward, with increasing northwesterly wind stress over the Labrador Sea in winter, and increasing ice cover between latitudes 45 and 55° N in February. Since water that is on the Labrador Shelf in winter flows in the Labrador current and reaches the Grand Banks in the following spring and summer, the effect is to produce a water column with lowered surface salinity and cold subsurface waters over the Grand Banks in spring and summer. Following the evidence assembled by Cushing (1988, 1990a, b) and documented for the

Norwegian herring stocks by Ellertsen *et al.* (1990), low temperatures are associated with late development of zooplankton populations and poor growth and survival of cod larvae and juveniles. Low summer salinity is also associated with poor recruitment in the Grand Banks cod (Sutcliffe *et al.* 1983, Myers *et al.* 1993a), probably because the stronger stratification inhibits the upwelling of nutrients in summer during storm events.

This interpretation has been challenged by Dr. R. A. Meyers and his associates (Meyers *et al.* 1993a, b; Mertz and Meyers 1994; Hutchings and Meyers 1994). They began by studying records of the period when the Great Salinity Anomaly was centered over the Grand Banks. They found that no low-temperature anomaly was associated with the salinity anomaly. Possibly, this lack of a correlation arises because the low-temperature signal of the Labrador current obliterated the signal of the Great Salinity Anomaly. It gave Meyers *et al.* (1994) reason to cast doubt on the mechanism proposed to account for poor cod recruitment in those years.

The authors also argued that for the match/mismatch hypothesis to be applicable to northern cod, some regularity should be discernible between the average spawning date of the cod at various locations and the average timing of the zooplankton peak. They found that as one moves north from the Scotian shelf, to St. Pierre Bank, and on to the Grand Banks, the average spawning date of cod becomes later; surprisingly, on the Labrador shelf the spawning date is earlier. The data on zooplankton are very inadequate, but the available evidence does not indicate a correspondence with the pattern of spawning dates of cod. There is a great need for good data on the biological oceanographic processes in the area. Until such data are collected, the matter remains unresolved.

Herring stocks of the North Sea

Between 1965 and 1977, the landings of herring from the North Sea fell from about 1.2 million tons to a very low level, and the fishery was closed from 1977 to 1980. As the stocks fell, so did the recruitment, but from 1980 onward there was greatly improved recruitment from a low biomass of spawning fish, giving reason to suspect that the change was associated with changing environmental factors (Bailey and Steele 1992).

There are two schools of thought about the nature of these environmental changes. As we have seen (section 9.5.1) Aebischer *et al.* (1990) drew attention to parallel changes in climate, plankton, herring, and birds in both the North Atlantic and the North Sea. In 1980, all of these factors changed. The climate of western Europe began to warm, the plankton biomass began to increase, and the herring populations showed improved recruitment. Munk and Christiansen (1990) documented improved survival in the herring spawned off southern Scotland and northern England, which would be consistent with these trends.

The alternative explanation for improved herring recruitment in the North Sea, proposed by Corten (1990), was that changing current patterns caused variation in the transport of larvae from the northwest spawning grounds to the southeast nursery areas. The period of expanding stocks was one in which these conditions became favorable after being unfavorable. Corten's proposal was based on surveys of larval distribution, but there were no hydrographic data to support it.

More recently, Lindeboom *et al.* (in press) has documented several changes that occurred in the North Sea at about the same time. In addition to the changes in plankton, herring, and kittiwake stocks reported by Aebischer *et al.* (1990), the numbers of dogfish caught sharply declined, starting in 1978 and continuing for a decade. Also in 1978, the velvet swimming crab disappeared from Dutch coastal waters. In the Wadden Sea, inshore of the chain of islands along the Dutch and German coasts, a sharp increase in phytoplankton and in benthic biomass occurred in the late 1970s; at the same time the number of young eider ducks increased by two orders of magnitude. The phytoplankton of this area showed a sharp increase in *Phaeocystis* blooms. All of this evidence strongly suggests a major change in the ecosystem of the North Sea and adjacent waters, which Lindeboom and colleagues tentatively characterize as a switch of emphasis from benthic production to predominantly pelagic production.

In general, the view is gaining ground that while overfishing of commercial stocks of the North Sea is undoubtedly an important influence, there are many changes that occurred at about the same time in the open Atlantic, the North Sea, and various coastal waters that seem to be associated with changing atmospheric influences.

9.5.3 Evidence from distribution of organisms

Let us recall atmospheric events in the first 25 years of the 20th century. The North Atlantic oscillation pressure index was predominantly positive, so that the track of the westerlies across the Atlantic was well to the south and relatively warm air was brought to the shores of western Europe. It is probable that there was a corresponding warming of coastal waters, or movement of warm water masses to more northerly locations. Cushing (1982) has documented the northward movement of animals in the eastern Atlantic during this period. There were three main groups of events. First, an unusually large number of surface-living and pelagic subtropical animals were carried to western France and the British Isles. These included the Portuguese man-of-war (*Physalia physalis*), two species of goose barnacle (*Lepas*), loggerhead turtles (*Caretta*), and the octopus. Second, Atlantic, as opposed to Arctic, species of fish and bottom-living animals reached further north than they had previously been recorded. Swordfish, pollack, and ray appeared in Icelandic waters, cod and haddock appeared in the Barents Sea, and the hermit crab *Eupagurus,* together with three species of benthic molluscs, appeared on the Murman coast for the first time. Third, in waters off Nova Scotia there were, during this period, an unusually large number of new records of subtropical species (borne, presumably, by the Gulf Stream) but also of Arctic fish species. We have seen that when the North Atlantic oscillation pressure index is positive, winds tend to blow from a northerly direction over Labrador, and the weather is colder. The presence of Arctic species off Nova Scotia may reflect a cooling of coastal waters of eastern Canada, with perhaps a strengthening of the Labrador current.

All these events reached their peak in the period 1925–1935, which Cushing has dubbed 'the dramatic decade.'

9.5.4 Events in the English Channel: the Russell cycle

The end of the dramatic decade marked the beginning of a major change in the biota of the English Channel. Most of the observations on this change were made by staff of the Marine Laboratory at Plymouth, and the most important evidence was collected and summarized by Sir Frederick Russell (1973), so the event is known as the Russell cycle.

The story begins with a decline in recruitment to the herring stock in the western part of the English Channel, beginning in 1925. The last recorded year-class entered the fishery

Fig. 9.17 Changes in the abundance of fish larvae in the English Channel 1926–1972. These changes are part of the Russell cycle. After Russell (1973).

in 1931, and the Plymouth herring stock collapsed in 1936. Throughout this period, macroplankton was being regularly monitored at International Station E1, off Plymouth, and samples were taken at the surface and near the sea bed for phosphate analysis. It had previously been established that two species of the arrow worm *Sagitta* tended to occur in different water masses and could be taken as indicators of those water masses. *S. setosa* characterized water that had been found to be resident in the English Channel while *S. elegans* characterized water occurring to the west of the channel, south of Ireland. In the autumn of 1931 the quantity of macroplankton declined by a factor of four, and *S. elegans* was replaced by *S. setosa*. During the next decade the number of nonclupeid fish larvae (e.g., flatfish, cod family) declined drastically (Fig. 9.17), but the number of pilchard eggs increased. During the same period, the amount of phosphorus present in the water column during winter also declined. It seemed as if there had been some shift in a key environmental parameter that had caused a total change in the structure of the ecosystem. The old system had indicator organisms that it shared with a water mass to the west. It had abundant larvae of flatfish and gadoids, and in the winter, when phytoplankton activity was minimal, it had a good supply of dissolved nutrients (as indexed by phosphorus). Suddenly, a persistent shift occurred, with different indicator organisms, less nutrients, less macroplankton, and far fewer fish larvae. The change was from a system dominated by herring, macroplankton, and bottom-living fishes to one dominated by pilchard and small plankton.

For the next 25 years only a few comparable observations were made, but these indicated that numbers of nonclupeid fish larvae remained low. However, between 1965 and 1979, the situation reversed itself. In 1965 the total number of fish larvae increased by an order of magnitude. Over the next 5 years the winter level of phosphorus rose to pre-1930 levels. In 1970 *S. elegans* came to predominate over *S. setosa* and by 1978 its abundance was back to earlier levels. During the 1970s the number of copepods returned to normal levels and the number of pilchard eggs decreased.

There were interesting changes along the shore lines during the same period. An unusual seaweed, *Laminaria ochroleuca,* appeared in the Plymouth area in the late 1940s, spread widely in the south and west of England by 1960, but declined sharply between 1965 and

1975. The barnacle *Chthalamus stellatus* is better adapted to higher temperatures than *Balanus balanoides*. The warm-water species spread eastward in the English Channel during the 1930s and 1940s, but in the 1960s retreated quite sharply.

All of these events point to a tendency for the English Channel to have been invaded by more southerly species during the pre-1940 period of positive values for the North Atlantic oscillation index, and for these to have retreated during the period when the index became negative. The special feature of the Russell cycle is that it appears to have involved a major flip from one fairly stable configuration of organisms in the ecosystem to another quite different mix of species. In ecosystem theory it has long been recognized that multiple stable states may exist for a particular community of organisms, and that relatively small perturbations may cause a flip from one stable state to the other. The point of separation between the two states is referred to as a bifurcation. Discussion of these ideas is to be found in May and Oster (1976), Allen (1985), and Mann (1988).

A full understanding of the mechanisms of change involved in the Russell cycle still eludes us. However, the English Channel is a well-studied area close to the zone of transition between the subtropical gyre and the subarctic gyre in the North Atlantic. When changes in the North Atlantic oscillation pressure index caused changes from colder, more northerly weather to warmer, more westerly weather off the coast of western Europe, the biological community of the English Channel was induced to flip from one stable state to another.

Two papers from the 1980s take a longer view. Southward *et al.* (1988), quoting Cunningham (1906), pointed out that in the English Channel herring (*Clupea herengus*) are near the southern end of their range and pilchard (*Sardina pilchardus*) are near the northern end of theirs. Reconstructing climate and fish landings from contemporary reports, Southward showed that regular oscillation has occurred between herring in cool periods and pilchard in warm periods, and the Russell cycle was but the most recent example of this phenomenon. Binet (1986) reconstructed a complementary story about the relative abundance of herring in the English Channel and the North Sea.

9.6 SUMMARY FOR THE NORTH ATLANTIC

The North Atlantic oscillation winter pressure index (Fig. 9.06) enables one to divide the years of the 20th century into four main groups: (i) before 1935 the index was predominantly positive, so that the prevailing winds approaching western Europe were from the south and west, and were warmer than average; (ii) between 1935 and 1950 the index oscillated about the long-term mean; and (iii) from 1950 to 1965 the index was falling, and remained negative until 1971. This was a period of increasing cold northerly winds over the waters of the northeast Atlantic. Finally, (iv) after 1971 there was a return to predominantly positive anomalies in the NAO index, similar to those seen in the early part of the 20th century, with a warming trend off western Europe but a cooling trend in Newfoundland and Labrador. These changes were reflected in changes in sea surface temperatures.

The biological events associated with the four periods are illustrated in Figs. 9.18 and 9.19. Beginning about 1917 cod spread north along the coast of western Greenland, probably as a result of receiving a good supply of larvae from Iceland in the Irminger current. Soon afterward, subtropical species began to appear off the coast of western Europe. The

Fig. 9.18 Diagram summarizing biological events associated with the warming trend in the eastern North Atlantic, when the North Atlantic oscillation winter pressure index tended to be positive and the westerlies took a southern route.

Fig. 9.19 Diagram summarizing biological events associated with the cooling trend in the eastern North Atlantic, when the NAO winter pressure index tended to be negative and northerly winds predominated.

peak of these incidents was in the period 1925–1935, the 'dramatic decade.' In the mouth of the English Channel, herring were replaced by pilchard, and the numbers of young stages of gadoid fish and flatfish dropped sharply. The whole balance of the plankton in the English Channel shifted in the early 1930s, and remained changed for more than 30 years. The decade 1940–1950 saw a marked improvement in recruitment of cod in many parts of the North Atlantic, and between 1960 and 1970 the cod landings from West Greenland reached a high level, then collapsed.

It is interesting to note the lag between the onset of the cooling trend, in about 1950, and the downturn in biological events. Cod recruitment in the North Atlantic did not begin to decline until well into the 1950s, the pilchard-dominated community in the English Channel did not begin to decline until about 1965, and the collapse of the West Greenland cod fishery took place between 1968 and 1970.

Once the cooling trend was well established, with weakening of westerlies and a marked increase in northerly winds along the east coast of Greenland and across to the coast of western Europe, a well-documented decline in the biomass of phytoplankton and zooplankton was seen in the North Sea and the northeast Atlantic. Interestingly, the delay in the onset of the spring bloom that accompanied this decline seemed to be advantageous to young cod in the North Sea, for their numbers increased between 1960 and 1970. Herring, on the other hand, were declining during this period, and the decline continued until the fishery was closed in 1977.

The northerly winds blowing down the east coast of Greenland created a major slug of cold, low-salinity water that travelled all the way around the sub-Arctic gyre, taking about 15 years. Most fish stocks in its path showed poor recruitment while it was present, possibly because the cold water delayed the surface warming required to establish the spring bloom of plankton.

In about 1980 the phytoplankton and zooplankton stocks of the northeast Atlantic began an upward trend in biomass that seems to have been maintained. On the opposite side of the Atlantic, however, the winters on the Labrador shelf were characterized by low temperatures and strong winds. It was a period in which the cod stocks of the Labrador shelf and northern Grand Banks succumbed to heavy fishing pressure, so that the fishery had to be closed in 1992. It is strongly suspected that the flux of cold, low-salinity water onto the Grand Banks in the Labrador current contributed to the collapse.

There is a long history of biological observations off the coasts of western Europe, and there is little doubt that many of the events summarized here are linked to large-scale changes in weather patterns. However, these patterns are not as clearly delineated as the southern oscillation in the Pacific Ocean, so the links between physical and biological phenomena in the Atlantic are rather more tenuous, and much remains to be done to establish the mechanisms linking the two.

9.7 VARIABILITY IN THE SOUTHERN OCEAN

As discussed in section 3.3.8, there is a zone of westerly winds and an associated Antarctic circumpolar current girdling the southern hemisphere around 50° S. Interesting observations on the variability in strength and position of these winds, and on associated changes in the ocean around Tasmania, have been made by Dr. Graham P. Harris and colleagues. An example (Harris *et al.* 1992) concerns the strength and position of the westerly winds, and their effect on the food web supporting jack mackerel *Trachurus declivis.* The weather-patterns of Tasmania region are dominated by periods of high zonal westerly wind interspersed with calmer periods. The alternation is caused by the passage of Rossby waves in the upper-level circulation of the atmosphere south of 40° S, which in turn cause the zone of westerly winds to shift north and south over Tasmania. As the waves pass over Tasmania, the westerlies strengthen and weaken. The pattern shows (i) short-term, (ii) seasonal, and (iii) interannual fluctuations.

High westerly wind stress causes advection of colder, nutrient-rich sub-Antarctic water up the eastern side of Tasmania and reduces water column stability. This condition implies more frequent overturn of the water column, increased nutrient upwelling, and increased new production in the phytoplankton. Spring blooms tend to be later but stronger. In periods of low wind stress, subtropical water moves down from the north and replaces the cool, nutrient-rich water. Less upwelling of nutrients occurs and spring blooms tend to be earlier but weaker. There is a short-term periodicity of about 40 days in these events, a cycle that is superimposed on interannual variability related to ENSO; this variability is, in turn, superimposed on a decadal-scale pattern of change that may be related to global warming.

An intensive study in Storm Bay, southeastern Tasmania, from 1985 to 1989 took place during the aftermath of the 1983–1984 El Niño, when wind stress showed the usual 40-day oscillation but an overall declining trend. The authors compared data from 1985 and 1986, a more windy period, with data from 1988 and 1989, a relatively calm period. In any given wind event, nutrient-rich waters rose to the surface, leading to a bloom of diatoms. As the wind died down, stratification arose, nutrients became depleted, and diatoms were replaced by flagellates (see Chapter 3). Copepods and euphausids, which have longer life cycles than the phytoplankton, rose and fell in numbers on a seasonal rather than a 40-day basis. In years when the wind events were more frequent, the larger zooplankton populations grew rapidly, providing food for the jack mackerel. In years when wind events and diatom blooms were much less frequent, the numbers of krill, for example, decreased by three or four orders of magnitude, and the fishery for jack mackerel collapsed. It is believed that the mackerel moved south as the zonal westerly winds moved south.

Harris *et al.* (1988) have documented a number of other changes in the fishery that are correlated with the oceanographic changes described above. Catches of Tasmanian spiny lobsters, *Jasus novaehollandiae,* show a clear correlation with high air temperature and pressure, attributable to the intensification of fishing effort that takes place during favorable weather, but also show a significant correlation with the same environmental conditions with a time lag of 7 years. It appears that recruitment is influenced by the favorable conditions, which leads to enhanced catches 7 years later when the lobsters reach catchable size. The importance of the oceanographic conditions is reinforced by the finding of a similar correlation with landings of New Zealand lobsters and a correlation of opposite sign with landings of *Jasus verreauxi* in New South Wales, Australia. Years of high winds also favor good recruitment of Tasmanian scallops, bring more southern bluefin tuna into Tasmanian waters, and influence the numbers of leopard seals migrating northward from Antarctica to Macquarie Island (about 55° S).

Use of satellite data has made it possible to anticipate some of the changes described above. It seems reasonable to suppose that as the sequence of events connecting atmospheric phenomena with changes in ocean productivity becomes more widely understood in other parts of the world, our ability to forecast trends in stocks of fish and shellfish will show a corresponding improvement.

10

The Oceans and Global Climate Change: Physical and Biological Aspects

10.1 INTRODUCTION

The oceans have a deep circulation, the thermohaline circulation, which we have so far mentioned very little. Water is heated in equatorial regions, then moves poleward in major currents, giving off heat to the atmosphere. In sub-Arctic regions cooling and ice formation cause water to become more dense. The water then sinks to form the 'deep water.' This sinking is the beginning of a long journey close to the ocean floor. Some of the deep water travels south in the Atlantic basin, moves across to the Pacific basin, and there moves slowly northward in a journey that may take a thousand years.

At the regions of deep water formation large quantities of carbon dioxide dissolved in the water sink to great depth and are removed from contact with the atmosphere. Conversely, at regions of upwelling, especially the large upwellings at the tropical divergence, heating of the cold upwelled water causes it to give off billions of tons of carbon dioxide. These are the major physical mechanisms by which the oceans exchange carbon dioxide with the atmosphere, and there is no particular reason to think that they are not approximately in balance. However, there are in addition important biological processes that remove carbon dioxide from the atmosphere and transfer it to the deep ocean. Over 99% of the carbon dioxide added to the earth's atmosphere throughout its history has been taken up by phytoplankton and sedimented to the sea floor to form the calcareous rocks and the fossil fuels. This biological mechanism is known as the biological pump.

For at least 150 years the carbon dioxide concentration of the atmosphere has been rising as a result of man's activities in cutting down the forests and burning fossil fuel. It is now expected that there will be a global rise in atmospheric temperature—the 'greenhouse' effect. Before the magnitude of this important change can be predicted, it will be necessary to understand the extent to which the oceans may be a reservoir for excess carbon dioxide. In this chapter we seek first of all to describe the mechanism of global warming and the present-day global carbon cycle. We shall then explore the relative importance of oceanic sources and sinks compared with industrial activities and terrestrial biota. Finally, we shall attempt to peer into the future, and consider what changes might be expected to occur in physical and biological mechanisms for circulating carbon in the ocean if the expected rise in global atmospheric temperature occurs.

10.2 PHYSICAL ASPECTS

10.2.1 The greenhouse effect

(a) The process

The average surface temperature of the earth and the lower atmosphere is about 15 °C. If there were no water vapor, carbon dioxide, or methane in the atmosphere, the surface temperature would be below freezing by $\simeq 18$ °C and all the rivers, lakes, and oceans would be frozen solid. The reason for the higher, more habitable temperature is the fact that these greenhouse gases delay heat from leaving the earth by trapping it in the lower atmosphere.

As pointed out in Chapter 3 all the heat received on the earth comes originally from the sun's surface ($\simeq 6000$ °C) via electromagnetic radiation with wavelengths between 0.2 and 2.4 μm, often called the short-wave radiation. Of this incoming radiation $\simeq 31\%$ is reflected back into space, $\simeq 23\%$ is absorbed by the ozone, water vapor, clouds, and dust in the atmosphere, and 46% is absorbed by the land and water at the earth's surface (Mitchell 1989). All these absorbers in turn radiate heat in the form of electromagnetic radiation. The wavelength of this radiation is much longer than the incoming radiation in accordance with Planck's radiation law, which dictates that cooler bodies radiate at longer wavelengths. The radiation from the surface of the earth and the atmosphere is thus at wavelengths between 5 and 100 μm, the long-wave radiation.

The atmosphere is quite transparent to the short-wave radiation from the sun as indicated above by the fact that only 23% of the incoming radiation is absorbed by the atmosphere. The long-wave radiation is another story. Roughly 90% of the long-wave radiation leaving the earth's surface is absorbed in the atmosphere by the greenhouse gases. This heat eventually reaches the upper layers of the lower atmosphere through convection and is lost to outer space, but the absorption by the greenhouse gases delays the loss and keeps the lower atmosphere warmer than if the atmosphere were transparent to the long-wave radiation. The amount of heat trapped and the resulting temperature of the atmosphere clearly vary directly with the concentrations of these gases. If they are very concentrated, as on the planet Venus, the temperature is very high ($+400$ °C) and if they are at low concentrations, as on Mars, the temperature is very low (-50 °C).

On the earth there are seven or so gases that contribute to warming the atmosphere, ranging from the naturally occurring carbon dioxide, water vapor, methane, ozone, and nitrous oxide to the human-produced chlorofluorocarbons (Freon). The warming effect of each of these gases is different because their concentrations are different and because they absorb radiation with different efficiencies at different wavelengths. At present, about 65%

of the warming effect is due to water vapor and 32% to carbon dioxide. Between them they absorb almost all the long-wave radiation with wavelengths longer than 15 μm and shorter than 8 μm. The other gases each contribute ≃1% or less to the total effect but an increase of one molecule of chlorofluorocarbon is about 10,000 times more effective at trapping heat than one molecule of carbon dioxide (Mitchell 1989). This difference arises because the chlorofluorocarbons strongly absorb energy in the 'window' between 8 and 12 μm, where water vapor and carbon dioxide do not absorb. A small increase in the chlorofluorocarbons thus makes a big difference in the total absorption because the band is still full of unabsorbed radiation, whereas a similar increase in carbon dioxide can produce only a small increase in absorption in a band where most of the radiation is already absorbed.

For the future of life on the earth, the important point is that the concentrations of these gases in the atmosphere are slowly increasing. Carbon dioxide, for example, has been monitored since 1958 and these data along with determinations of the concentrations in the 'old' air trapped in the polar ice sheets clearly indicate that the concentration in the atmosphere has increased by about 25% since 1850 (Schneider 1989). Most of this increase is due to the burning of wood, coal, oil, and gas. Deforestation in the northern hemisphere also plays an important role as the reduced biomass takes up less carbon dioxide from the atmosphere and the decomposing litter gives off carbon dioxide. It is many years before new growth begins to take up carbon dioxide faster than the old forest. The trace gases such as methane and nitrous oxide have also been increasing. Although their contribution to the total budget is small, at the present time they together produce about the same *increase* in the greenhouse effect as does the increased level of carbon dioxide (Hansen *et al.* 1988). These well-documented increases raise a question that is today of great concern throughout the world—that is, what effects are these increases having on the world's temperature, precipitation, ice cover, and biological processes?

(b) Trends in air temperature

Finding the answers, which is, of course, not easy, involves two main parts: first, the monitoring of the important variables to determine if and how they are changing with time, and second, the construction of a sophisticated model of the climate to determine which processes are causing the observed changes. The main variable to monitor is the surface air temperature, which has been measured accurately throughout the past 150 years over land and sea. As can be imagined, the temperature records are very uneven in quality and coverage and creating a useful data set for the necessary length of time is prone to some unusual problems. For instance, the air in a city can be ≃0.2 °C higher because of the activity in the city and the shape of the buildings. This 'heat island' effect must be removed to get a true picture of the long-term temperature trends (Hansen and Lebedeff 1988). At sea other problems arise. Folland *et al.* (1984) analyzed the available sea surface temperature data but used only the night-time values because the sun's warming of the ships during the day produces falsely high values. They also point out that the temperatures obtained from ships during the second World War appear to be high because the thermometers had to be taken inside the ship to be read, as no lights were allowed on the open deck. These and other spurious signals have to be carefully weeded out of the record as they can create systematic errors that are as large as the signal that is being sought.

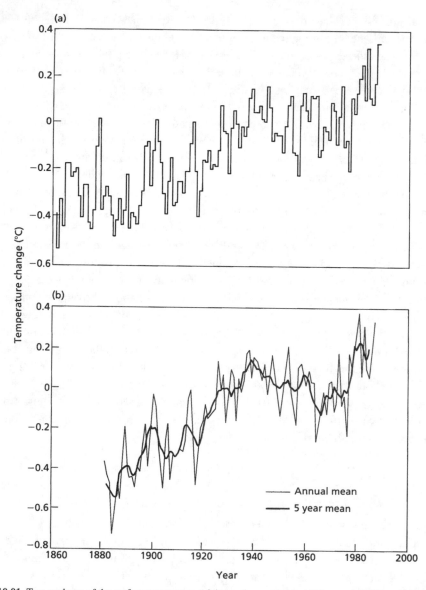

Fig. 10.01 Two analyses of the surface temperature of the earth over the past 100 years relative to the average temperature between 1951 and 1981, adapted from Schneider (1989). (a) From Jones *et al.* (1988); includes temperatures from continents, islands, and the sea surface; and (b) from Hansen and Lebedeff (1988); approximately the same data set as in (a) except no sea surface temperatures are included.

A clean data set of the world's temperature is now available and updated analyses appear regularly. Two of these, Jones *et al.* (1988) and Hansen and Lebedeff (1988), are reviewed by Schneider (1989) (Fig. 10.01). Each analysis uses a different technique but both use roughly the same raw data except that Jones *et al.* (1988) include temperatures over the sea. The two graphs agree in that they both indicate that the temperature of the earth's surface as a whole has increased by ≃0.5 °C over the past 100 years, with the decade of the 1980s being the warmest on record. For a long time, the existence of a temperature increase has been regarded as unproven, but the evidence becomes more convincing with each new analysis. Even now, the cause of the rise is uncertain. Some invoke a change in solar radiation (the solar constant) but no one doubts that increasing atmospheric carbon dioxide eventually leads to a rise in temperature at the earth's surface, so the most probable cause of the present rise in temperature is the observed increase in atmospheric carbon dioxide.

As indicated above, it has been possible to determine from direct measurement the increases in the concentrations of the greenhouse gases. If their concentrations are known it is a relatively straightforward calculation to determine the consequent atmospheric heating rates (Mitchell 1989). The difficult issue to determine is the temperature change due to the changed heating rate. The atmosphere and/or ocean may react to the increased heating with processes that produce positive or negative feedback. For example, the content of water vapor in the atmosphere is usually assumed to remain in equilibrium with the oceans and varies with the air temperature. If the atmosphere warms up because of an increase in the greenhouse gases, the amount of water vapor in the air will automatically increase because of the higher temperature. This process leads to a further increase in the absorption of long-wave radiation and causes a further increase in temperature, or a positive feedback. Another positive feedback is related to the snow and ice cover over land. If the atmosphere warms up, the area of the earth's surface covered with snow and ice decreases while the area of soil and water increases. These darker-colored surfaces absorb more of the sun's heat than the white snow and ice, causing a further increase in temperature. On the other hand, an increase in temperature and water vapor results in more clouds. The increased cloud cover has two opposing effects. The clouds reflect back into space a portion of the incoming solar radiation, resulting in lower temperatures. The clouds also cut down the amount of back-radiation escaping from the earth to outer space. This effect is particularly prominent at night, and has a warming effect. It is the total effect of all such processes that results in the observed temperature variations. One way of integrating all the processes is by mathematical models.

(c) Climate models and predictions

Climate models are computer programs in which the atmosphere is represented by a three-dimensional grid (see Mitchell 1989, for review). One cell in the grid typically represents 500–1000 km in the east–west and north–south directions and 10–20 km in the vertical. At the grid points forming corners of each cell variables such as temperature, humidity, pressure, and wind are specified by the average conditions thought to exist at some time in the past. Changes in these variables over a short time interval (hours) are then calculated according to the equations governing thermodynamics, momentum balances, and mass conservation. The calculated changes are then used to update the values at the grid points. The process continues until the program has modelled the required time interval under study (decades).

The aim is to create a model sophisticated enough to predict accurately the consequences of the increases in the greenhouse gases. It is improved by comparing the results to the observations until it can predict the observed changes accurately, after which it can be used to estimate future trends with some confidence. This process, of course, is no different from the usual scientific method of comparing observations to a model, but the climate system of the earth is too complicated to be modelled with a few linear equations and it is only with the largest and fastest computers that even rudimentary models are possible. Even then, when variables have been changed to make a model fit observations, it is often impossible to know whether the right variables have been adjusted.

All of the recent climate models predict that the increases in the greenhouse gases will cause the temperature of the earth to rise. The amount varies for each model but lies in the range of 1.5–5.5 °C for a doubling of the concentration of carbon dioxide that is expected by about 2030. As the models improve, they become increasingly complicated as they include more of the relevant processes. In the most recent model (Mitchell *et al.* 1995), a model atmosphere is coupled to a model ocean and sulfate aerosols are included in addition to the greenhouse gases. Sulfate aerosols are, like CO_2, derived from combustion of fossil fuels but they generate a cooling effect by increasing the amount of short-wave radiation reflected back into space. The comparisons between this model's results and the observed temperature over the past 133 years show a very realistic agreement, especially when the sulfate aerosols are included.

The effects of a worldwide increase in temperature will depend, of course, on the size of the increase and will be beneficial for some areas and processes and disastrous for others. If the water in the oceans increases in temperature it will expand and cause a rise in sea level that will be added to the rise in sea level caused by the partial melting of the polar ice caps. In the extreme case the oceans will rise by a meter or so and inundate heavily populated low-lying areas, which also tend to be rich agricultural areas. Millions of people will be affected. The higher temperatures are also expected to change patterns of agriculture as some areas will become too hot or too dry to grow the present crops, while other areas that are now too cold become warm enough to produce crops.

One of the biggest problems with the present climate models is the fact that the oceans are not completely included. This situation arises partly because the circulation of the oceans is less well known than the circulation of the atmosphere and partly because the computers are not yet sophisticated enough to include both the atmosphere and the oceans in a coupled model. A typical method of including the oceans in the models (Hansen *et al.* 1988) is to assume the oceans do not change appreciably during the time period being modelled. Thus the oceanic heat transport, the seasonal changes in temperature, the mixed layer depth, and other factors are assumed to remain the same for each year. The global climate models, therefore, are not yet capable of predicting or incorporating large-scale sea surface temperature anomalies as were found to be so important in the discussion of the El Niño in Chapter 9.

In the next sections our purpose is first to indicate why the oceans are important in the global carbon dioxide budget and then to describe the deep ocean currents that are responsible for moving the carbon dioxide from the surface layers down into the deeper layers of the various oceans.

10.2.2 The carbon cycle

The main source of carbon throughout earth's history has been in the form of carbon dioxide ejected into the atmosphere by volcanoes. Revelle (1982, 1983) estimates that over time volcanoes have added $\simeq 50 \times 10^6$ gigatons (1 gigaton = 1 Gt = $10^9 \times 10^3$ kg = 10^{12} kg) of carbon dioxide to the atmosphere and are still adding a further 0.04 Gt y^{-1}.

Fig. 10.02 Quantities of carbon in the various reservoirs of the world and the annual flux due to volcanoes and burning of fossil fuels. Units are gigatons.

Estimates of the magnitude of the reservoirs and fluxes in the global carbon cycle are constantly being updated. The following estimate is from Sundquist (1993). The burning of fossil fuel, together with deforestation and changes in land use, adds about 7 Gt C per year (Fig. 10.02). Some of the carbon dioxide in the atmosphere becomes incorporated into the structure of terrestrial plants, while some enters the ocean by dissolving in the surface layers. At present the atmosphere holds about 750 Gt C and an annual variation of about 10 Gt, reflecting the summertime transfer of CO_2 from the atmosphere to the land plants by an excess of photosynthesis over respiration and decomposition, and the wintertime transfer back to the atmosphere through continuing respiration and decomposition. The terrestrial plants and the soil contain about 2160 Gt, the mixed layer of the ocean 900 Gt, and the deep ocean 36,400 Gt. Rocks formed from ocean sediment, together with the present sediments, contain more than 2000 times the total amount present in all of these reservoirs (about 90 million Gt), of which 20 million Gt is organic matter, and 5000 Gt of that takes the form of recoverable fossil fuels. The sediments, therefore, contain the vast majority (99.5%) of all of the carbon ever issued into the atmosphere by volcanoes. If no oceans existed to form the sediments, the concentration of carbon dioxide in the atmosphere would be far higher than it is and the average temperature on the earth would be closer to that found on Venus, which does have a high concentration of CO_2 and where it is too hot for water to remain liquid.

From direct measurements it is observed that the concentration of CO_2 in the atmosphere is over 353 ppm and is increasing by about 0.35%, or 1.2 ppm y^{-1}. In absolute terms, the increase is about 3.4 Gt y^{-1}, which is much less than the estimated 7 Gt that is being added each year to the atmosphere through human activities. The results of modelling physical and biological processes in the oceans suggest that they take up about 2.0 Gt per year. The fate of the missing 1.6 Gt of CO_2 is not yet fully understood. Some have argued that increasing atmospheric CO_2 must have the effect of causing an acceleration in the biological uptake of the gas both on land and in the oceans, but the evidence is inconclusive.

Since the oceans and their sedimentary rocks contain almost all the CO_2 ever issued into the atmosphere and they possibly absorb a significant fraction of the additional CO_2 being

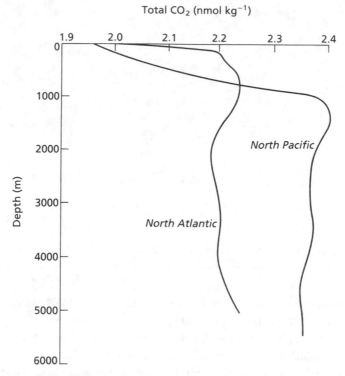

Fig. 10.03 Vertical profiles of total carbon dioxide concentration in the North Atlantic at GEOSECS station 37 (12° N 51° W) and in the North Pacific at GEOSECS station 214 (32° N 177° W), adapted from the GEOSECS atlases Vols. 2 and 4 (1981), respectively.

put into the atmosphere each year, it is obvious that the processes controlling the disposition of the gas are vital to understanding the climate system of the world.

Carbon dioxide enters the ocean by dissolving in the surface waters of the ocean at a rate determined by the difference in the partial pressure of the gas in the air and the water. Once in the ocean only a small amount of gas remains in the dissolved form, as most of it reacts with the water to form carbonic acid (H_2CO_3), bicarbonate ions (HCO_3^-), and carbonate ions (CO_3^{-2}) (Takahashi 1989). In the surface layer the carbon is incorporated into organic compounds, skeletons, and shells, especially during the spring bloom, but when the organisms die the remains sink. Some decompose, releasing the carbon back into the water column, but some are buried in sediments. This flux of carbon out of the euphotic zone into the deeper layers is called the 'biological carbon pump' and is explored in more detail in section 10.3. The result of the downward flux is an increase of the concentration of carbon dioxide from ≈2.0 nmol kg^{-1} (n = nano = 10^{-9}) in the surface layer to about 2.2–2.4 nmol kg^{-1} at 1000 m (Fig. 10.03). From 1000 m to the bottom the concentration increases or decreases only slightly. As is evident in Fig. 10.03 the concentration of CO_2 below 1000 m is about 10% higher in the North Pacific Ocean than in the North Atlantic. This difference is due to the differences in the rates of the deep circulation between oceans. Deep water passes more rapidly through the Atlantic than the Pacific (Takahashi 1989). As a result, the deep water in the Pacific Ocean has more time to accumulate the CO_2 being pumped down from the upper layer. As indicated in the next section the deep water originates in the North Atlantic and works its way into the Pacific, which suggests that the difference in the CO_2 concentrations between the two oceans of $\simeq 0.2$ nmol kg^{-1} indicates an increase above the base concentration of 2.2 nmol kg^{-1} contained in the North Atlantic deep water when it begins its journey. The formation and flow of these deep waters is part of the thermohaline circulation as distinct from the wind-driven circulation.

10.2.3 The deep thermohaline circulation

The fact that the deep and bottom waters of the oceans are colder than the near-surface waters is now common knowledge but when it was first discovered in 1791 the implications were far from understood (Warren 1981). Scientists at that time considered the ocean to be a relatively stagnant body of water in which the temperature at any location was in equilibrium with the atmosphere. In such an ocean heat absorbed from the sun would slowly diffuse downward into the deep waters, which would become much warmer than they are observed to be. The cold water, it was finally realized, must originate in the cold regions of the world and sink to the bottom under the warm water. This process was first suggested by Count von Rumford in 1797 as a consequence of laboratory experiments on thermally driven convection in fluids.

Rumford and subsequent authors envisaged large convectively driven circulations in which the surface waters in the polar regions cooled to be a high enough density to sink to the bottom. This new deep water would then flow away from the high latitudes and fill the deep basins of the oceans. As more cold water was added, the deep layer would become thicker, thus pushing cold water up toward the surface throughout the warmer regions of the oceans. The rate of this deep water upwelling has been calculated to be about 10^{-5} cm s^{-1}, which is too small to be measured, but high enough to have important effects on the structure of the ocean.

When the upward-moving cold water reaches a sufficiently shallow depth it is warmed by the downward diffusion of heat from the surface layer. Over the long term, a balance is achieved in which the downward diffusion of heat equals the rate that heat is being moved back up by the upward motion in the water. This advection–diffusion balance in the vertical flow of heat leads to the ocean's being divided into a warm upper layer and a cold lower layer separated by the permanent thermocline (as distinct from the seasonal thermocline). The rising deep water slowly becomes part of the warm upper layer and eventually flows back toward the poles, where it is again cooled to renew the deep water and complete the convection cell. This convection circulation, because it depends on changes in density related to changes in temperature and/or changes in salinity, is called the thermohaline circulation. It exists alongside and in addition to the wind-driven circulation, which interestingly is more or less confined to the upper layer of the ocean (Chapter 8) created by the thermohaline circulation (Gill 1982).

The reasons for the existence of the thermohaline circulation are not, however, as straightforward as originally perceived. The temperature decrease in the water between the equator and the polar regions creates horizontal density and pressure gradients. In an idealized world without irregularly shaped continents and ocean basins, the rotation of the earth would simply lead to geostrophically balanced eastward currents perpendicular to the poleward pressure gradient. In such a system, any decrease in the temperature of polar water or increase in the equatorial water would increase the pressure gradient and increase the speed of the eastward flow. The formation and sinking of dense waters occurs in particular locations where the conditions are quite different from those in the ideal world.

Because much of the deep water is formed near the surface in high latitudes, the formation of ice is often an important feature of the process. When ice is formed, there is a partitioning between the water and the salt. The ice is made from the water; the salt and other impurities are excluded from the ice and trapped in small channels or cavities within the ice. With time the salt slowly drains out of these channels into the water below and thereby increases the density of the underlying water.

One of the regions where dense water is formed is over the continental shelf around Antarctica, especially in the Weddell Sea. This sea is situated from 15 to 60° W and from 65 to 75° S, bounded on the west by the Antarctic Peninsula and on the east by Queen Maud Land. In the winter, new ice is formed on the western side of the sea. As we have seen, salt is excluded from the ice and added to the water below. Prevailing winds blow the ice offshore, leaving behind the more saline water. Thus the water flowing into the sea from the east is both cooled and made more saline by the formation of ice. The resulting water has a salinity of about 34.65 and a temperature of -0.7 °C and flows off the shelf down to the bottom of the ocean. This flow, along with similar processes on the other shelves around Antarctica, produces $2-5 \times 10^6$ m^3 s^{-1} (Warren 1981) of dense water that subsequently spreads northward at the bottom of the Atlantic, Pacific, and Indian Oceans as the Antarctic bottom water.

Another important source of dense water is found in the Greenland and Norwegian Seas where relatively saline water from the North Atlantic current is cooled to reach a density high enough to form bottom water. Clarke *et al.* (1990) describe the processes in detail and point out that one of the important features is the fact that the water circulates in a large well-defined cyclonic gyre in which the warmer saltier water from the Norwegian current and the colder fresher waters from the East Greenland current are mixed together to create the water that, when cooled, will form the denser deep water. Within this cyclonic flow, which is partly forced by the bathymetry, the water is confined to remain in a region of intense cooling for a considerable length of time. Consequently, the water continues to lose heat and gets progressively denser, with the result that the upper mixed layer becomes deeper and deeper until it reaches the bottom. The role of ice, if it is important in this area, is different than in the Weddell Sea. The extraction of freshwater to increase the salinity of the underlying water is not important because there is no net gain of salt in the formation area. This state exists because the amount of salt added to the water from the freezing process is balanced by the amount of freshwater added by melting ice. If ice is important, Clarke *et al.* (1990) suggest, it will be due to the leads between the ice floes, where the heat loss from the water is greatest and where the convection will begin. The spacing between the leads may, therefore, control the spacing of the convection cells.

The dense water formed in the Greenland and Norwegian Seas is important to the world's oceans because after it flows into the North Atlantic over the sills between Iceland and Scotland it forms one of the main components of the North Atlantic deep water (Harvey and Theodorou 1986). Similar processes in the sea to the north of Denmark Strait, between Greenland and Iceland, produce a slightly different form of dense water that flows over the sill in Denmark Strait to form the bottom water in the Labrador Sea (Swift *et al.* 1980).

There are other areas in the ocean where deep penetrative convection occurs in winter to form homogeneous dense water but the density is not generally as high as the water formed in the polar seas. In the Labrador Sea, for instance, there exists a cyclonic circulation, as in the Norwegian Sea, bringing together the relatively warm saline water from the Irminger current and the cold and fresher water from the Labrador current. The slow circular flow maintains the water in the region while it attains the properties that are just right to lead to deep convection in the depth of a cold winter. The depth of this convection has been observed to 2000 m so it does not form bottom water but remains at an intermediate depth. The depth of the convection in this region is also very dependent on the atmospheric conditions, and these can be very variable from one year to the next (Lazier 1980). When the winter winds are from the northwest they are exceptionally cold and the convection is the deepest, but in years when the wind patterns cause warmer air to flow persistently over the area the convection is quite shallow. Intermediate water masses are also formed in the northwest Pacific Ocean but no deep or bottom water is formed there (Warren 1983).

The densest water having a major influence on any of the major oceans is formed in the Mediterranean Sea and flows into the Atlantic through the Straits of Gibraltar (Armi and Farmer 1988). This water differs from the other dense waters because it is formed primarily by evaporation and the consequent increase in salinity, rather than by the decrease in temperature found in the dense polar waters. One of the interesting things about the Mediterranean water is that even though it is the densest water flowing into the Atlantic, having a σ_t of ≈ 29.0 as compared with 27.8 for the Antarctic bottom water, it does not descend to the bottom of the ocean. Killworth (1977) pointed out that this situation arises because the relatively warm Mediterranean water at 12–13 °C is less compressible than the cold (-0.7 °C) Antarctic bottom water. The more highly compressible cold water becomes even more dense than the surrounding water as it descends toward the bottom because of the increase in pressure. The less compressible 12 °C water does not increase in density as rapidly as the cold water as it descends to higher pressures, so it becomes entrained into the surrounding water more readily than the Antarctic bottom water.

The movement of the various intermediate, deep, and bottom waters is detected by tracking anomalous properties, a process known as the core layer method of water mass analysis. The dense water that flows off the shelves around Antarctica, for example, is followed by its characteristic low temperature. The North Atlantic deep water is traced by its high salinity and high dissolved oxygen content relative to the layers above and below. A suggestion of some of these water masses and their supposed flows is illustrated in Fig. 10.04. In the north the section begins in the Norwegian Sea and the flow of Norwegian Sea water into the North Atlantic over the Iceland–Faeroe rise. This flow is one of the main components of the North Atlantic deep water that flows into the western basin of the ocean through the Charlie Gibbs fracture zone in the mid-Atlantic ridge at 53° N. South of this point the diagram illustrates distributions west of the ridge.

The central waters of the subtropical gyres are outlined by the base of the permanent thermocline. Its depth is shallower at the equator than in the centers of the gyres and significantly deeper in the North Atlantic than in the South Atlantic. Beneath the permanent thermocline lie the intermediate waters that are formed in the high latitudes of both oceans. The Antarctic intermediate water renewed around Antarctica occupies a much larger volume than does the Labrador Sea water formed between Canada and Greenland in the Labrador Sea. The Mediterranean water (not shown) exhibits a strong salinity maximum at intermediate depths, especially on the eastern side of the mid-Atlantic ridge. Below the intermediate waters are the deep and bottom waters. The principal deep water is in the North

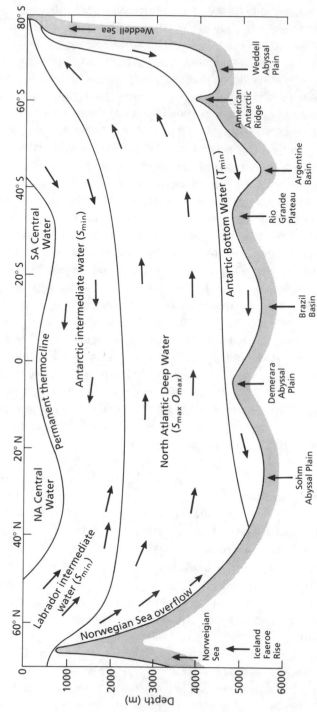

Fig. 10.04 Vertical section through the North and South Atlantic Oceans from the Norwegian Sea to Antarctica indicating the major water masses, their principal characteristics, and their directions of flow. S_{min} = low salinity, S_{max} = high salinity, O_{max} = high oxygen content, T_{min} = low temperature. Adapted from Dietrich *et al.* (1980).

Atlantic deep water characterized by relatively high salinity and oxygen values. It is observed throughout the length of the Atlantic Ocean between $\approx$2000 m and $\approx$4500 m. The water has also been shown to provide most of the deep water in the Indian and Pacific Oceans. The largest volume of bottom water originates on the shelves around Antarctica and flows north in all the major ocean basins. Its northern limit in the Atlantic is about 40° N, where it gives way to the Denmark Strait overflow water (not shown), which lies at the bottom in the Labrador Basin.

The flow of water depicted in Fig. 10.04 assumes that the north–south convection cells driven by the thermohaline circulation are broad ocean-wide features. A north–south section at any other longitude should show the same pattern of flow. There are, however, good dynamical reasons to doubt that such a simple pattern is realistic. As pointed out by Stommel (1958) and discussed by Warren (1981), the deep slow flows associated with the thermohaline circulation must, with very little error, be in geostrophic balance. Also, because the deep waters are being supplied continuously, there must be a slow upward movement in the upper layers of the ocean. This upward velocity is like the Ekman pumping we discussed in Chapter 8 and the balance between the vertical motion and the north–south flow is the same as described by Eq. 8.10 in association with the discussion of the Ekman pumping and the Sverdrup flow. That equation is

$$V_S = \frac{f}{\beta H} W_E \tag{10.01}$$

where β is the rate of change with latitude in the Coriolis parameter f, and V_S and W_E are the north–south and vertical velocities, respectively. In Chapter 8 the vertical velocity W_E was the Ekman pumping and was downward or negative leading to a negative or equatorward Sverdrup velocity V_S. In the present case the vertical velocity is upward or positive leading to a positive or poleward interior flow. This strange result leads to the conclusion that the deep waters formed in the polar regions are flowing toward their sources in the interior of the oceans. Because of this restriction Stommel (1958) postulated that the deep waters must flow from the north to the south, or from south to north in the case of the Antarctic bottom water in narrow western boundary currents much like those associated with the wind-driven circulation in the upper layer. These boundary currents have indeed been found and Warren (1981) gives a good description of them.

The dense waters thus flow away from their sources in narrow western boundary currents and move laterally into the interior of the oceans through eddy diffusion. An illustration of the flow of one of the most important deep waters—that is, the flow of the North Atlantic deep water (NADW)—is illustrated in Fig. 10.05 (Warren 1981, Gordon 1986). The flow begins at the sill between Iceland and Scotland that separates the North Atlantic from the Norwegian Sea. The overflow water mixes with the resident deep water in the eastern basin of the North Atlantic, then moves into the western basin of the North Atlantic through the gap in the mid-Atlantic ridge at 53° N known as the Charlie Gibbs fracture zone. The flow continues in a counter-clockwise fashion around the Irminger and Labrador Seas and then moves south toward the equator and the South Atlantic Ocean along the western boundary of the ocean. In the southern ocean the NADW is incorporated into the eastward-flowing Antarctic circumpolar current, from which it makes its way into the Indian and Pacific Oceans. The water enters the remainder of each ocean basin from the western boundary

Fig. 10.05 The flow of North Atlantic deep water (solid arrows) from the northeastern North Atlantic south to the Antarctic circumpolar current, and thence to the Indian and Pacific Oceans. The return flow (open arrows), after Gordon (1986), is in the thermocline and mixed layers. Slow upwelling of deep water is assumed to occur where the permanent thermocline exists.

currents by eddy diffusion. The regions where convection renews the deep and intermediate waters of the Norwegian and Labrador Seas are indicated in Fig. 10.05 by the circled crosses. Water must slowly upwell throughout all the ocean basins to provide for the continuous supply of deep and bottom water.

The return flow to the North Atlantic starts in the western Pacific Ocean north of Australia (Gordon 1986) and follows the path indicated by open arrows in Fig. 10.05. The flow out of the Pacific into the Indian Ocean transports about 8.5×10^6 m^3 s^{-3} and continues across the southern Indian Ocean into the Aguhlas current reaching the South Atlantic by flowing around the southern tip of Africa. At this point the transport is $\simeq 13.5 \times 10^6$ m^3 s^{-3}. The flow then becomes part of the equatorial current system in the Atlantic and eventually joins the Gulf Stream system, which delivers $\simeq 18.5 \times 10^6$ m^3 s^{-1} of upper layer water back to the Norwegian Sea, ready to be turned into deep water. This apparently continuous circulation connecting the deep and surface flows throughout the world's oceans is sometimes loosely referred to, and depicted as, a conveyor belt (Steele 1989).

10.3 THE BIOLOGICAL PUMP

Given the existence of a very large, slowly moving deep ocean circulation that may take a thousand years to traverse the major oceans, we need to look at the mechanisms that will take carbon from the atmosphere and transfer it either to the sea floor or to the deep parts of the ocean, for in either event it will be unable to return to the atmosphere for hundreds of years. We saw earlier that, at the limited areas that are sites of deep water formation, carbon dioxide dissolved in the water is carried to depth. We will now examine mechanisms operating over the entire surface of the oceans by which carbon is fixed in photosynthesis and introduced into the deep ocean by the sinking of dead organisms. It is called the biological pump.

10.3.1 The magnitude of carbon fixation in organic matter

The biological fixation of carbon is taking place continuously in the world ocean, but the pattern is extremely complex as may be seen in the satellite images (Plate 1). At any one place it changes with the alternation of light and dark. In most places this cycle lasts 24 hours, but within the Arctic and Antarctic circles there is a period of continuous darkness in winter and light in summer. It also changes seasonally, especially in temperate waters where there is an alternation of shallow mixed layer in summer and deep mixed layer in winter. Waters of the continental shelves tend to be more productive than the open ocean, and sites of upwelling more productive than sites where upwelling is absent. These phenomena have been reviewed in previous chapters.

Until recently, all estimates of global rates of carbon fixation by phytoplankton have been based on compilations of the world's literature on field measurements made at particular points at particular times. They range from 20 to 55 Gt y^{-1} of carbon. That literature has been reviewed by Walsh and Dieterle (1988), who suggest a figure of about 30 Gt y^{-1} of

Table 10.1 Aquatic photosynthesis and sedimentation of organic carbon. From Walsh and Dieterle (1988).

Region	Area (km^2)	Net primary production ($\times\ 10^{12}$ kg C y^{-1})	Sediment organic carbon sink ($\times\ 10^{12}$ kg C y^{-1})
Open ocean	3.1×10^8	18.6	0.19
Continental shelf	2.7×10^7	5.40	0
Continental slope	3.2×10^7	2.24	0.50
Freshwater marshes	1.6×10^6	1.51	0.15
Estuaries/deltas	1.4×10^6	0.92	0.20
Salt marshes	3.5×10^5	0.49	0.05
Rivers/lakes	2.0×10^6	0.40	0.13
Coral reefs	1.1×10^5	0.30	0.01
Seaweed beds	2.0×10^4	0.03	0
Totals	3.75×10^8	29.89 (C input)	1.23 (C output)

carbon for total photosynthesis by phytoplankton less respiration—the net fixation (Table 10.1). Of this amount, they estimate that more than half is fixed in the open ocean, which has the lowest rate of fixation per unit area, but has an enormous area compared with the other habitats listed. As we discussed in Chapter 8, there is still a great deal of uncertainty about the magnitude of primary production in the major ocean gyres. Within the past decade estimates have been put forward that vary by an order of magnitude (see Chapter 8; Platt *et al.* 1989 and references therein). With our newly acquired ability to monitor the whole ocean by satellite sensors, it is hoped that the confidence limits of estimates will be greatly decreased.

Satellite color images now make it possible to make estimates of the distribution of chlorophyll in the surface waters of the world ocean, but this estimate gives, at best, an indication of the distribution of phytoplankton biomass. To convert from this figure to rates of carbon fixation is no easy task. Platt and Sathyendranath (1988) put forward an algorithm that combines a spectral and angular model of submarine light with a model of the spectral response of algal photosynthesis. They pointed out that one needs to know from field measurements the geographical and temporal variation in the physiological rate parameters of the algae and the biological structure of the water column. They also suggest that such data be collected, after which it should be possible to use satellite data to routinely estimate rates of carbon fixation in any location that has the appropriate satellite coverage. As we showed in Chapter 3, an estimate of global primary production using satellite data has now been made (Longhurst *et al.* in press). According to the assumptions made about the overestimation of chlorophyll in turbid coastal area, global net primary production is estimated at 45–50 Gt C y^{-1}. This figure is considerably higher than that arrived at by Walsh and Dieterle (1988) but similar to the estimate obtained by Martin *et al.* (1987) by extrapolation from a small number of ^{14}C incubations made under ultra-clean conditions (see section 10.3.2). The estimates in Longhurst *et al.* (1995) of the distribution of primary production between coastal waters and the open ocean are not very different from those of Walsh and Dieterle (1988).

In Chapter 8, we also discussed two categories of primary production: regenerated production, in which the algae use as a nitrogen source the ammonia excreted by the consumers in the system, and new production, in which the algae use nitrate upwelled from below the nutricline. The ratio of new production to total production is called the *f* ratio. In an idealized situation in which the pelagic zone of the ocean is in a biochemical steady state, regenerated production can be considered a closed loop with no net gain or loss of carbon or nitrogen. In this situation the flux of nitrogen leaving the pelagic zone in sedimenting organic particles is balanced by the upward flux of nitrate–N into the zone. Similarly, the downward flux of carbon in organic matter is balanced by uptake from the atmosphere. (There is also an upward flux of carbon by eddy diffusion, but the vertical gradient in carbon is so slight that this component can be ignored.) This fact brings us to the realization that when considering the magnitude of the biological pumping of carbon from the mixed layer down to the interior of the ocean, it is not the total production that we need to know, but the new production. Hence, even if a routine method is developed for using satellites to determine total net production of phytoplankton over wide areas, it will also be necessary to know the f ratio for each area under study. Fortunately, there is less disagreement among different observers about the levels of new production than about the levels of total production.

One of the most difficult problems facing those trying to determine the true magnitude of new production is that much of it is believed to occur as a result of intermittent bursts of upwelling of nitrate, resulting from eddy activity, or internal waves (Goldman 1988). Any method of observation that is not continuous is likely to miss a proportion of these intermittent events. In a permanently stratified tropical ocean with a deep nutricline, the new production may take place at considerable depth and escape detection by satellites. In places where this production occurs, the lower part of the euphotic zone will have a relatively high proportion of new production (high f ratio) while the upper part may function mainly on regenerated nitrogen (low f ratio). Hence there is the added difficulty of a nonuniform f ratio at a given geographic location.

In spite of all these complications, the role of the phytoplankton in pumping carbon down into the interior of the ocean is of such importance in the context of rising atmospheric CO_2 that a great deal of effort will be made to evaluate the magnitude of the new production.

10.3.2 Sinking of organic matter into the interior of the ocean

Primary production may be by organisms of a wide range of sizes, from large diatoms down to picoplankton, and their consumers may be small ciliates, copepods in the mid-size range, larger euphausids, or large gelatinous zooplankton that nevertheless filter very fine particles. These consumers produce fecal pellets with a great range of sizes and sinking rates. Calculation of the downward transport of carbon by evaluating all the terms in such a complicated food web may be a nearly impossible task, and it is not surprising that there have been a number of attempts to obtain the value of the sinking flux by direct measurement, using particle traps. For example, Martin *et al.* (1987) reported on the material intercepted in traps placed at depths ranging from 100 to 2000 m in the northeast Pacific off California. They estimated that at 100 m depth a station close to the coast had a downward flux of 7.1 mol C m^{-2} y^{-1} while at the same depth a station far out in the subtropical gyre had a downward flux of only 1.2 mol C m^{-2} y^{-1}. These figures yielded f ratios estimated at 0.14 for the open ocean and 0.17 for the coastal zone. In an area of active upwelling it was 0.20. Applying these percentages to some of their own primary production data, Martin *et al.* (1987) tentatively extrapolated to the world ocean, concluding that global total production was on the order of 50 Gt of carbon, of which 7.4 Gt was new production. Rather larger estimates of f ratios have been arrived at by Sambrotto *et al.*(1993). They presented data

Fig. 10.06 Downward flux of carbon in the northeast Pacific as determined by sediment traps. Composite of six open ocean stations. From Martin *et al.* (1987).

showing that the ratio of carbon fixed to nitrate taken up by the phytoplankton is often very different from the commonly used value of 6.6 (the Redfield ratio). More carbon is fixed per molecule of nitrate taken up than is commonly supposed. Thus, the ratio of new production to total production, the *f* ratio, may be 0.25–0.35 rather than 0.15–0.20, which in turn suggests that the carbon export from surface waters may be on the order of 15 Gt y^{-1}.

Martin *et al.* (1987) were also able to reconstruct the vertical profiles of sinking organic matter, down to 2000 m. As expected, the deeper a trap was placed, the less material it collected. This study gave a measure of the rate at which organic matter was being decomposed during the journey downward. Flux data from six open ocean stations were combined and fitted rather well the expression

$$F = 1.53 \, (z/100)^{-0.858}, \qquad (r^2 = 0.81) \tag{10.02}$$

where z is depth in meters.

As Fig. 10.06 shows, the flux at 500 m is only about 0.4 mol C m^{-2} y^{-1} and at 2000 m it is on the order of 0.1 mol C m^{-2} y^{-1}. The difference is accounted for by the oxidation of the organic carbon to CO_2 during its downward journey. It works out that 50% of the carbon removed from the surface is regenerated in less than 300 m, 75% by 500 m, and 90% by 1500 m. The authors conclude that this rapid regeneration in near-surface waters means that the majority of the CO_2 removed from the upper 100 m will be available for exchange with the atmosphere within 10 years. That conclusion may be true for the area studied (between northern California and Hawaii) but in areas where the permanent thermocline is normally less than 100 m from the surface, much of the sinking carbon would be removed to the interior of the ocean and cut off from the atmosphere for at least several decades.

The question then arises, is this situation typical for oligotrophic gyres in the world ocean? Jenkins (1982) and Jenkins and Goldman (1985) studied oxygen utilization rates at a range of depths in the North Atlantic subtropical gyre. Martin *et al.* (1987) converted their carbon oxidation rates to oxygen utilization rates and compared the two sets of results. While the rates at 100 m in the Atlantic were somewhat lower than in the Pacific, the rates measured from 200 to 500 m were all higher in the Atlantic, indicating that less carbon was oxidized above 200 m and more was oxidized at 200–500 m than in the Pacific.

As part of the Joint Global Ocean Flux Study (JGOFS), a cooperative study was undertaken along a transect in the eastern half of the North Atlantic, known as the North Atlantic Bloom Experiment (NABE). As part of this study, Honjo and Manganini (1993) deployed automated time-series sediment traps for 1 year at two sites. They found that at a depth of 2 km the annual mass flux was 22 and 27 g m^{-2} at the southern (34° N) and northern (48° N) stations, respectively. At the southern station, material from the spring bloom was identifiable from January to May, constituting 62% of the annual total, and at the northern station from March to May, contributing 50% of the total. When converted to fluxes of particulate organic carbon at 2000 m, the values were 1.03 and 1.38 gC m^{-2} y^{-1} at the southern and northern stations, respectively. These values are close to the 0.1 mol C m^{-2} y^{-1} reported for the northeast Pacific by Martin *et al.* (1987).

Interpretation of differing patterns of sinking fluxes is difficult, complex, and far from resolution. Trapping efficiency depends on the design of the trap (Butman *et al.* 1986) and as yet it has not been possible to reconcile fluxes measured in traps with fluxes determined from process studies in the water column above. Moreover, one quickly gets into a mass of biological detail. Consider, for example, the timing of the spring bloom of phytoplankton in relation to the ability of zooplankton to consume it. If zooplankton are present in good numbers in surface waters and are able to keep up with the consumption of phytoplankton as fast as it is produced, more phytoplankton carbon will be oxidized in the near-surface waters. We saw in section 8.6 that the sub-Arctic gyres of the North Atlantic and North Pacific differ in this respect. On the other hand, common forms of zooplankton such as copepods produce large quantities of feces that sink rapidly and would have the effect of accelerating downward movement of carbon.

Bacteria have the opposite effect. If a dead organism begins to break into smaller particles and release dissolved organic matter, both the particulate and the dissolved organic matter provide food for bacteria, which then multiply rapidly. Bacteria are so small, however, that they have a negligible sinking rate, and large quantities of carbon fixed in surface waters may remain in suspension and be converted back to CO_2 by bacterial respiration. The whole question of the operation of the biological pump is a matter of active debate. See, for example, Karl *et al.* (1988) and Cho and Azam (1988).

10.3.3 Sedimentation on the sea floor

The last stage in this biological pumping is sedimentation on the sea floor. As far as the major ocean gyres are concerned, the rate of carbon fixation by phytoplankton is relatively low. A large proportion of carbon that sinks is regenerated to CO_2 in the water column, so that deep ocean sediments receive only a few g m^{-2} of organic carbon annually. As Table 10.1 shows, the whole of the deep ocean floor accounts for only about 15% of the world's carbon storage.

The situation on the continental shelves is quite different. They have much higher rates of primary production than the open ocean, and budgets show that perhaps half of this production is not grazed by the zooplankton, but sinks to the bottom (Walsh *et al.* 1981, Walsh 1983). One might therefore expect large accumulations of sedimentary carbon on the shelves. Instead, one finds large accumulations on the continental slopes, and the authors postulate that there is cross-shelf transport and export to the continental slopes that worldwide amounts to 2.7 Gt C y^{-1}. Perhaps two-thirds of this carbon is oxidized to CO_2 and returned to shelf and slope waters, but from studies of the organic carbon content of the sediments, their porosity, and ^{210}Pb measurements of sediment accumulation rates, it is concluded that 0.9 Gt of carbon is stored annually on the shelves and slopes, along with another 0.2 Gt in the estuaries and deltas.

10.3.4 The significance of biological fixation of carbon in calcium carbonate skeletons

Reef-building activities by corals and formation of tests by such planktonic forms as foraminifera are clearly mechanisms for the removal of carbon from the mixed layer of the ocean. Worldwide geological deposits of calcium carbonate attest to the long-term and pervasive influence of this activity by marine organisms. The question arises, should this downward flux of inorganic carbon be counted as a term of the biological pump, when considering ocean–atmospheric interactions? The short answer is no. Formation of calcium carbonate removes CO_2 from the mixed layer (and indirectly from the atmosphere) but it also removes calcium. The removal of calcium affects the total alkalinity by the liberation of hydrogen ions:

$$Ca^{2+} + HCO_3^- \Rightarrow CaCO_3 + H^+$$

This process shifts the equilibrium between bicarbonate and carbon dioxide in the water, leading to the liberation of carbon dioxide to the water and eventually back to the atmosphere:

$$H^+ + HCO_3^- \Rightarrow CO_2 + H_2O$$

The addition of CO_2 by this process more than compensates for its removal in the formation of carbonate, so there is no net downward pumping from the atmosphere to the interior of the ocean. One may then ask, 'But surely the sinking of calcium carbonate skeletons represents a net downward flux of carbon, a biological pumping?' The answer is, 'This flux is compensated for by a change in the alkalinity of the surface waters, but not by a flux of carbon from the atmosphere. In the steady-state situation, the change in alkalinity is compensated by advection of some neutralizing material from other sources, such as the land, and not by a flux of carbon dioxide from the atmosphere.'

10.3.5 Summary of the biological pump

Among those attempting to model carbon dioxide fluxes between the atmosphere and the ocean, there is agreement that biological processes lead to a net flux of carbon into the ocean. Part of the primary production in the mixed layer is consumed and respired in that layer so that the carbon dioxide released by respiration is returned to the atmosphere, and therefore leads to no net loss to the atmosphere. Another part (equivalent to the 'new production') sinks out of the mixed layer, carrying with it fixed carbon that is withdrawn from contact with the atmosphere for some considerable period. At present, the details of the biological pump are very uncertain. We do not know the total amount of carbon fixed world-wide on an annual basis, we do not know what proportion of that total is new production, and we do not know the details of the sinking process or the regeneration of carbon that takes place in the water column. Finally, we have only the roughest of estimates for the amount of carbon that is buried in the sediments.

Nevertheless, it is frequently useful to construct models that represent a synthesis of knowledge at a particular time about the total flux of carbon dioxide between the atmosphere and the ocean. Takahashi (1989) has reviewed a number of models and arrived at the following synthesis. Carbon dioxide transfer across the sea surface is dependent on the difference between the partial pressure of carbon dioxide in surface waters and in the air, multiplied by a gas transfer coefficient. The partial-pressure difference is strongly influenced by biological carbon fixation, since vigorous photosynthesis by algae reduces the carbon dioxide concentration in surface waters. On the other hand, if cold water is upwelled and warmed at the surface, the change in temperature will increase the partial pressure, because warm water saturates at a lower gas concentration than cold water.

The gas transfer coefficient is dependent on such factors as wind strength, degree of surface turbulence, and the shape of the waves. Attempts to quantify these variables in wind tunnel experiments have led to widely differing results, but some cross-checking against the values used to stimulate conditions in the ocean can be carried out by checking the rate at which radon gas and radioactive carbon have moved between the ocean and the atmosphere.

With the use of the best estimates for the values of partial-pressure differences and gas transfer coefficients in different parts of the world ocean, calculations have been made of net carbon flux between the ocean and the atmosphere, which Takahashi (1989) claims are accurate within ± 0.3 Gt y^{-1}. The greatest flux from the ocean to the atmosphere is believed to result from warming of upwelled water at the equatorial divergences. Although strong biological pumping occurs there, its magnitude is believed to be substantially less than the outgassing resulting from warming of upwelled water. The major sinks for carbon dioxide are found in polar and subpolar regions. Cooling of water as it flows from the equator toward the poles causes it to take up CO_2 from the atmosphere, and deep water formation, mentioned earlier, leads to the downward transport of large amounts of the absorbed CO_2. In addition, the biological pumping that occurs in temperate waters transports large quantities of carbon to the interior of the oceans. The overall estimate according to Takahashi (1989) is a net flux of 1.6 ± 0.3 Gt y^{-1} of carbon from the atmosphere to the oceans.

In the 1990 IPCC report on climate change (Houghton *et al.* 1990) it was concluded that, based on a number of carbon models, the oceanic uptake of CO_2 for the decade 1980–1989 was 2.0 ± 0.8 Gt C yr^{-1}. The best estimate of the carbon budget for that decade follows:

	Gt C yr^{-1}
Emissions from fossil fuels into the atmosphere	5.4 ± 0.5
Emissions from deforestation and land use	1.6 ± 1.0
Accumulation in the atmosphere	3.4 ± 0.2
Uptake by the ocean	2.0 ± 0.8
Net imbalance	1.6 ± 1.4

Fig. 10.07 Data on carbon dioxide content and deuterium δD levels (indicative of atmospheric temperature) of gas bubbles in the Vostok ice core at various dates back to 160,000 years BP. From Barnola *et al.* (1987).

The IPCC report concludes that the biological pump does not contribute to the sequestering of excess CO_2, since the marine biota do not respond directly to the CO_2 increase, and atmospheric CO_2 was apparently constant for at least 1000 years before the Industrial Revolution (Sundquist 1993). On the other hand, future climate changes that might cause changes in the amount of upwelling of nutrients, and hence marked changes in the biological fixation of CO_2, could be expected to influence atmospheric CO_2.

10.4 EVIDENCE FROM PALEOCLIMATE STUDIES

The magnitude of the recent rise in atmospheric CO_2 from about 280 ppm in the 17th century to over 350 ppm at present should be viewed in the context of long-term changes. From studies of the carbon dioxide content of bubbles formed in the ice caps thousands of years ago, we now know that during the last ice age CO_2 was as low as 180 ppm. Some of the most complete evidence comes from a Vostok ice core (Lorius *et al.* 1988). Vostok station is located on the central part of east Antarctica, where the present mean annual temperature is −55.5 °C. The ice core was drilled to a depth of 2200 m and the deepest part is 160,000 years old (Fig. 10.07). Ambient temperature through the life of the core (expressed as the difference from present surface temperatures) has been estimated from the isotope composition (both deuterium and ^{18}O, which give good agreement). There was a rise of about 10 °C about 15,000 years ago, marking the end of the last ice age. The previous interglacial period was from about 140,000 to 116,000 years BP and its beginning was also signalled by a relatively rapid rise in mean temperature of about 10 °C. Temperatures during the glacial period fluctuated considerably, and this is often attributed to variations in the earth's orbit (Hays *et al.* 1976). As we turn our attention to the record of CO_2 in the bubbles trapped in the ice, there is a remarkable correspondence with the temperature, each rise in temperature being accompanied by a corresponding rise in atmospheric CO_2. At the end of each of the last two glacial periods, when the temperature rose about 10 °C, the carbon dioxide content of the atmosphere rose by about 40%. Interpretation of this result is controversial, analogous to the classic 'chicken and egg' situation. Was the temperature change brought about by changing CO_2 levels or did the changing CO_2 levels result from the climate change?

This text is not the place to debate theories of the causes of ice ages, but it seems to be agreed that a glacial period is brought on by a triggering factor that does not in itself cause the large drop in temperature. The triggering factor leads to a series of climatic feedbacks (such as increasing albedo as a result of the increasing ice cover) that lead to the observed drop in mean temperature, formation of greatly enlarged ice caps, drop in sea level, and so on.

Many regard changes in the earth's orbit ('orbital forcing') as the prime initiating factor (but see Broecker [1992] for a review of arguments against that view). The direction that the earth's axis points, in relation to the stars, cycles every 23,000 years. The obliquity of the earth's axis in relation to its tilted position changes on a cycle of 41,000 years, and its orbital eccentricity has a period of 100,000 years. Evidence that past ice ages have corresponded with these periodicities comes from ocean sediment cores going back 800,000 years (Hays *et al.* 1976; Kerr 1983). Three parameters were measured: (i) oxygen isotopic composition of planktonic foraminifera (a measure of temperature); (ii) a statistical analysis of radiolarian assemblages, previously shown to be correlated with temperature; and (iii) percentages of another radiolarian, not used in the previous data, thought to be an indicator of stratification in the water column. Spectral analysis showed that climatic variation in these records peaked every 23,000, 42,000, and 100,000 years, strongly supporting the theory that orbital forcing is a fundamental cause of quaternary ice ages.

Glacial periods can be satisfactorily accounted for by various models involving orbital forcing, climatic feedbacks, and atmospheric carbon dioxide (reviewed in Lorius *et al.* 1988), but falling CO_2 levels are responsible for more than half of the effect. This result would not be expected from the direct radiative effect of the CO_2 changes, which require amplification. Nevertheless, the CO_2 changes are essential to the models. This consideration brings us to the question: What caused them?

10.4.1 Theories concerning rapid changes in atmospheric CO_2

We have seen that the ending of the last two ice ages was marked by a roughly 40% rise in atmospheric CO_2. If we could understand the causes and consequences of these changes, we should be in a better position to attempt to forecast the effect of further increases in atmospheric CO_2 as a result of man's activities. Obviously, two classes of factors are involved, physical and biological. The ocean is divided into two concentric spheres: the surface mixed layer whose CO_2 partial pressure is a little above or below atmospheric value, and the deep ocean which is supersaturated with CO_2 and contains a huge reservoir, 50–60 times that of the atmosphere. Any physical mechanism that increases the transport from the deep ocean to the surface mixed layer will cause a rise in atmospheric CO_2. Alternatively, since the biological pump is responsible for the downward transport of carbon dioxide into the deep ocean, any factor that decreases biological activity while upward transport from the deep ocean continues at its normal rate will lead to increasing atmospheric CO_2. Most explanations put forward to date have both physical and biological elements.

Mix (1989) proposed a mainly biological model, and reviewed a number of preceding models. Several early workers in the field had suggested that during glacial periods the supply of nutrients to the phytoplankton was increased, especially at areas of upwelling, thus stimulating biological productivity and drawing down the atmospheric CO_2. At the end of the glaciation the flux of nutrients reverted to something approaching present-day levels and the atmospheric CO_2 rose sharply. Fortunately, it is possible to test this hypothesis. The concentrations of dissolved cadmium and phosphate in the sea are closely correlated, and from the cadmium-to-calcium ratio in the shells of benthic foraminifera it is possible to calculate the ocean's phosphate content in times past. There is no evidence for high nutrient concentrations in upwelling areas in glacial periods, or of a sharp change at the end of those periods.

Several modellers have concentrated on processes in the Antarctic. As we discussed earlier, deep water enriched in nutrients rises to the surface, but extensive biological pumping also occurs and the net result is that the southern ocean is a major sink for carbon. It has been suggested that iron is a limiting factor in the biological use of nutrients, and that a higher input of iron-rich dust during the last ice age could have stimulated biological productivity, reducing atmospheric CO_2 (see section 3.2.8). Another suite of models has invoked increased nutrient upwelling at the tropical divergence, with increased biological production and a decrease in atmospheric CO_2 (even though outgassing of CO_2 at the equator would presumably continue).

Attempts have been made to calculate changes in ocean productivity between glacial and interglacial times by comparing the organic content and the carbon isotope ratios of the appropriate parts of ocean bottom cores. A recent example is Sarnthein *et al.* (1988). They concluded that there was a massive decrease in new production at low- and mid-latitude upwelling areas at the termination of the last ice age with a consequent increase in atmospheric CO_2. They attributed it to orbital forcing, leading to an increase in high-latitude isolation, reduction in ice cover, and a reduction in the strength of meridional trades.

Mix (1989), while not disagreeing with the conclusions of Sarnthein *et al.* (1988), cited numerous difficulties with interpretation of organic matter data from cores. He proposed instead to carry out a major reconstruction of ocean productivity in times past, using the community structure of foraminifera assemblages as his index. Beginning with a global ocean productivity map, he developed for the North and South Atlantic Ocean a relationship between foraminifera assemblages on the contemporary sea floor ('core tops') and the level of productivity of the waters above. This analysis was done using standard transfer function techniques (Imbrie and Kipp 1971). The result was a regression of measured productivity against estimated productivity with an error envelope of ± 12 g C m^{-2} y^{-1}. Then, turning to the foraminiferan assemblages present at the glacial maximum 18,000 years BP, he constructed surface productivity maps for that time. They were markedly different from contemporary productivity maps. The largest increases were seen at the equator and in subtropical regions, the least in eastern boundary currents. Overall, Mix (1989) estimated an increase of 18% in total primary productivity of the North Atlantic. Using the arguments of Eppley and Peterson (1979) that show that higher rates of production tend to have a higher proportion of new production, Mix (1989) estimated that new production would have been 38% higher overall, and 87% higher in equatorial regions during the coldest period of the last ice age.

It seems that during the last ice age atmospheric CO_2 was reduced by stronger biological pumping, particularly at the equator. There is evidence for this decrease both in the foraminiferan communities from that period, and from the amounts and types of organic carbon in the sediments. The earlier suggestion that this CO_2 reduction was caused by an increase in dissolved nutrients in the deep ocean is rejected. Instead, modifications of the present-day atmosphere–ocean interactions are invoked. During the ice ages, it is suggested, a stronger thermal gradient between the poles and the equator resulted in stronger trade winds and stronger upwelling of nutrient-rich water, particularly at the equator. There may well have been enhanced upwelling in Antarctic waters, but the evidence from paleobiology is less convincing.

Sundquist (1993) reviewed these arguments and added a further important consideration. Because the amount of CO_2 in the deep ocean is 50–60 times that of the atmosphere, and a delicate balance exists between carbonate, bicarbonate, and dissolved CO_2, any mechanism that changes this balance in the deep ocean can liberate large quantities of CO_2, a portion of which will find its way into the atmosphere at sites of upwelling. Hence, the mechanism delivering biologically fixed carbonate to the deep ocean ('the carbonate pump') is intimately related to events during deglaciation.

The current scenario for the life of an ice age now appears to have the following components:

1 Changes in the earth's orbit led to reduced solar radiation at high latitudes, lower temperatures, and spread of the polar ice caps.

2 Increase in the thermal gradient between the poles and the equator led to strengthening of trade winds and to complex changes in the patterns of primary production in the oceans including increased upwelling of nutrient-rich water at the equator.

3 Increased upwelling caused enhanced biological pumping, and a decrease in atmospheric CO_2, which in turn led to further cooling.

4 After an appropriate period the orbital position of the earth changed enough to begin a reversal of the process.

10.4.2 Postulated changes in the rate of deep water formation

We saw in section 10.2.3 that the world's major source of deep water formation is in the North Atlantic. Broecker *et al.* (1985) pointed out that as polar water sinks, water from the south moves north to take its place. While doing so, it cools and about 5×10^{21} calories of heat are released into the atmosphere each year. This energy modifies the climate of the North Atlantic basin and adjacent land areas. The water that sinks in the Norwegian and Greenland Seas is relatively deficient in nitrogen and phosphorus. Foraminifera in surface sediments taken from the floor of the North Atlantic have a low cadmium-to-calcium ratio, indicative of this low nutrient status.

Broecker *et al.* (1985) reviewed the evidence that the formation of deep water in the North Atlantic was severely reduced in glacial times. It has been found that the cadmium-to-calcium ratio in benthic foraminifera was higher in glacial times than today, indicating a more nutrient-rich water. It has also been shown that their stable carbon isotope signature in glacial times was indicative of deep water much more like that found in the Pacific during the same period. All this evidence suggests that the organisms living deep in the North Atlantic were in water that had a different origin than today, and this finding opens the possibility that in glacial times the thermohaline circulation of the world was radically different than it is today. Even the possibility of reversal, with deep water formation in the Pacific and upwelling in the Atlantic, should not be ruled out. Just how this would relate to the scenario reviewed above in 10.4.1 remains unclear.

Evidence that radical changes have occurred in the operation of the deep circulation of the North Atlantic has been reviewed by Bond (1995). From interpretations of foraminiferan communities and stable isotopes of carbon and oxygen in cores, it has been shown that major shifts in the operation of the 'Atlantic Conveyor' arose during the last glaciation. Confirmation of major changes in bottom current speed were obtained by studying grain size distribution in cores. Details of the various findings are not in agreement, but it now seems certain that major changes in deep ocean circulation accompanied the climate fluctuations during the last ice age.

10.5 PHYTOPLANKTON AND DIMETHYLSULFIDE

In addition to their role in fixation of carbon dioxide, phytoplankton are thought by some workers to influence the earth's climate through the production of dimethylsulfide, or DMS (Charlson *et al.* 1987). Most species of phytoplankton excrete DMS, and some of it escapes to the atmosphere where it reacts to form a sulfate and a methane sulfonate aerosol, MSA. The latter substance stimulates the formation of clouds over the oceans, thus increasing albedo and reducing temperature at the earth's surface.

Dimethylsulfide is the only volatile sulfur compound excreted by phytoplankton and its biological function is unclear. The concentration of DMS in surface waters is only weakly correlated with the biomass or productivity of the phytoplankton. Once released into the water, DMS may be broken down photochemically, removed by bacteria, or released into the atmosphere. Rates of emission from the sea to the atmosphere are in the range of 2.2–5.7 mmol m^{-2} y^{-1}, with the highest values being found in warm, highly productive coastal regions. Coccolithophorids, which are most abundant in tropical, oligotrophic waters, have a high rate of DMS production per unit biomass, while diatoms have a low rate of production.

Cloud condensation nuclei (CCN) are essential to the formation of clouds of liquid water droplets. It is thought that at the present time the air over continents has an abundance of particles that could act as CCN, but that air over the oceans remote from land has a relative shortage of suitable particles. Water-soluble sulfates are believed to be the natural CCN in the marine air. Sea-salt particles are possible candidates, but concentrations at cloud height are much too low for them to be important. Attention therefore focused on 'non-sea-salt-sulfate,' and DMS was identified as the most likely precursor of such sulfate particles.

Charlson *et al.* (1987) then developed at considerable length the technical arguments that led them to believe that an increase in DMS in the atmosphere (with water content held constant) would lead to a larger number of smaller droplets, and that this development would increase the earth's albedo, which, in turn, would have a cooling effect. Since DMS is released into the atmosphere at the highest rate in tropical waters, the authors argued that it serves a useful function in moderating the extremes of temperature over the tropical oceans. They see this condition as part of the global feedback control to which Lovelock (1979) drew attention in his book *Gaia*.

The importance of DMS as a climatic influence was challenged by Schwartz (1988). He argued that if DMS emissions have a strong influence on global albedo, an even stronger effect should be detectable from man-made sulfur dioxide emissions, for these are thought to release twice as much sulfur into the atmosphere as the phytoplankton. Moreover, the release is mostly in the northern hemisphere, and has increased to its present level during the past 100 years. Schwartz searched the records for any evidence of higher albedo in areas most under the influence of industrial sulfur emissions, and found none. He also found no evidence of reduced temperatures in the northern hemisphere compared with the southern. His conclusion was that anthropogenic sulfates do not have an important effect on the cloud component of albedo. By extension, he suggested that neither do the products of DMS emission. On the other hand, Legrand *et al.* (1991) showed that in the Vostok ice core (Fig. 10.07) the levels of DMS and non-sea-salt-sulfate (NSSS) in the atmosphere was much greater during the later stages of the last glacial period than they are today, suggesting that this condition was a factor in global cooling.

An important body of data pertinent to this subject was published by Prospero and Savoie (1989). They reported on analyses of methane sulfonate aerosol (MSA) and NSSS, on eight Pacific islands. They took Fanning and American Samoa as sites totally remote from industrial activity and found that the ratio MSA/NSSS was 0.065. They then used this ratio to calculate what was the proportion of non-sea-salt sulfates derived from DMS at other Pacific sites. They concluded that about 80% of all cloud condensation nuclei over the North Pacific were attributable to DMS from phytoplankton. They claimed that most of the industrial sulfur emissions emphasized by Schwartz are deposited on land, locally or regionally, so that, for example, only 20–25% of those produced in North America are exported from the continent. In light of these conclusions, the original hypothesis of Charleson *et al.* (1987) may still be valid. The possibility that, in the context of global warming, DMS production by phytoplankton could be a negative feedback, moderating some of the effects of increasing atmospheric CO_2, cannot be ruled out.

Reviewing the role of natural and man-made aerosols, Charlson and Wigley (1994) concluded that anthropogenic sulfur emissions far exceed those from phytoplankton, and that it is highly probable that global warming would have been greater, especially in the northern hemisphere, if there had not been an increase in albedo associated with the sulfur aerosols. In the elaborate coupled ocean-atmosphere model of Mitchell *et al.* (1995), the addition of a term for sulfate aerosols considerably improved the agreement with global means and large-scale patterns of temperature in recent decades.

10.6 SOME SPECULATIONS ABOUT THE FUTURE

Almost every topic discussed in this chapter is marked by large areas of uncertainty. Yet we are confronted with urgent global problems, and to meet these challenges we need all the insight we can muster about the interactions of the atmosphere, the terrestrial biota, and the oceans. While the first edition of this book was in press, there appeared an authoritative assessment of the climate change problem (Houghton *et al.* 1990) issued by the Intergovernmental Panel on Climate Change (IPCC). An update of the conclusions was published as an IPCC supplement (Houghton *et al.* 1992).

The following is a summary of current understanding of the global warming problem, taken from Hare (1988) and the IPCC reports. About half of the predicted warming is due to increased carbon dioxide levels in the atmosphere, the other half to increasing chloro-fluorocarbons, nitrous oxide, methane, and ozone (the latter in the lower layers only). In combination they are expected to lead to the equivalent of a doubling of carbon dioxide by the year 2030. According to existing ocean–atmosphere models, this condition should lead to an increase of between 1.5 and 4.5 °C in mean global surface temperatures, with the high latitudes warming the most, especially in autumn and winter. Sea level may rise between 8 cm and 29 cm, with a best estimate of 18 cm (Houghton *et al.* 1990), chiefly because of an expansion of the ocean's water column and only secondarily from melting of glacial ice.

If sea-level rise is too rapid, natural succession of coastal ecosystems cannot occur because the rate of change is faster than anything that has occurred since the last glaciation. The effect will be exacerbated where developed areas behind the coastal systems are physically protected. For example, under a scenario of half a meter rise by the year 2100, it is estimated that more than 4000 square miles of coastal wetland could be lost in the contiguous United States. In the short term, as marshes flood, die, and decompose, the release of nutrients would stimulate productivity, but the loss of detritus-based food webs and of habitat for seabirds, spawning fish, shrimp, and shellfish would likely lead to reduced fisheries yields. If coastal protection is absent, it may be possible for some coastal communities to migrate inland with the rising waters.

At publication time, a second major IPCC assessment was in preparation, and only the general findings had been made public. One of these findings is that, on the basis of the observations in nature and the results of the most sophisticated models, the majority of scientists now believe that the changes in global temperature are unlikely to be entirely natural and are at least in part the result of anthropogenic emissions.

What, then, are the expected responses of the oceans to such a change? A preliminary study commissioned by the Department of Fisheries and Oceans of the Canadian Government (Wright *et al.* 1986) came to the following conclusions (some of which are general while others are particular to the North Atlantic).

1   Sea surface temperatures generally will rise, with the largest changes expected near 60° N. This increase will lead to greater amounts of evaporation and hence a more vigorous hydrological cycle.

2   The expected balance will be that mid-latitudes will have the greatest increase in evaporation, with a corresponding increase in precipitation in polar and tropical regions. As a result, meridional (i.e., north–south) gradients in surface salinity will increase, creating fresher northern conditions and saltier subtropical conditions.

3   The thickness, areal extent, and duration of ice cover will decrease.

4   Enhanced surface warming at high latitudes will mean a reduced meridional temperature gradient and probably a reduction in wind stress of the order of 10% over the entire North Atlantic. This decline, in turn, will lead to reduced flow in the major wind-driven gyres so that the Gulf Stream and the Kuroshio current, for example, will be less strong. This conclusion may be contrasted with that of Bakun (1990), which states that warming of the land will lead to local intensification of the temperature gradient between land and sea, which will lead to more coastal upwelling.

5   Increased precipitation in the higher latitudes should lead to more river run-off, greater stability in coastal water columns, and a strengthening of buoyancy-driven currents. For the western North Atlantic, this situation should mean that while the Gulf Stream will relax, the Labrador current will be enhanced.

6 Increased buoyancy in surface waters, combined with reduced wind stress should lead to a reduction in rate of intermediate water formation in the Labrador Sea. In the Norwegian and Greenland Seas the inflow of freshwater is expected to be less, and the amount of wind mixing may increase as a result of the retreat of the ice. For this area it is difficult to predict whether, in the short term, deep water formation will increase or decrease.

7 In the long term, global warming should lead to increased evaporation in the North and South Atlantic basins, thus increasing average salinity. According to the theories of Broecker *et al.* (1985) this scenario will probably decrease the deep water formation and thermohaline circulation. However, at the same time, less cooling is expected at the surface, which will tend to reduce deep water formation. The balance between these two tendencies cannot yet be predicted.

What are the likely biological consequences of these changes? This question is difficult, if not almost impossible, to answer. If, as suggested, there is a general reduction in wind stress there will probably be less upwelling of new nitrogen at the equator and other major upwelling areas, leading to less biological production. Reduced biological pumping would be a positive feedback toward increasing atmospheric CO_2. Until we are in a position to make predictive models of ocean circulation, we really do not know whether the upwelling of nutrient-rich water will increase or decrease overall in the world ocean. Planktonic organisms, with their relatively short life histories, may adapt much more rapidly to changing climate than would longer-lived organisms like trees and mammals. After a few years of changed environmental conditions, the planktonic community may change its species composition and adapt well to the new regime.

For coastal waters, there is a suggestion that global warming may already have led to a decrease in primary productivity and zooplankton biomass. Roemmich and McGowan (1995) showed that in the California current the annual average temperature of the surface layer (0–100 m) has increased by 1.5 °C in some places since 1951, while the biomass of zooplankton has decreased by 80%. They attribute the effect to an increase in buoyancy of the surface layer, which tends to inhibit wind-driven upwelling, thereby decreasing the supply of nutrients to the planktonic food web. There is speculation that these declines may be linked to the decline in populations of sooty shearwaters and other marine birds, and to the roughly 35% reduction in commercial fish landings in the California current. It is not clear, however, whether the warming trend is part of an interdecadal fluctuation, or whether it is a long-term trend resulting from climate forcing (Hill 1995).

In Monterey Bay, California, the annual mean shoreline water temperatures have increased by 0.75 °C during the past 60 years, while decadal mean summer maximum temperatures have increased by 2.2 °C. In the intertidal fauna, the abundance of southern species has increased while the abundance of northern species has decreased, indicating a northward shift in species ranges (Barry *et al.* 1995).

A second report commissioned by the Canadian Government (Frank *et al.* 1988) called for a preliminary assessment of the changes to be expected in the fisheries of Atlantic Canada. The authors reiterated many times that their report was no more than informed speculation. Nevertheless, their ideas have interesting links with some of the material dealt with in earlier chapters in this book. For example, it has been suggested that global warming will lead to increased precipitation in high latitudes, which means increased freshwater run-off and greater buoyancy in coastal waters. This change can be expected to lead to more rapid warming and stabilization of coastal waters in spring. Among the species that might be expected to benefit from this change are lobsters and scallops. Lobster landings have shown interesting positive correlations with river run-off in the first year of the lobster's life (section 4.7.2), and scallop year-class success has been positively correlated with the water temperature in the year in which they were recruited.

Species such as cod and haddock, on the other hand, might be adversely affected. It was found in the Gulf of Maine area that the ctenophore *Pleurobrachia* (a jellyfish-like animal), when present in its peak abundance (>1 m^{-3}), can graze down the zooplankton on which larval cod and haddock rely. Normally, the larval fish occur in spring and the ctenophores in mid-summer so competition for food is minimal. However, in 1983 the sea surface was unusually warm between January and April and the time of peak ctenophore abundance advanced from summer to spring. As a result, the ctenophores and larval haddock coincided, and the 1983 haddock year-class was the lowest on record (Frank 1986). Hence, there is a possibility that freshening and warming of the coastal waters off Nova Scotia and in the Gulf of Maine might lead to reduced success of cod and haddock recruitment.

It would naturally be expected that if there was a general warming of the ocean, the geographic range of fish species would shift northward. There is a certain amount of observational data from previous warming events to indicate which fish might be most affected. For example, during the warming trend of the 1940s, there was a northward shift in the abundance and distribution of mackerel, lobster, and menhaden, and green crab established a resident population north of any previously recorded location. During the cooling trend of the 1960s, American plaice and butterfish retracted their range southward, and capelin and spiny dogfish extended their migrations southward. Other species, such as haddock, yellowtail flounder, winter flounder, and winter skate appeared not to change their range in response to temperature changes, and it is suggested that their distribution is determined more by the presence of an appropriate bottom type. From all of this evidence it is concluded that in the Gulf of Maine, for example, many species might be displaced northward by global warming, and their place taken by an assemblage of fishes more characteristic of the mid-Atlantic bight. This population would include menhaden, butterfish, red hake, silver hake, and herring.

From Chapter 9 we may recall that the warming event of the 1940s was associated with a strong northward extension of the range of cod in West Greenland, and we may speculate that a similar event might be precipitated by global warming in the 1990s. Similarly, the changes in the whole biological community in the English Channel, which took place in association with the warming of the 1940s, might be expected to recur in the years to come.

In section 8.4.4 it was shown that when a warm-core ring interacts with the continental shelf, there may be adverse effects on year-classes of fish because the larvae are swept off the shelf and dispersed into the open ocean. One prediction that could be made is that this situation will occur less frequently because reduction in strength of the Gulf Stream as a result of global warming will lead to the formation of fewer warm-core rings.

In the discussion of tidal currents (section 7.3.2) it was shown that the distribution of herring spawning grounds in the North Atlantic often coincides with tidally mixed areas, and that the young stages often seem to be retained in those areas for a considerable period. It was further suggested that the upper limit of the size of a spawning stock might be determined by the limits of the tidally mixed area. Under the conditions of global warming, with a freshening of coastal waters, a small decrease in the extent of tidally mixed areas might be expected, which might have some consequence for the herring stocks.

To turn from species of commercial importance to more broad ecological principles, the freshening of coastal waters might be expected to lead to earlier onset of stratification in areas that are not tidally mixed, and hence a lengthening of the period of stratification. This change will shift the growth advantage from larger phytoplankton such as diatoms to smaller species such as dinoflagellates. In general, it should lead to a lengthening of the food chain between the primary producer and the larger consumer species, with a consequent loss of secondary productivity.

One might have hoped that it would be possible to offer an opinion as to whether ocean productivity would be higher or lower as a consequence of global warming. Certainly many features, such as the reduced wind stress and greater water column stability in north temperate coastal waters, suggest less upwelling of nutrient-rich water and hence reduced productivity. On the other hand, waters that are now ice-covered for a substantial part of the year will become ice-free and have increased productivity. We can say for certain that things will be different, and that those who exploit the ocean resources in a particular place will need to pass through a painful period of adjustment, but it is not possible to say what the overall effect will be on global ocean productivity.

PART D
DISCUSSION AND CONCLUSIONS

11

Questions for the Future

11.1 INTRODUCTION

We have reviewed the dynamics of marine ecosystems from three perspectives. In Chapter 2 we concentrated on the turbulent flow of the water surrounding the organisms to understand better how algae obtain inorganic nutrients and microscopic animals their algal food. Chapters 3–8 reviewed the biological and physical processes associated with formation of phytoplankton concentrations in spring blooms, at fronts, in coastal upwelling areas, at shelf edges, in rings, and in estuaries. The final perspective was of long-term variations in phytoplankton productivity due to climate changes.

In all of these discussions we reviewed information on consumer dynamics where it was available, but by far the greater part of the text concerns phytoplankton. The reason for this bias is that chlorophyll can be measured continuously by means of a fluorometer, and this serves as a useful index of phytoplankton biomass. Instruments for the continuous measurement of zooplankton biomass are still in the developmental stage. Future advances of marine ecology require major improvements in automated methods for collection of data on zooplankton.

As marine ecology emerges as an integrated discipline concerned with the physics, chemistry, and biology of the oceans, there is a search for generalities around which to organize the multiplicity of observations. We see this search revolving around three questions.

1 Is there a common mechanism to account for the occurrence of high biological productivity in a variety of physical environments?

2 To what extent are events in marine ecosystems determined by the physical processes? To paraphrase Hartline (1980), do physical factors feed fish?

3 How can we develop concepts and models that span the enormous range of scales in marine ecology, from the microscopic to the global and from seconds to geological ages?

11.2 A BASIC MECHANISM OF PHYTOPLANKTON PRODUCTIVITY

It was pointed out by Legendre (1981), and is amply confirmed by our review, that there is one sequence of events that occurs in a variety of physical settings and on time scales ranging from a few hours to a year, that normally leads to an increase in phytoplankton

351

production. The essence of it is strong vertical mixing followed by stratification of the water column. As first described by Gran (1931) and presented as a quantitative model by Sverdrup (1953), it was offered as the explanation of the spring bloom in temperate waters. The vertical mixing brings nutrients from depth to surface waters, and the formation of stratification confines the phytoplankton to a well-lit zone in which daily photosynthesis exceeds daily respiration. The driving force for vertical mixing is wind stress at the surface and the chief agent of stratification is solar heating.

In estuaries and parts of the continental shelf vertical mixing may be driven by tidal currents interacting with the bottom, and stratification may be mainly a function of freshwater run-off, but the effect on the phytoplankton is the same. The major difference is the time scale. The strength of tidal currents varies with a diurnal rhythm and with the fortnightly cycle of spring and neap tides. When tidal currents are strong, vertical mixing is also strong, and when tidal currents are weak stratification is more marked. We have reviewed examples of high phytoplankton productivity corresponding with each period of stratification.

At tidally mixed fronts, the area of tidal mixing increases with the spring tides and decreases on the neap tides, causing the front to move back and forth horizontally. At some geographical locations there is an alternation of tidal mixing and stratification that leads to enhanced phytoplankton production and biomass.

Similarly, in areas noted for their coastal upwelling, the prevailing wind brings nutrient-rich water to the surface and a relaxation of that wind permits surface warming and stratification. The high productivity is associated with the relaxation of the winds, as the phytoplankton utilize the upwelled nutrients. Even in areas of more or less continuous winds, a horizontal succession of events can be detected. Close to the coast, nutrients are abundant in the freshly upwelled water, but as it streams away from the area of upwelling stratification sets in and phytoplankton biomass increases. Still further from the coast is a zone in which the zooplankton become more abundant as they feed on the phytoplankton. A similar zonation is found in relation to equatorial upwelling.

Here, then, is a mechanism of biological productivity that is found in situations as diverse as a temperate estuary and a mid-ocean band of upwelling running parallel with the equator. The alternation of vertical mixing and stratification is surely one of the most important sequences in marine ecology.

In Chapter 3, we referred to examples that appeared to contradict the Gran/Sverdrup model—namely, instances where a phytoplankton bloom was initiated in the apparent absence of stratification. A bloom (i.e., an accumulation of phytoplankton biomass) is the result of the growth term for a phytoplankton population being significantly larger than the loss term. Platt *et al.* (1994) pointed out that the loss term is usually either grazing or sinking, so if these processes were unusually low, biomass could accumulate and a bloom could form. If light increased rapidly due to increasing day length, or if the actual depth of the mixed layer was less than that indicated by profiles of conductivity or temperature, the growth term could increase while the loss terms remained unchanged. The last situation is thought to occur when a body of water is subject to an unusually low level of wind mixing for a protracted period. The phytoplankton are moved vertically at such a slow rate that they are able to grow and multiply in the absence of a clearly defined shallow thermocline. Because in the examples on record of blooms forming before the onset of stratification, the critical parameters for the test of these mechanisms were not measured, no conclusive evidence has been gathered for violation of the Gran/Sverdrup model.

Making a clear distinction between changes in phytoplankton productivity and the accumulation of biomass (i.e., bloom formation) has enabled oceanographers to begin the development of a global classification of pelagic ecosystems (Banse and English 1944; Longhurst 1995) that embodies the Gran/Sverdrup model in conjunction with other physical and biological mechanisms. For example, Longhurst calls the areas exhibiting the classical pattern of spring bloom and subsidiary autumn bloom the Westerly Winds Domain. A variant of this pattern is found in the North Pacific, where the formation of a deep mixed layer in winter is limited by the low salinity of the surface waters (see section 3.3.8). Poleward of this area, where spring comes very late but the days of summer are very long, the main bloom takes place in summer. A subsidiary fall bloom in that locale may be mediated by reduced grazing pressure as the zooplankton sink to overwintering depths. Areas exhibiting this pattern are called the Polar Domain. The Trade Wind Domain is the area that is permanently stratified and in which the biomass of phytoplankton fluctuates with small amplitude in response to seasonal variations in trade winds. Finally, Longhurst (1995) recognizes a Coastal Domain in which a wide range of mechanisms, including tidal mixing, internal waves, wind stress, river run-off, or topography, cause periodic upwelling that is interspersed with stratification caused by either surface warming or reduced salinity. These mechanisms are analyzed in Chapters 4–7 of this textbook. All are consistent with the Gran/Sverdrup model.

Now that good satellite coverage of the ocean surface is available, it seems likely that by combining this information with shipboard data on the structure of the water column, rapid progress will be made toward an understanding of primary productivity, its magnitude, and its variability, on a global scale.

11.3 THE RELATIONSHIP BETWEEN PHYSICAL AND BIOLOGICAL PROCESSES

In the majority of situations we have reviewed, biological processes appear to be strongly influenced by the physics, while the physical processes are largely independent of the biology. This perception led Hartline (1980) to write an article entitled 'Coastal Upwelling: Physical Factors Feed Fish.' In estuaries, on continental shelves, and in the open ocean, the physical processes appear to set the stage (albeit a constantly changing stage) on which the biological play is enacted. The organisms are adapted to the physical conditions that occur in a given place, but, in the short term, the physical conditions are very little affected by the biology.

There are some exceptions. Lewis *et al.* (1983) documented an example of short-term feedback from biological to physical processes. They showed that at an oligotrophic ocean station the chlorophyll maximum was sufficiently strong that the heat absorbed from the downwelling radiation caused local heating of the water, which caused increased vertical mixing and deepening of the mixed layer. The chlorophyll maximum also decreased the rate of heating of the water immediately below it, thereby increasing the stability of that part of the thermocline. The phenomenon is probably widespread in the ocean.

Stramska and Dickey (1993) summarized several examples of local heating caused by absorption of light in phytoplankton cells and provided new data from the North Atlantic. They modelled the situation and found that such local heating was a factor in the stabilization of the water column and formation of a bloom that preceded the seasonal thermocline.

When we turn to a time scale of thousands of years, the feedback from biology to physics is far from trivial. The present composition of the earth's atmosphere and the present average temperature at the surface of the earth have been profoundly influenced by living organisms. Volcanoes have been constantly adding to the carbon dioxide content of the atmosphere, but 99.5% of that carbon dioxide has been transported to the ocean sediments by marine organisms. They fixed carbon dioxide in their calcareous skeletons and incorporated carbon dioxide in their organic matter. As they sank to the bottom of the ocean, they gave rise to the carbonaceous rocks and to the deposits of fossil fuel. Without these processes it is likely that the carbon dioxide content of the atmosphere would be very high and the temperature of the surface of the earth would be about 400 °C. Life as we know it would be impossible.

Biological processes are also thought to have played a major part in starting and ending the ice ages. It has been found that the rise in mean temperature (about 10 °C) associated with the end of the last glaciation was accompanied by a rise in atmospheric carbon dioxide. There is not yet agreement about the mechanisms involved. Changes in the earth's orbit are believed to have had a triggering effect, but it seems that there was also a sharp decrease in the fixation of atmospheric carbon dioxide by the phytoplankton. If this hypothesis turns out to be correct, biological processes are responsible for profoundly influencing the composition of the atmosphere, the temperature of the earth's surface, and as a consequence, the temperature and circulation of the whole global ocean.

This fact should not blind us to the limitations of the biological pump as a factor for offsetting anthropogenic increases in atmospheric carbon dioxide (section 10.3.5). For about 1000 years before the Industrial Revolution, atmospheric CO_2 was relatively stable, so that the biological pump must have been in equilibrium with processes returning CO_2 from the ocean to the atmosphere. No evidence exists to suggest that the biological pump responds to changing atmospheric CO_2 directly, so it would not be a factor in sequestering anthropogenic carbon (Houghton *et al.* 1990). On the other hand, any global change that modified ocean circulation and materially altered the upwelling of nutrients would be likely to change the operation of the biological pump and the exchange of CO_2 between ocean and atmosphere.

11.3.1 The concept of auxiliary energy

In our review we have seen that physical processes in the ocean are driven by solar energy that is transmitted to the water mass as heat, wind energy, or freshwater run-off, and result in currents, turbulence, and patterns of stratification. Solar energy is also captured by phytoplankton, passed from one organism to the other through the feeding process, and finally dissipated in the heat of respiration. Physical oceanography deals with the first of these pathways and biological oceanography with the second. In marine ecology we become concerned with both, and we then notice that the two pathways are interrelated. Water movement breaks down the boundary layers around organisms, transports nutrients and waste products, and influences the rate of encounter between planktonic predators and their prey. Organisms use currents to assist their migrations and sessile organisms occupy sites with a range of current speeds appropriate to their way of life. These physical energies assist the transfer of energy in the biological food web without themselves taking part in the process. From a biological point of view they have been labelled auxiliary energy.

The concept was first elucidated by Odum (1967a, b) who drew attention to the way in which the ebb and flow of the tide assists the productivity of a salt marsh, and the way in which agricultural productivity (as measured by energy flow) is enhanced by the use of auxiliary energy to prepare the ground, remove weeds, spread fertilizers, and so on. In the history of agriculture the auxiliary energy has changed from being human labor (the farmer and his family) through the use of domestic animals to the present-day use of fossil fuels. Odum drew attention to the fact that the auxiliary energy for a salt marsh is provided free by nature, while the auxiliary energy of a farm is costly.

Margalef (1974, 1978b) pointed out that the physical energy that upwells nutrient-rich water in the ocean is an energy subsidy to the phytoplankton. He showed that a model relating photosynthesis to energy of upwelling is, in many situations, a better predictor of phytoplankton production than the more complex models involving light, nutrients, and temperature. He also developed detailed explanations of the morphology and physiology of various phytoplankton species in terms of their adaptations to high or low levels of auxiliary energy. For example, diatoms tend to be characteristic of turbulent water in which nutrients are being upwelled vigorously, while flagellates are better suited to low-turbulence, low-nutrient waters.

The topic was further explored by Legendre (1981), Legendre and Demers (1984, 1985), and Legendre *et al.* (1986). He and his colleagues pointed out that high biological productivity often occurs at places where there is a sudden change of auxiliary energy, such as fronts, thermoclines, sediment–water interfaces, and the underside of ice. They also pointed out that for there to be an effective physical–biological interaction there must be a matching of scales in time and space. For example, since the doubling time for phytoplankton cells is in the range of hours to days, favorable physical conditions must persist for at least a few days for there to be a marked increase in phytoplankton populations. A good example of the importance of matching time scales was reviewed in Chapter 6. High phytoplankton productivity on tidally mixed fronts is stimulated by changes in the strength of tidal mixing and has a periodicity of 14 days, whereas high productivity at the shelf break is stimulated by internal waves that form on the ebbing tides with a semi-diurnal rhythm. Since the generation time of meso-zooplankton is of the order of weeks, they cannot reproduce fast enough to use the pulses of primary production occurring at fortnightly intervals on the tidally mixed fronts, but they are able to build their populations by using the more continuous supply of phytoplankton generated in association with the internal lee waves at the shelf break.

One of the problems inherent in trying to build the concept of auxiliary energy into numerical models is that it is difficult to determine what proportion of the energy of physical processes impinges on biological processes. Margalef (1978a) made some rough estimates and developed comparative plots of the regressions of primary production in the plankton and in agriculture on auxiliary energy (Fig. 11.01). The diagram shows that a given amount of primary production in the plankton requires about 10 times more auxiliary energy than the same production on land. This finding is not surprising when we recall that in the ocean nutrient regeneration often occurs at great depth, far removed from the euphotic zone, whereas on land nutrients are regenerated in the soil, close to the roots of the plants.

Fig. 11.01 Auxiliary energy (W m^{-2}) used to enhance primary production in agricultural systems and in the phytoplankton. From Margalef (1978a).

It will be interesting to see whether the concept of auxiliary energy leads to better models of primary production. Now that it is possible to obtain vertical profiles of turbulence (see section 3.3.3), data on primary production as a function of turbulent energy may begin to accumulate. Lande *et al.* (1989) used data on the vertical distribution of cell concentrations and turbulence to calculate the population growth rates at different depths of those phytoplankton cells that could be regarded as nonmotile and neutrally buoyant. These cells were all in the size range known as ultraphytoplankton, and consisted of prochlorophytes (<1 μm), cyanobacteria (≈1 μm), small eukaryotes (<2 μm), and larger eukaryotes (2–4 μm). Taking the dispersal due to turbulent diffusion as the loss term, they calculated the population growth rate necessary to compensate for this loss, at various depths. For all groups, the highest population growth rates occurred near the bottom of the euphotic zone, near the top of the nutracline. Cyanobacteria grew most rapidly at 85–95 m and eukaryotes at 100–115 m depth. Below 150 m growth rates were all negative and cell numbers decreased exponentially with depth.

11.4 THE QUESTION OF SCALE

The drawing in Fig. 11.02 helps to illustrate some problems of scale that confront us in the future. The figure represents the 1000 km domain of a fictitious fish that is shown moving from place to place on its annual migration. From its spawning grounds in the early spring the fish swims to the upper layer of the open ocean to feed on the zooplankton that are feeding on the early phytoplankton blooms. Later they move to a tidal front over the shelf before swimming on to feed on a small spawning fish. They next move on to the outer region of the continental shelf where internal tidal waves have enhanced the productivity and finally they end up back at their own spawning grounds.

The fish swim thousands of kilometers over the year to feed at a series of areas of enhanced productivity that are each about 50 km wide. From the size of territory associated with each species it is clear that complete studies of fish will need to cover a much larger area than studies of phytoplankton. The time scales associated with the two organisms are also much different. Individual phytoplankton live for a few days and the population of a patch can change significantly over the same length of time. Fish live for 5–10 years and

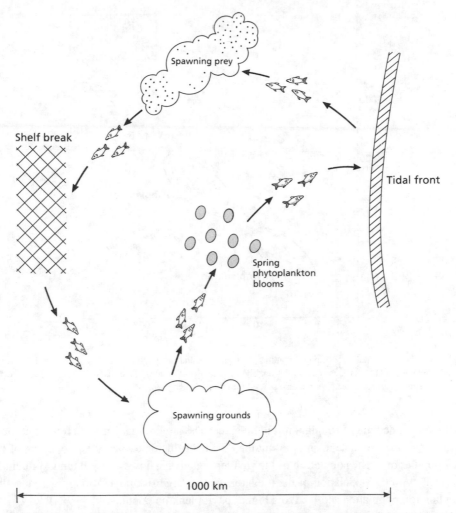

Fig. 11.02 A schematic diagram to illustrate the contrast between the 1000 km scale of the fish's domain and the 50 km scale of the various features, such as fronts and plankton blooms, upon which the fish depend.

measurable changes in their population occur over a year or so. Thus, besides covering a large area, the study of fish will also have to take a long time to understand the causes of any variations in time. The length and time scales associated with the populations are sometimes plotted jointly on a chart such as the one shown in Fig. 11.03 (Steele 1981).

The fish's domain, in the upper right of the figure, is thousands of kilometers across and the population takes years to change significantly. Phytoplankton, at the other extreme, change their numbers quickly and are concentrated over small scales. The length and time scales of the zooplankton lie in between those of the phytoplankton and the fish. They also lie in between the other two as prey to the fish and predators to the phytoplankton. The health of the fish population is obviously dependent on the health and availability of the zooplankton, which in turn is dependent on the phytoplankton, but these links between the groups can be rather 'elastic.' That is, the zooplankton population covers a sufficiently larger area and is sufficiently mobile to survive short-term fluctuations in the phytoplankton population at any one location. The fish in turn are mobile enough not to be strongly

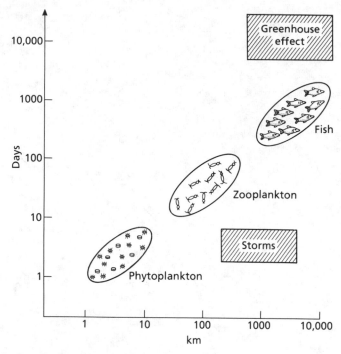

Fig. 11.03 The horizontal scale of phytoplankton, zooplankton, and fish populations plotted against the time required for a significant change in the population. The time and length scales of storms and climatic effects such as a change in the greenhouse effect are also plotted for comparison.

affected by a decline of zooplankton in one area. The larger organisms, therefore, because of their mobility are able to smooth out the effects of fluctuations in the population of the prey populations. The population of the smaller organisms, however, will tend to fluctuate inversely with the population of the predator. Thus a large fish population may reduce the zooplankton population and lead to a higher phytoplankton population.

Changes in the environment will also cause large variations in the populations of the smaller organisms. Violent storms can disrupt and disperse spawning fish and increase the turbulence in the upper layer to disperse plankton patches. Longer-term changes in temperature such as those expected from the greenhouse effect will change the phytoplankton species of a given region and possibly the locations of phytoplankton concentrations will change. A plankton population is said to absorb the variations due to storms and adapt to the multiyear variations (Steele 1988).

Not shown in Figs. 11.02 and 11.03 are important relationships between the fish and its predators, and between the fish and its prey. It is also necessary to understand fluctuations in the other fish populations inhabiting the same space as the one illustrated because they may strongly interact. The diagram in Fig. 11.03 may then sprout other axes to depict these other relationships such as the different trophic levels and the different levels of hierarchy of each particular species. Four axes are too many to show on one diagram, so for simplicity the time/length scale plane is sometimes collapsed to a line with the assumption that time scales increase directly in proportion to the length scales. Such a figure is given in Fig. 11.04. It must be remembered that the simplification comes at the cost of ignoring some important effects such as short- and long-term meteorological forcing.

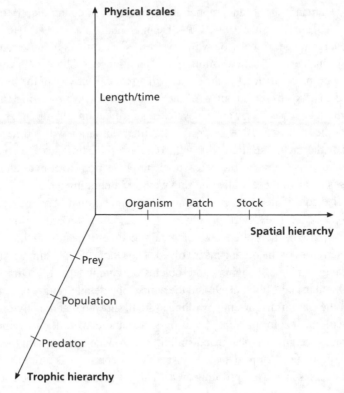

Fig. 11.04 The three main ecological variables plotted on a three-dimensional grid as suggested by Steele (1988).

Diagrams like those in Figs. 11.03 and 11.04 help to organize marine ecology and aid in separating the important and tractable relationships that have to be understood. One initiative aimed at understanding the interactions between the passively floating phytoplankton, the actively swimming zooplankton, and the turbulence in the water was discussed at a workshop entitled 'Small-scale Bio-physical Coupling' (Bucklin *et al.* 1989). The participants in this workshop were experts in acoustics, numerical modelling, physical oceanography, and biological oceanography. They agreed that progress in the field is most likely to be made through study of small-scale physical and biological processes with the object of understanding how the small organisms, both passive and active, interact with one another and with the small-scale physical processes.

To achieve this understanding they propose to obtain observations over a series of length and time scales. At the 1-m scale, optical and acoustic instruments will obtain high-resolution time-series observations of currents and the movements of particles and organisms from 5 μm to 5 mm in size. Results from these fine-scale observations will lead to the development of accurate parameterizations of the fine-scale physical and biological processes that can then be used in the models connecting the fine-scale observations to the large-scale ones.

The observations over the larger scales will include the use of submersibles to observe zooplankton, moored instruments to record temperature, velocity, and light intensity, temperature and velocity microstructure profilers, drifters, and an acoustic imaging array—in short, a high-resolution multidiscipline experiment using the most advanced instrumentation.

Kenneth Denman is an oceanographer at the forefront of contemporary concerns with the problems of scale (Denman 1993, 1994; Denman and Abbott 1994). He has drawn attention to the large range of scales on which processes occur that are important to the question of climate change and global warming. As we mentioned in Chapter 10, any process that affects the upwelling of nutrients in the ocean will affect the activity of the biological pump and, in turn, the carbon dioxide content of the atmosphere. This upwelling can be affected by storms (time scale 1 week), by annual cycles of stratification and mixing, by ENSO-type events with a time scale of 5–10 years, by thermocline ventilation with a time scale of about 30 years, and by thermohaline circulation with a characteristic time scale of about 500 years. If we wish to understand how the biological pump might respond to climate change, we must study processes on all of these scales and find ways of integrating them.

Denman also contributed to an important symposium entitled 'Aquatic Ecology: Scale, Pattern and Process.' He reviewed the physical processes at work at various scales in the ocean, and showed how their effect on biological processes depends on the time and space scales of the activities of the organisms involved. The whole symposium volume (see Denman 1994) is a goldmine of insights into problems of scale in aquatic environments.

Finally, Denman and Abbott (1994), in the course of a detailed study of a study of satellite images of the California current upwelling system, showed that the phytoplankton cells (identified by pigment color patterns) act as passive markers of water movement. Their biological dynamics of growth, death, and sinking are slow compared with the physical dynamics. Thus, pigment color as detected by satellite can be used as a marker for the study of physical processes in active upwelling areas. This system is likely to be a useful tool for future studies.

Hats off to the dedicated, hard-working individuals who will write the next chapters in this vital and fascinating field!

Appendix

Table A.1 Variables, constants, and nondimensional numbers

Symbol	Quantity or variable	Typical value and/or units
A	Area of cross-section	m^2
α	Thermal expansion coefficient (water)	$10^{-4}\ °C^{-1}$
α, α_λ	Absorption coefficient	$\simeq 0.1\ m^{-1}$ for red light
		$\simeq 0.001\ m^{-1}$ for blue light
β	Variation of Coriolis parameter with latitude	$2 \times 10^{-11}\ m^{-1}\ s^{-1}\ (\phi = 45°)$
C	Concentration of a substance	$kg\ m^{-3}$
C_D	Drag coefficient	varies with shape and flow
c	Specific heat of water	$4.2\ kJ\ kg^{-1}\ °C^{-1}$
D	Molecular diffusivity	$1.5 \times 10^{-7}\ m^2\ s^{-1}$ (heat)
		$1.5 \times 10^{-9}\ m^2\ s^{-1}$ (salt)
D_c	Compensation depth	m
D_E	Ekman depth	$\simeq 80\ m$
D_t	Tidal energy dissipation	$W\ m^{-2}$
d	Typical diameter	m
δ	Boundary layer thickness	m
ε	Turbulent energy dissipation	$W\ kg^{-1}$
F	Flux	$kg\ m^{-2}\ s^{-1}$
F_D	Drag force	N
f	Coriolis parameter $(2\Omega \sin \phi)$	$10^{-4}\ s^{-1}\ (\phi = 45°)$
ϕ	Latitude	degrees
g	Gravitational acceleration	$9.98\ m\ s^{-2}$
g'	Reduced gravity $\left(\dfrac{\Delta\rho}{\rho}g\right)$	$0.01\ m\ s^{-2}$
h	Height of water column	m
I	Intensity of light	$W\ m^{-2}$
J	Energy, in Joules	$N\ m$
K	Half-saturation constant	$kg\ m^{-3}$
K_h	Horizontal eddy diffusivity	$500\ m^2\ s^{-1}$
K_v	Vertical eddy diffusivity	$10^{-4}\ m^2\ s^{-1}$
l	Typical length	m
L_b	Buoyancy length scale $(\varepsilon/N^3)^{1/2}$	$1–60\ m$
L_d	Diffusive length scale $2\pi\ (\nu D^2/\varepsilon)^{1/4}$ (Batchelor scale)	$0.2–1.0\ mm$
L_v	Viscous length scale $2\pi(\nu^3/\varepsilon)^{1/4}$ (Kolmogoroff scale)	$6–35\ mm$
M	Molar concentration	$mol\ 1^{-1} = g\text{-at}\ 1^{-1}$
μM	Micromolar concentration	$\mu g\text{-at}\ 1^{-1}$

Table A.1 Variables, constants, and nondimensional numbers (*Continued*)

Symbol	Quantity or variable	Typical value and/or units
M_E	Ekman transport	kg m s^{-1}
N	Force, in newtons	kg m s^{-2}
N	Brunt–Väisälä frequency $(g/\rho \, d\rho/dz)^{1/2}$	10^{-3}–10^{-2} s^{-1}
P_a	Pressure, in Pascals	N m^{-2}
R_e	External Rossby radius $(gh)^{1/2}/f$	1000 km
R_i	Internal Rossby radius $(g'h)^{1/2}/f$	30 km
Re	Reynolds number ud/ν	
ρ	Density	1000 kg m^{-3} (freshwater)
		1026 kg m^{-3} (salt water)
σ_t	Sigma-t (density anomaly)	$(\rho - 1000)$ kg m^{-3}
τ	Stress	N m^{-2}
u, v, w	Components of velocity along x, y, z	m s^{-1}
$\bar{u}, \bar{v}, \bar{w}$	Average velocity	m s^{-1}
u', v', w'	Fluctuating component of velocity	m s^{-1}
u_*	Friction velocity $\left(\dfrac{\tau}{\rho}\right)^{1/2}$	m s^{-1}
ν	Kinematic viscosity	10^{-6} m^2 s^{-1} (water)
W	Power, in Watts	J s^{-1}
Ω	Angular velocity	rad s^{-1}
x, y, z	Position coordinates	m
ζ	Relative vorticity	s^{-1}

Table A.2 Basic units

Quantity	Name	Symbol
mass	gram	g
length	meter	m
time	second	s
amount of substance	mole	mol
temperature	degree Celsius	deg C, °C

Table A.3 Prefixes

Prefix	Factor	Symbol
pico	10^{-12}	p
nano	10^{-9}	n
micro	10^{-6}	μ
milli	10^{-3}	m
centi	10^{-2}	c
deci	10^{-1}	d
deca	10	da
hecto	10^2	h
kilo	10^3	k
mega	10^6	M
giga	10^9	G

References

Abelson, A., Miloh, T. and Loya, Y. (1993) Flow patterns induced by substrata and body morphologies of benthic organisms, and their roles in determining availability of food particles. *Limnol. Oceanogr.* 1116–1124.

Aebischer, N.J., Coulson, J.C. and Colebrook, J. M. (1990) Parallel long-term trends across four marine trophic levels and weather. *Nature* 347: 753–755.

Aguilera, J., Jiménez, C., Rodriguez-Maroto, J.M. and Niell, F.X. (1994) Influence of subsidiary energy on growth of *Dunialiella viridis* Teodoresco: the role of extra energy in algal growth. *J. Appl. Phycol.* 6: 323–330.

Alber, M. and Valiela, I. (1995) Organic aggregates in detrital food webs: incorporation in bay scallops (*Argopecten irradians*). *Mar. Ecol. Prog. Ser.* 121: 117–124.

Alldredge, A.L. and Hamner, W. M. (1980) Recurring aggregations of zooplankton by a tidal current. *Estuar. Coast. Mar. Sci.* 10: 31–37.

Alldredge, A. L. and Cohen, Y. (1987) Can microscale chemical patches persist in the sea? Microelectrode study of marine snow, fecal pellets. *Science* 235: 689–691.

Allen, P. M. (1985) Ecology, thermodynamics and self-organization: towards a new understanding of complexity, pp. 3–26. In: R.E. Ulanowicz and T. Platt (Eds) *Ecosystem Theory for Biological Oceanography. Can. Bull. Fish. Aquat. Sci.* 213 pp.

Anderson, S.M. and Charters, A. C. (1982) A fluid dynamic study of water flow through *Gelidium nudifrons. Limnol. Oceanogr.* 27: 399–412.

Andrews, W. R.H. and Hutchings, L. (1980) Upwelling in the southern Benguela current. *Prog. Oceanogr.* 9: 1–81.

Anon. (1992) *Our Living Oceans.* Report on the status of U.S. living marine resources. US Dept. Commerce. Washington, DC. 148 pp.

Ansa-Emmin, M. (1982) Fisheries in the CINECA region. *Rapp. P.-v. Reun. Cons. Int. Explor. Mer.* 180: 405–422.

Aoki, I. and Inagaki, T. (1992) Acoustic observations of fish schools and scattering layers in a Kuroshio warm core ring and its environs. *Fisheries Oceanogr.* 1: 137–142.

Apel, J. R. (1987) *Principles of Ocean Physics.* Academic Press, New York. 634 pp.

Apel, J. R., Byrne, H.M., Proni, J.R. and Charnell, R.L. (1975) Observations of oceanic internal and surface waves from the Earth Resources Technology Satellite. *J. Geophys. Res.* 80: 865–881.

Armi, L. and Farmer, D.M. (1988) The flow of Mediterranean water through the Strait of Gibraltar. *Prog. Oceanogr.* 21: 1–105.

Armstrong, D.A., Mitchell-Innes, B.A., Verhaye-Dua, F., Waldron, H. and Hutchings, L. (1987) Physical and biological features across the upwelling front in the southern Benguela. In: A.I.L. Payne, J.A. Gulland and K.H. Brink (Eds) *The Benguela and Comparable Ecosystems. South African J. Mar. Sci.* 5: 171–190. Sea Fisheries Research Institute, Cape Town.

Astthorsson, O.S., Hallgrimsson, I. and Jonsson, G.S. (1983) Variations in zooplankton densities in Icelandic waters in spring during the years 1961–1982. *Rit. Fiskideildar* (*J. Mar. Res. Inst. Reykjavik*) 7: 73–113.

Atkins, W.R.G. (1928) Seasonal variation in the phosphate and silicate content of sea water during 1926 and 1927 in relation to the phytoplankton crop. *J. Mar. Biol. Assoc.* 15: 191–205.

Atkinson, L.P., O'Malley, P. G., Yoder, J.A. and Paffenhöfer, G.A. (1984) The effect of summertime shelf break upwelling on nutrient flux in southeastern United States continental shelf waters. *J. Mar. Res.* 42:969–993.

Atkinson, L.P. and Targett, T.E. (1983) Upwelling along the 60-m isobath from Cape Canaveral to Cape Hatteras and its relationship to fish distribution. *Deep-Sea Res.* 30: 221–226.

Avise, J.C., Nelson, W.S., Arnold, J., Koehn, R.K., Williams, G.C. and Thorsteinsson, V. (1990) The evolutionary genetic status of Icelandic eels. *Evolution* 44: 1254–1262.

Backus, R.H. and Bourne, D.W. (Eds) (1987) *Georges Bank.* M.I.T. Press, Cambridge, Mass. 593 pp.

Backus, R.H. and Craddock, J.E. (1982) Mesopelagic fishes in Gulf Stream cold core rings. *J. Mar. Res.* 40 (suppl.): 1–20.

Bailey, K.M. (1981) Larval transport and recruitment of Pacific hake *Merluccius productus. Mar. Ecol. Prog. Ser.* 6: 1–9.

Bailey, R.A. and Steele, J.H. (1992) North Sea herring fluctuations, pp. 213–230. In: M.H. Glantz (Ed), *Climate Variability, Climate Change and Fisheries.* Cambridge University Press, Cambridge.

Bakun, A. (1973) Coastal upwelling indices, west coast of North America 1946–71. NOAA Tech. Rep. NMFS SSRF-671. US Dept. Commerce, Seattle.

Bakun, A. (1990) Global climate change and intensification of coastal ocean upwelling. *Science* 247: 198–201.

Bakun, A. and Parrish, R.H. (1982) Turbulence, transport and pelagic fish in the California and Peru current systems. *CalCOFI Rep.* 23: 99–112.

Banse, K. (1987) Clouds, deep chlorophyll maxima and the nutrient supply to the mixed layer of stratified water bodies. *J. Plankt. Res.* 9: 1031–1036.

Banse, K. and English, D.C. (1994) Seasonality of coastal zone color scanner phytoplankton pigment in the offshore oceans. *J. Geophys. Res.* 99 (C4): 7323–7345.

Barber, R.T., Chavez, F.P. and Kogelschatz, J.E. (1985) Biological effects of El Niño. pp. 399–438. In: M. Vegas (Ed) *Seminario Regional Ciencas Tecnologia y Agression Ambiental: El Fenomeno 'El Niño.'* Contec Press, Lima, Peru.

Barber, R.T. and Smith, R.L. (1981) Coastal upwelling ecosystems, pp. 31–68. In: A.R. Longhurst (Ed) *Analysis of Marine Ecosystems.* Academic Press, New York. 741 pp.

Barnett, A.M. and Jahn, A.E. (1987) Pattern and persistence of a nearshore planktonic ecosystem off Southern California. *Continental Shelf Res.* 7: 1–25.

Barnola, J.M., Raynaud, D., Korotkevich, Y.S. and Lorius, C. (1987) Vostok ice core provides 160,000-year record of atmospheric CO_2. *Nature* 329: 408–412.

Barry, J.P., Baxter, C.H., Sagarin, R.D. and Gilman, S.E. (1995) Climate-related, long-term faunal changes in a California rocky intertidal community. *Science* 267: 672–675.

Barton, E.D., Huyer, A. and Smith, R.L. (1977) Temporal variation observed in the hydrographic regime near Cabo Corbeiro in the northwest African upwelling region, February to April 1974. *Deep-Sea Res.* 24: 7–24.

Beamish, R.J. (1993) Climate and exceptional fish production off the west coast of North America. *Can. J. Fish. Aquat. Sci.* 50: 2270–2291.

Beamish, R.J. and Bouillon, D.R. (1993) Pacific salmon production trends in relation to climate. *Can. J. Fish. Aquat. Sci.* 50: 1002–1016.

Beamish, R.J., Neville, C.M., Thomson, B.L., Harrison, P.J. and St. John, M. (In press) The effect of Fraser River discharge on interannual production of Pacific salmon and herring in the Strait of Georgia. *Can. J. Fish. Aquat. Sci.*

Berg, H.C. and Purcell, E.M. (1977) Physics of chemoreception. *Biophys. J.* 20: 193–215.

Bernal, P. and McGowan, J.A. (1981) Advection and upwelling in the California current. *Coast. Estuar. Sci.* 1: 381–399.

Bienfang, P.K., Syper, J. and Laws, E. (1983) Sinking rate and pigment responses to light-limitation of a marine diatom: implications to dynamics of chlorophyll maximum layers. *Oceanol. Acta* 6: 55–62.

Bigelow, H.B. (1927) Physical oceanography of the Gulf of Maine. *Bull. U.S. Bureau Fisheries* 40: 511–1027.

Biggs, D.C. (1982) Zooplankton excretion and NH_4^+ cycling in near-surface waters of the Southern Ocean. 1. Ross Sea, Austral Summer 1977–78. *Polar Biol.* 1: 55–67.

Binet, D. (1986) French sardine and herring fisheries: a tentative description of their fluctuations since the eighteenth century. *Int. Symp. Long Term Changes Mar. Fish. Pop.* 253–272.

Blackburn, M. (1977) Studies on pelagic animal biomasses, pp. 283–299. In: N. R. Andersen and B. J. Zuharec (Eds) *Oceanic Sound Scattering Prediction.* Plenum Press, New York.

Blackburn, M. (1981) Low latitude gyral regions, pp. 3–30. In: A. R. Longhurst (Ed) *Analysis of Marine Ecosystems.* Academic Press, New York. 741 pp.

Blanton, J.O., Atkinson, L.P., de Castillejo, F.F. and Montero, A.L. (1984) Coastal upwelling off the Rias Bajas, Galicia, Northwest Spain. 1. Hydrographic studies. *Rapp. P.-v. Réun. Cons. Int. Explor. Mer.* 183: 79–90.

Blanton, J.O., Tenore, K.R., Castillejo, R., Atkinson, L.P., Schwing, F.B. and Lavin, A. (1987) The relationship of upwelling to mussel production in the rias on the western coast of Spain. *J. Mar. Res.* 45: 497–511.

Boëtius, J. and Harding E.F. (1985) A re-examination of Johannes Schmidt's Atlantic eel investigations. *Dana* 4: 129–162.

Boje, R. and Tomczak, M. (Eds) (1978) *Upwelling Ecosystems.* Springer-Verlag, Berlin. 303 pp.

Bond, G.C. (1995) Climate and the conveyor. *Nature* 377: 1995.

Boudreau, B., Simard, Y. and Bourget, E. (1991) Behavioural response of the planktonic stages of the American lobster *Homarus americanus* to thermal gradients, and ecological implications. *Mar. Ecol. Prog. Ser.* 76: 13–23.

Boudreau, B., Simard, Y. and Bourget, E. (1992) Influence of a thermocline on vertical distribution and settlement of post-larvae of the American lobster *Homarus americanus* Milne-Edwards. *J. Exp. Mar. Biol. Ecol.* 162: 35–49.

Bousfield, E.L. (1955) Ecological control of the occurrence of barnacles in the Miramichi Estuary. *Bull. Nat. Mus. Can.* 137: 1–69.

Bowden, K.F. (1983) *Physical Oceanography of Coastal Waters.* Ellis Horwood, New York, 302 pp.

Bowden, K.F. (1977) Heat budget considerations in the study of upwelling, pp. 277–290. In: M. Angel (Ed) *A Voyage of Discovery.* Pergamon Press, New York.

Bowen, J.D., Stolzenbach, K.D. and Chisholm, S.W. (1993) Simulating bacterial clustering around phytoplankton cells in a turbulent ocean. *Limnol. Oceanogr.* 38: 36–51.

Bowers, D.G. and Simpson, J.H. (1987) Mean position of tidal fronts in European-shelf seas. *Continental Shelf Res.* 7: 35–44.

Bowman, M.J. and Iverson, R.L. (1978) Estuarine and plume fronts, pp. 87–104. In: M.J. Bowman and W.E. Esaias (Eds) *Oceanic Fronts in Coastal Processes.* Springer-Verlag, New York.

Bowman, M.J., Yentsch, C.M. and Peterson, W.J. (Eds) (1986) *Tidal Mixing and Plankton Dynamics.* Springer-Verlag, Berlin.

Boyd, P.W., Strom, S., Whitney, F.A., Doherty, S., Wen, M.E., Harrison, P.J., Wong, C.S. and Varela, D.E. (1995a) The N.E. subarctic Pacific in winter: 1. Biological standing stocks. *Mar. Ecol. Prog. Ser* 128:11–24.

Boyd, P.W., Whitney, F.A., Harrison, P.J. and Wong, C.S. (1995b) The N.E. subarctic Pacific in winter: 2. Biological rate processes. *Mar. Ecol. Prog. Ser.* 128: 25–34.

Brander, K. (Ed.) (1994) Spawning and life history information for North Atlantic cod stocks. ICES Coop. Res. Rep. No. 205.

Brandt, A., Sarabun, C.C., Seliger, H.H. and Tyler, M.A. (1986) The effects of a broad spectrum of physical activity on the biological processes in the Chesapeake Bay, pp. 361–384. In: J.C.L. Nihoul (Ed) *Marine Interfaces Ecohydrodynamics.* Elsevier, Amsterdam.

Brickman, D. and Loder, J.W. (1993) Energetics of the internal tide on northern Georges Bank. *J. Phys. Oceanogr.* 23: 409–424.

Brink, K.H. (1983) The near-surface dynamics of coastal upwelling. *Prog. Oceanogr.* 12: 223–257.

Brink, K.H., Halpern, D. and Smith, R.L. (1980) Circulation in the Peru upwelling system near 15 deg. S. *J. Geophys. Res.* 85: 4036–4048.

Broecker, W.S. (1992) Upset for Milankovitch theory. *Nature* 359: 779–780.

Broecker, W.S., Peteet, D.M. and Rind, D. (1985) Does the ocean–atmosphere system have more than one stable mode of operation? *Nature* 315: 21–26.

Broenkow, W.W. (1965) The distribution of nutrients in the Costa Rica dome in the eastern tropical Pacific Ocean. *Limnol. Oceanogr.* 10: 40–52.

Bucklin, A., Brandt, A. and Orr, M. (1989) Small-scale biophysical coupling. Preliminary report of workshop sponsored by the Office of Naval Research. Washington, DC.

Budyko, M.I. (1974) *Climate and Life.* Academic Press, New York. 508 pp.

Bugden, G.L., Hargrave, B.T., Sinclair, M.M., Tang, C.L., Therriault, J.-C. and Yeats, P.A. (1982) Fresh water runoff effects in the marine environment: the Gulf of St. Lawrence example. *Can. Tech. Rep. Fish. Aquat. Sci.* 1078, 89 pp.

Bumpus, D.F. (1976) Review of the physical oceanography of George's Bank. ICNAF Res. Bull. No. 12: 119–134.

Butman, C.A. (1986) Larval settlement of soft-sediment invertebrates: some predictions based on an analysis of near-bottom profiles, pp. 487–513. In: J. C. L. Nihoul (Ed) *Marine Interfaces Ecohydrodynamics.* Elsevier, Amsterdam.

Butman, C.A., Grant, W.D. and Stolzenbach, K.D. (1986) Predictions of sediment trap biases in turbulent flows: a theoretical analysis based on observations from the literature. *J. Mar. Res.* 44: 601–644.

Cairns J.L. (1967) Asymmetry of internal tidal waves in shallow coastal waters. *J. Geophys. Res.* 72: 3563–3565.

Canby, T.Y. (1984) El Niño's ill wind. *National Geographic* 165: 144–183.

Cane, M.A., Eshel, G. and Buckland, R.W. (1994) Forecasting Zimbabwean maize yield using eastern equatorial Pacific sea surface temperature. *Nature* 370: 204–205.

Carpenter, R.C., Hackney, J.M. and Adey, W.H. (1991) Measurements of primary productivity and nitrogenase activity of coral reef algae in a chamber incorporating oscillatory flow. *Limnol. Oceanogr.* 36: 40–49.

Carpenter, R.C. and Williams, S.L. (1993) Effects of algal turf canopy height and microscale topography on profiles of flow speed in a coral forereef environment. *Limnol. Oceanogr.* 38: 687–694.

Chapman, P. and Shannon, L.V. (1985) The Benguela ecosystem, Part 2. Chemistry and related processes. *Oceanogr. Mar. Biol. Ann. Rev.* 23: 183–251.

Chapman, R. and Nof, D. (1988) The sinking of warm-core rings. *J. Phys. Oceanogr.* 18: 565–583.

Charlson, R.J., Lovelock, J.E., Andreae, M.O. and Warren, S.G. (1987) Oceanic phytoplankton, atmospheric sulphur, cloud albedo and climate. *Nature* 326: 655–661.

Charlson, R.J. and Wigley, T.M.L. (1994) Sulfate aerosols and climate change. *Sci. Am.* February 1994: 48–57.

Charney, J.G. (1955) The Gulf Stream as an inertial boundary layer. *Proc. Natl. Acad. Sci.* 41: 731–740.

Chavez, F.P. and Barber, R.T. (1987) An estimate of new production in the equatorial Pacific. *Deep-Sea Res.* 34: 1229–1243.

Chelton, D.B., Bernal, P.A. and McGowan, J.A. (1982) Large-scale physical and biological interaction in the California current. *J. Mar. Res.* 40: 1095–1125.

Chenoweth, S.B., Libby, D.A., Stephenson, R.L. and Power, M.J. (1989) Origin and dispersal of larval herring (*Clupea harengus* L.) in coastal waters of eastern Maine and southwestern New Brunswick. *Can. J. Fish. Aquat. Sci.* 46: 624–632.

Chereskin, T.K. (1983) Generation of internal waves in Massachusetts Bay. *J. Geophys. Res.* 88: 2649–2661.

Cho, B.C. and Azam, F. (1988) Major role of bacteria in biogeochemical fluxes in the ocean's interior. *Nature* 332: 441–443.

Chriss, T.M. and Caldwell, D.R. (1984) Universal similarity and thickness of the viscous sublayer at the ocean floor. *J. Geophys. Res.* 89: 6403–6414.

Clarke, R.A., Hill, H.W., Reiniger, R.F. and Warren, B.A. (1980) Current system south and east of the Grand Banks of Newfoundland. *J. Phys. Oceanogr.* 10: 25–65.

Clarke, R.A., Swift, J.H., Reid, J.L. and Koltermann, K.P. (1991). The formation of the Greenland Sea deep water: double diffusion or deep convection? *Deep-Sea Res.* 37: 1385–1424.

Cloern, J.E. (1991) Tidal stirring and phytoplankton bloom dynamics in an estuary. *J. Mar. Res.* 49: 203–221.

Cloern, J.E., Alpine, A.E., Cole, B.E., Wong, R.L.J., Arthur, J.F. and Ball, M.D. (1983) River discharge controls phytoplankton dynamics in the north San Francisco Bay estuary. *Estuar. Coast. Shelf Sci.* 16: 415–429.

Codispoti, L.A., Dugdale, R.C. and Minas, H.J. (1982) A comparison of the nutrient regimes off northwest Africa, Peru and Baja California. *Rapp. P.-v. Réun. Cons. Int. Explor. Mer.* 180: 184–201.

Codispoti, L.A. and Friederich, G.E. (1978) Local and mesoscale influences on nutrient variability in the northwest African upwelling region near Cabo Corbeiro. *Deep-Sea Res.* 25: 751–770.

Coelho, M.L. (1985) Review of the influence of oceanographic factors on cephalopod distribution and life history. *NAFO Sci. Coun. Studies* 9: 47–57.

Colebrook, J.M. (1979) Continuous plankton records: seasonal cycles of phytoplankton and copepods in the North Atlantic Ocean and the North Sea. *Mar. Biol.* 51: 23–32.

Colebrook, J.M. (1985) Continuous plankton records: overwintering and annual fluctuations in the abundance of zooplankton. *Mar. Biol.* 84: 261–265.

Colebrook, J.M. (1986) Environmental influences on long-term variability in marine plankton. *Hydrobiologia* 142: 309–325.

Conte, M.H., Bishop, J.K.B. and Backus, R.H. (1986) Nonmigratory, 12kHz, deep scattering layers of Sargasso Sea origin in warm-core rings. *Deep-Sea Res.* 33: 1869–1884.

Corner, E.D.S., Head, R.N. and Kilvington, C.C. (1972) On the nutrition and metabolism of zooplankton. VIII. The grazing of *Biddulphia* cells by *Calanus helgolandicus*. *J. Mar. Biol. Ass. U.K.* 52: 847–861.

Corten, A. (1990) Long-term trends in pelagic fish stocks of the North Sea and adjacent waters, and their possible connection to hydrographic changes. *Neth. J. Sea Res.* 25: 227–235.

Cowles, T.J., Roman, M.R., Gauzens, A.L. and Copley, N.J. (1987) Short-term changes in the biology of a warm-core ring: zooplankton biomass and grazing. *Limnol. Oceanogr.* 32: 653–664.

Creutzberg, F. (1985) A persistent chlorophyll-a maximum coinciding with an enriched benthic zone, pp. 97–108. In: P.E. Gibbs (Ed) *Proceedings of the Nineteenth European Marine Biology Symposium.* Cambridge University Press, Cambridge.

Crisp, D.J. (1960) Factors influencing growth rate in *Balanus balanoides*. *J. Anim. Ecol.* 29: 95–116.

Csanady, G.T. (1979) The life and death of a warm-core ring. *J. Geophys. Res.* 84: (C2): 777–780.

Csanady, G.T. (1981) Circulation in the coastal ocean. *Adv. Geophys.* 23: 101–183.

Cullen, J.J. (1991) Hypotheses to explain high-nutrient conditions in the open sea. *Limnol. Oceanogr.* 36: 1578–1599.

Cunningham, J.T. (1906) Fishes, pp. 291–306. In: W. Page (Ed) *The Victoria History of the County of Cornwall,* Vol. 1. Archibald Constable, London.

Currie, J.T. (1984) Microscale nutrient patches: do they matter to the plankton? *Limnol. Oceanogr.* 29: 211–214.

Cury, P. and Roy, C. (1989) Optimal environmental window and pelagic fish recruitment success in upwelling areas. *Can. J. Fish. Aquat. Sci.* 46: 670–680.

Cushing, D.H. (1968) Grazing by herbivorous copepods in the sea. *J. Cons. Int. Explor. Mer.* 32: 70–82.

Cushing, D.H. (1969) Upwelling and fish production. F.A.O. Fish. Tech. Paper 84: 38 pp.

Cushing, D.H. (1971) Upwelling and the production of fish. *Adv. Mar. Biol.* 9: 255–334.

Cushing, D.H. (1975) *Marine Ecology and Fisheries.* Cambridge University Press, Cambridge, UK. 278 pp.

Cushing, D.H. (1978) Upper trophic levels in upwelling areas, pp. 101–110. In: R. Boje and M. Tomczak (Eds) *Upwelling Ecosystems.* Springer-Verlag, Berlin.

Cushing, D.H. (1979) Climatic variation and marine fisheries, pp. 608–627. In: *Proceedings of the World Climate Conference, Geneva 1979.* World Meteorological Organization, Geneva.

Cushing, D.H. (1982) *Climate and Fisheries*. Academic Press, London. 373 pp.

Cushing, D.H. (1986) The migration of larval and juvenile fish from spawning ground to nursery ground. *J. Cons. Int. Explor. Mer.* 43: 43–49.

Cushing, D.H. (1988) The northerly wind, pp. 235–244. In: B.J. Rothschild (Ed) *Towards a Theory on Biological–Physical Interactions in the World Ocean*. Kluwer, Amsterdam.

Cushing, D.H. (1990a) Plankton production and year-class strength in fish populations: an update of the match/mismatch hypothesis. *Adv. Mar. Biol.* 26: 249–293.

Cushing, D.H. (1990b) Recent studies on long-term changes in the sea. *Freshwater Biol.* 23: 71–84.

Cushing, D.H. and Dickson, R.R. (1976) The biological response in the sea to climate change. *Adv. Mar. Biol.* 14: 1–122.

da Silva, A.J. (1986) River run-off and shrimp abundance in a tropical coastal ecosystem: the example of the Sofala Bank (Central Mozambique), pp. 329–344. In: S. Skreslet (Ed) *The Role of Freshwater Outflow in Coastal Marine Ecosystems*. Springer-Verlag, Berlin.

Dade, W.B. (1993) Near-bed turbulence and hydrodynamic control of diffusional mass transfer at the sea floor. *Limnol. Oceanogr.* 38: 52–69.

Dandonneau, Y. (1988) Seasonal or aperiodic cessation of oligotrophy in the tropical Pacific Ocean, pp. 137–156. In: B.J. Rothschild (Ed) *Towards a Theory on Biological–Physical Interactions in the World Ocean*. Kluwer, Dordrecht.

Davies, J.M. and Payne, R. (1984) Supply of organic matter to the sediment in the North Sea during a spring phytoplankton bloom. *Mar. Biol.* 78: 313–324.

Davis, C.S., Flierl, G.R., Wiebe, P.H. and Franks, P.J.S. (1991) Micropatchiness, turbulence and recruitment in plankton. *J. Mar. Res.* 49: 109–151.

Dawe, E.G. and Warren, W.G. (1992) Recruitment of short-finned squid in the Northwest Atlantic Ocean and some environmental relationships. *J. Cephal. Biol.* 2: 1–21.

de Baar, H.J.W., de Jong, J.T.M., Bakker, D.C.E., Loscher, B.M., Veth, C., Bathmann, U. and Smetacek, V. (1995) Importance of iron for plankton blooms and carbon dioxide drawdown in the Southern Ocean. *Nature* 373: 412–415.

de Young, B. and Pond, S. (1988) The deepwater exchange cycle in Indian Arm, British Columbia. *Estuar. Coast. Shelf Sci.* 26: 285–308.

Deardorff, J.W., Willis, G.E. and Lilly, D.K. (1969) Laboratory investigation of non-steady penetrative convection. *J. Fluid Mech.* 35: 7–31.

Defant, A. (1958) *Ebb and Flow*. University of Michigan Press, Ann Arbor, Mich. 121 pp.

Denman, K.L. (1973) A time-dependent model of the upper ocean. *J. Phys. Oceanogr.* 3: 173–184.

Denman, K.L. (1993) The ocean carbon cycle and climate change: an analysis of interconnected scales, pp. 213–223. In: S.A. Levin, T.M. Powell and J.H. Steele (Eds) *Patch Dynamics*. Springer-Verlag, Berlin. 455 pp.

Denman, K.L. (1994) Scale-determining biological–physical interactions in oceanic food webs, pp. 377–402. In: P.S. Giller, A.G. Hildrew and D.G. Rafaelli (Eds) *Aquatic Ecology: Scale, Pattern and Process*. Blackwell Science, Oxford.

Denman, K.L. and Abbott, M.R. (1994) Time scales of pattern evolution from cross-spectrum analysis of advanced very high resolution radiometer and coastal zone color scanner imagery. *J. Geophys. Res.* 99: 7433–7442.

Denman, K.L. and Miyake, M. (1973) Upper layer modification at ocean station Papa: observations and simulation. *J. Phys. Oceanogr.* 3: 185–196.

Dickson, R.R. and Brander, K.M. (1993) The effects of a changing windfield on cod stocks of the North Atlantic. *Fish. Oceanogr.* 2: 124–153.

Dickson, R.R., Kelly, P.M., Colebrook, J.M., Wooster, W.S. and Cushing, D.H. (1988a) North winds and production in the eastern North Atlantic. *J. Plankt. Res.* 10: 151–169.

Dickson, R.R., Lamb, H.H., Malmberg, S.-A. and Colebrook, J.M. (1975) Climatic reversal in the northern North Atlantic. *Nature* 256: 479–482.

Dickson, R.R., Meincke, J., Malmberg, S.-A. and Lee, A.J. (1988b) The 'great salinity anomaly' in the northern North Atlantic 1968–1982. *Prog. Oceanogr.* 20: 103–151.

Dietrich, G. (1950) Die anomale jahresschwankung des wärmeinhalts im Englischen Kanal, ihre Ursachen und auswirkungen. *Deutsche Hydrograph Zeit.* 3: 184–201.

Dietrich, G., Kalle, K., Krauss, W. and Siedler, G. (1980) *General Oceanography: An Introduction*, 2nd Ed. Wiley, New York. 626 pp.

Doty, M.S. and Oguri, M. (1956) The island mass effect. *J. Cons. Int. Explor. Mer.* 22: 33–37.

Dower, K.M. and Lucas, M.I. (1993) Photosynthesis–irradiance relationships and production associated with a warm-core ring shed from the Agulhas retroflection south of Africa. *Mar. Ecol. Prog. Ser.* 95: 141–154.

Drinkwater, K.F. (1986) On the role of freshwater outflow on coastal ecosystems—A workshop summary, pp. 429–438. In: S. Skreslet (Ed) *The Role of Freshwater Outflow in Coastal Marine Ecosystems*. Springer-Verlag, Berlin.

Drinkwater, K.F. (1987) 'Sutcliffe revisited': Previously published correlations between fish stocks and environmental indices and their recent performance, pp. 41–61. In: R.I. Perry and K.T. Frank (Eds) *Environmental Effects on Recruitment to Canadian Atlantic Fish Stocks.* Can. Tech. Rep. Fish. Aquat. Sci. 1556.

Drinkwater, K.F. (1988) The effect of freshwater discharge on the marine environment, pp. 415–430. In: W. Nicholaichuk and F. Quinn (Eds) *Proceedings of a Symposium on the Interbasin Transfer of Water: Impact and Research Needs for Canada.* Department of the Environment, National Hydrology Research Centre, Saskatoon, Canada. 504 pp.

Drinkwater, K.F. and Frank, K.T. (1994) Effects of river regulation and diversion on marine fish and invertebrates. *Aquatic Conservation: Freshwater and Marine Ecosystems* 4: 135–151.

Dugdale, R.C. and Goering, J.J. (1967) Uptake of new and regenerated forms of nitrogen in primary productivity. *Limnol. Oceanogr.* 12: 196–206.

Duncombe Rae, C.M., Boyd, A.J. and Crawford, R.J.M. (1992) 'Predation' of anchovy by an Agulhas ring: a possible contributory cause of a very poor year class of 1989. *S. Afr. J. Mar. Sci.* 12: 167–173.

Dyer, K. (1973) *Estuaries: A Physical Introduction.* Wiley, London.

Eckman, J.E. and Duggins, D.O. (1991) Life and death beneath macrophytic canopies: effects of understory growth rates and survival of marine benthic suspension feeders. *Oecologia* 87: 473–487.

Eilertsen, H.C. (1993) Spring blooms and stratification. *Nature* 363: 24.

Eldridge, P.M. and Sieracki, M.E. (1993) Biological and hydrodynamic regulation of the microbial food web in a periodically mixed estuary. *Limnol. Oceanogr.* 38: 1666–1679.

Ellertsen, B., Fossum, P., Solemdal, P., Sundby, S. and Tilseth, S. (1990) Environmental influence on recruitment and biomass yields in the Norwegian Sea ecosystem, pp. 19–35. In: K. Sherman, L.M. Alexander and B.D. Gold (Eds) *Large Marine Ecosystems: Patterns, Processes and Yields.* American Association for the Advancement of Science, Washington, DC.

El-Sayed, S.L. (1978) Primary productivity and estimates of potential yields of the Southern Ocean, pp. 141–160. In: M.A. McWhinnie (Ed) *Polar Research to the Present, and the Future.* Westview Press, Boulder, Colorado.

Enfield, D. B. (1989) El Niño past and present. *Rev. Geophys.* 27: 159–187.

Epifanio, C. E. (1987) The role of tidal fronts in maintaining patches of brachyuran zoeae in estuarine waters. *J. Crustacean Biol.* 7: 513–517.

Eppley, R.W. (1980) Estimating phytoplankton growth rates in the central oligotrophic oceans, pp. 231–242. In: P. G. Falkowski (Ed) *Primary Productivity in the Sea.* Plenum, New York.

Eppley, R.W. and Peterson, B.J. (1979) Particulate organic matter flux and planktonic new production in the deep ocean. *Nature* 282: 677–680.

Eppley, R.W., Renger, E.H. and Harrison, W.G. (1979) Nitrate and phytoplankton production in southern California waters. *Limnol. Oceanogr.* 24: 483–494.

Ewing, G. (1950) Slicks, surface films and internal waves. *J. Mar. Res.* 9: 161–187.

Evans, G.T. and Parslow, J.S. (1985) A model of annual plankton cycles. *Biol. Oceanogr.* 3: 327–347.

Falkowski, P.G. (1995) Ironing out what controls primary production in the nutrient rich waters of the open ocean. *Global Change Biol.* 1: 161–163.

Falkowski, P.G. and Wirick, C.D. (1981) A simulation model of the effects of vertical mixing on primary productivity. *Mar. Biol.* 65: 69–75.

Fang, Z. and Wallace, J.M. (1994) Arctic Sea ice variability on a time scale of weeks and its relation to atmospheric forcing. *J. Climate* 7: 1897–1914.

Farmer, D.M. and Freeland, H.J. (1983) The physical oceanography of fjords. *Prog. Oceanogr.* 12: 147–219.

Fasham, M.J.R. and Pugh, P.R. (1976) Observations on the horizontal coherence of chlorophyll-a and temperature. *Deep-Sea Res.* 23: 527–538.

Fasham, M.J.R., Sarmiento, J.L., Slater, D.J., Ducklow, H.W. and Williams, R. (1993) Ecosystem behaviour at Bermuda station 'S' and weather station 'I'. *Global Biogeochem. Cycles* 7: 379–415.

Flagg, C.N. and Beardsley, R.C. (1975) The 1974 M.I.T. New England shelf dynamics experiment (March 1974). Part 1. Hydrography. *M.I.T. Report* 75–1.

Flierl, G.R. and Davis, C.S. (1993) Biological effects of Gulf Stream meandering. *J. Mar. Res.* 51: 529–560.

Flierl, G.R. and Wroblewski, J.S. (1985) The possible influence of warm core Gulf Stream rings upon shelf water larval fish distribution. *Fish. Bull.* 83: 313–330.

Fofonoff, N.P. (1981) The Gulf Stream system, pp. 112–139. In: B.A. Warren and C. Wunsch (Eds) *Evolution of Physical Oceanography.* M.I.T. Press, Cambridge, Mass.

Fogg, G.E. (1985) Biological activities at a front in the western Irish Sea, pp. 87–96. In: P.E. Gibbs (Ed) *Proceedings of the Nineteenth European Marine Biology Symposium.* Cambridge University Press, Cambridge.

Folland, C.K., Parker, D.E. and Kates, F.E. (1984) Worldwide marine temperature fluctuations 1856–1981. *Nature* 310: 670–673.

Fortier, L. and Leggett, W.C. (1982) Fickian transport and the dispersal of fish larvae in estuaries. *Can. J. Fish. Aquat. Sci.* 39: 1150–1163.

Fortier, L. and Leggett, W.C. (1983) Vertical migrations and transport of larval fish in a partially mixed estuary. *Can. J. Fish. Aquat. Sci.* 40: 1543–1555.

Fortier, L. and Leggett, W.C. (1984) Small-scale covariability in the abundance of fish larvae and their prey. *Can. J. Fish. Aquat. Sci.* 41: 502–512.

Fournier, R.O. (1978) Biological aspects of the Nova Scotian shelf break fronts, pp. 69–77. In: M.J. Bowman and W.E. Esaias. *Oceanic Fronts in Coastal Processes.* Springer-Verlag, New York.

Frank, K.T. (1986) Ecological significance of the ctenophore *Pleurobrachia pileus* off southwestern Nova Scotia. *Can. J. Fish. Aquat. Sci.* 43: 211–222.

Frank, K.T., Perry, R.I., Drinkwater, K.F. and Lear, W.H. (1988) Changes in the fisheries of Atlantic Canada associated with global increases in atmospheric carbon dioxide: a preliminary report. *Can. Tech. Rep. Fish. Aquat. Sci.* 1652. 52 pp.

Franks, P.J.S., Wroblewski, J.S. and Flierl, G.R. (1986) Prediction of phytoplankton growth in response to the frictional decay of a warm-core ring. *J. Geophys. Res.* 91 (C6): 7603–7610.

Fransz, H.G. and Gieskes, W.W C. (1984) The unbalance of phytoplankton and copepods in the North Sea. *Rapp. P. -v. Réun. Cons. Int. Explor. Mer.* 183: 218–225.

Frechette, M. and Bourget, E. (1985a) Food-limited growth of *Mytilus edulis* L. in relation to the benthic boundary layer. *Can. J. Fish. Aquat. Sci.* 42: 1166–1170.

Frechette, M. and Bourget, E. (1985b) Energy flow between the pelagic and benthic zones: factors controlling particulate organic matter available to an intertidal mussel bed. *Can. J. Fish. Aquat. Sci.* 42: 1158–1165.

Frechette, M., Butman, C.A. and Geyer, W.R. (1989) The importance of boundary-layer flows in supplying phytoplankton to the benthic suspension feeder *Mytilus edulis* L. *Limnol. Oceanogr.* 34: 19–36.

Friedlander, A. and Smith, D. (1983) Sand Lance larvae found in entrainment feature associated with a warm core ring off Hudson Canyon. *Coast. Oceanogr. Climatol. News* 2: 3–4. (Obtainable from Center for Ocean Management Studies, University of Rhode Island, Kingston, RI 02881.)

Friedman, M.M. and Strickler, J.R. (1975) Chemoreceptors and feeding in calanoid copepods (Arthropoda: Crustacea). *Proc. Natl. Acad. Sci.* 72: 4185–4188.

Fuglister, F.C. (1960) *Atlantic Ocean Atlas,* Vol. 1. Woods Hole Oceanographic Institution.

Fulton, J.D. and LeBrasseur, R.J. (1985) Interannual shifting of the subarctic boundary and some of the biotic effects on juvenile salmonids, pp. 237–252. In: W.S. Wooster and D.L. Fluharty (Eds) *El Niño North.* Washington Sea Grant Program. University of Washington, Seattle.

Gade, H.G. and Edwards, A. (1980) Deep water renewal in fjords, pp. 453–490. In: H. J. Freeland, D.M. Farmer and C.D. Levings (Eds) *Fjord Oceanography.* Plenum Press, New York.

Gallegos, C.L. and Platt, T. (1982) Phytoplankton production and water motion in surface mixed layers. *Deep-Sea Res.* 29: 65–76.

Gammelsrød, T. (1992) Variation in shrimp abundance on the Sofala Bank, Mozambique, and its relation to the Zambesi River runoff. *Estuar. Coast. and Shelf. Sci.* 35: 91–103.

Gargett, A.E. (1984) Vertical eddy diffusivity in the ocean interior. *J. Mar. Res.* 42: 359–395.

Garrett, C.J.R. and Loder, J.W. (1981) Dynamical aspects of shallow-sea fronts. *Phil. Trans. Roy. Soc. Lond. A* 302: 562–581.

Garrett, C.J.R. and Loucks, R.H. (1976) Upwelling along the Yarmouth Shore of Nova Scotia. *J. Fish. Res. Bd. Can.* 33: 116–117.

Garvine, R.W. (1986) The role of brackish plumes in open shelf waters, pp. 47–65. In: S. Skreslet (Ed) *The Role of Freshwater Outflow in Coastal Marine Ecosystems.* Springer-Verlag, Berlin.

Garvine, R.W. and Monk, J.D. (1974) Frontal structure of a river plume. *J. Geophys. Res.* 79: 2251–2259.

Gavis, J. (1976) Munk and Riley revisited: nutrient diffusion transport and rates of phytoplankton growth. *J. Mar. Res.* 34: 161–179.

GEOSECS (1981) *Atlantic Expedition,* Vol. 2. Sections and Profiles. *Pacific Expedition,* Vol. 4, Sections and Profiles. National Science Foundation, Washington.

Gerard, V.A. (1982) In situ water motion and nutrient uptake by the giant kelp *Macrocystis pyrifera. Mar. Biol.* 69: 51–54.

Gerard, V.A. (1987) Hydrodynamic streamlining of *Laminaria saccharina* Lamour in response to mechanical stress. *J. Exp. Mar. Biol. Ecol.* 107: 237–244.

Gerard, V.A. and Mann, K.H. (1979) Growth and production of *Laminaria longicruris* (Phaeophyta) populations exposed to different intensities of water movement. *J. Phycol.* 15: 33–41.

Gerritsen, J. and Strickler, J.R. (1977) Encounter probabilities and community structure in zooplankton: a mathematical model. *J. Fish. Res. Board Can.* 34: 73–82.

Gill, A.E. (1982) *Atmosphere–Ocean Dynamics.* International Geophysics Series 30. Academic Press, New York. 662 pp.

Glantz, M.H. (1984) Floods, fires and famine: is El Niño to blame? *Oceanus* 27 (2): 14–19.

Glantz, M.H. (1985) Climate and fisheries: a Peruvian case study. CPPS Boletin ERFEN 15: 13–31.

Glémarec, M. (1973) The benthic community of the European North Atlantic continental shelf. *Oceanogr. Mar. Biol. Ann. Rev.* 11: 263–289.

Glover, R.S. (1967) The continuous plankton recorder survey of the North Atlantic. *Symp. Zool. Soc. Lond.* 19: 189–210.

Glynn, P.W. (1985) El Niño-associated disturbance to coral reefs and post-disturbance mortality by *Acanthaster planci. Mar. Ecol. Prog. Ser.* 26: 295–300.

Glynn, P.W. (1990) Coral mortality and disturbances to coral reefs in the tropical eastern Pacific, pp. 55–126. In: P.W. Glynn (Ed) *Global Ecological Consequences of the 1982–83 El Niño–Southern Oscillation.* Elsevier, Amsterdam.

Goldman, J.C. (1984a) Oceanic nutrient cycles, pp. 137–170. In: M.J.R. Fasham (Ed) *Flows of Energy and Materials in Marine Ecosystems: Theory and Practice.* Plenum Press, New York.

Goldman, J.C. (1984b) Conceptual role for microaggregates in pelagic waters. *Bull. Mar. Sci.* 35: 462–476.

Goldman, J.C. (1988) Spatial and temporal discontinuities of biological processes in surface waters, pp. 273–296. In: B.J. Rothschild (Ed) *Towards a Theory on Biological–Physical Interactions in the World Ocean.* Kluwer Academic, Amsterdam.

Goodrich, D.M. (1988) On meteorologically-induced flushing in three U.S. east-coast estuaries. *Estuar. Coast. Shelf Sci.* 26: 111–121.

Gordon, A.L. (1986) Interocean exchange of thermocline water. *J. Geophys. Res.* 91: 5037–5046.

Gower, J.F.R., Denman, K.L. and Holyer, R.J. (1980) Phytoplankton patchiness indicates the fluctuations spectrum of mesoscale oceanic structure. *Nature* 288: 157–159.

Graham, J.J. (1972) Retention of larval herring within the Sheepscot estuary of Maine. *Fish. Bull.* 70: 299–305.

Graham, N.E. and White, W.B. (1988) The El Niño cycle: a natural oscillator of the Pacific ocean–atmosphere system. *Science* 240: 1293–1302.

Gran, H.H. (1931) On the conditions for the production of plankton in the sea. *Rapp. P.-v. Réun. Cons. Int. Explor. Mer.* 75: 37–46.

Gran, H.H. and Braarud, T. (1935) A quantitative study of the phytoplankton in the Bay of Fundy and the Gulf of Maine (including observations on hydrography, chemistry and turbidity). *J. Biol. Bd. Canada* 1: 279–433.

Graybill, M.R. and Hodder, J. (1985) Effects of the 1982–83 El Niño on reproduction of six species of seabirds in Oregon. pp. 205–210. In: W.S. Wooster and D.L. Fluharty (Eds) *El Niño North: Niño Effects in the Eastern Subarctic Pacific Ocean.* Washington Sea Grant Program, University of Washington, Seattle

Greenberg, D.A. (1983) Modeling the mean barotropic circulation in the Bay of Fundy and Gulf of Maine. *J. Phys. Oceanogr.* 13: 886–904.

Grindley, J.R. (1964) On the effect of low-salinity water on the vertical migration of estuarine zooplankton. *Nature (Lond.)* 203: 781–782.

Gundersen, J.K. and Jørgensen, B.B. (1990) Microstructure of diffusive boundary layers and the oxygen uptake of the sea floor. *Nature* 345: 604–607.

Hachey, H.B. (1935) The effect of a storm on an inshore area with markedly stratified waters. *J. Biol. Board Can.* 1: 227–237.

Halpern, D. (1971) Observations on short-period internal waves in Massachusetts Bay. *J. Mar. Res.* 29: 116–132.

Hamilton, K. and Mysak, L.A. (1986) Possible effect of the Sitka Eddy on sockeye (*Onchorhynchus nerka*) and pink salmon (*Onchorhynchus gorbusha*) migration off southeast Alaska. *Can. J. Fish. Aquat. Sci.* 43: 498–504.

Haney, J.C. (1986) Seabird segregation at Gulf Stream frontal eddies. *Mar. Ecol. Prog. Ser.* 28: 279–285.

Hansen, J. and Lebedeff, S. (1988) Global surface air temperature: update through 1987. *Geophys. Res. Lett.* 15: 323–326.

Hansen, J., Fung, I., Lacis, A., Rind, D., Lebedeff, S., Ruedy, R., Russell, G. and Stone, P. (1988) Global climate changes as forecast by the Goddard Institute of Space Studies three dimensional model. *J. Geophys. Res.* 93: 9341–9364.

Hare, F.K. (1988) The global greenhouse effect. *Conference Proceedings: The Changing Atmosphere: Implications for Global Security.* WMO No. 710. World Meteorological Organization, Geneva. 483 pp.

Harris, G.P., Davies, P., Nunez, M. and Meyers, G. (1988) Interannual variability in climate and fisheries in Tasmania. *Nature* 333: 754–757.

Harris, G.P., Griffiths, F.B. and Clementson, L.A. (1992) Climate and fisheries off Tasmania—interactions of physics, food chains, and fish, pp. 585–597. In: A.I.L. Payne, K.H. Brink, K.H. Mann, and R. Hillborn (Eds). Benguela trophic functioning. *S. Afr. J. Mar. Sci.* 12.

Hartline, B.K. (1980) Coastal upwelling: physical factors feed fish. *Science* 208: 38–40.

Harvey, J. and Arhan, M. (1988) The water masses of the central North Atlantic in 1983–84. *J. Phys. Oceanogr.* 18: 1855–1875.

Harvey, J.G. and Theodorou, A. (1986) The circulation of the Norwegian Sea overflow in the eastern North Atlantic. *Oceanologica Acta* 9: 393–402.

Hays, J.D., Imbrie, J. and Shackelton, N.J. (1976) Variations in the earth orbit: pacemaker of the ice ages. *Science* 194: 1121–1131.

Hayward, T.L. and Venrick, E.L. (1982) Relation between surface chlorophyll, integrated chlorophyll and integrated primary production. *Marine Biology* 69: 247–252.

Heaps, N.S. (1980) A mechanism for local upwelling along the European continental slope. *Oceanologica Acta* 3: 449–454.

Hellerman, S. and Rosenstein, M. (1983) Normal monthly wind stress over the world ocean with error estimates. *J. Phys. Oceanogr.* 13: 1093–1104.

Hempel, G. (1982) The Canary current: studies of an upwelling system. *Rapp. P.-v. Réun. Cons. Int. Explor. Mer.* 180: 1–455.

Hendershott, M.C. (1981) Long waves and ocean tides, pp. 292–341. In: B.A. Warren and C. Wunch (Eds) *Evolution of Physical Oceanography*. M.I.T. Press, Cambridge, Mass.

Herbland, A. and Voituriez, B. (1979) Hydrological structure analysis for estimating the primary production in the tropical Atlantic Ocean. *J. Mar. Res.* 37: 87–101.

Herman, A.W., Sameoto, D.D. and Longhurst, A.R. (1981) Vertical and horizontal distribution patterns of copepods near the shelf-break south of Nova Scotia. *Can. J. Fish. Aquat. Sci.* 38: 1065–1076.

Hernández-León S. (1991) Accumulation of mesozooplankton in the wake area as a causative mechanism of the 'island-mass effect.' *Mar. Biol.* 109: 141–147.

Hill, D.K. (1995) Pacific warming unsettles ecosystems. *Science* 267: 1911–1912.

Hitchcock, G.L., Langdon, C. and Smayda, T.J. (1985) Seasonal variations in the phytoplankton biomass and productivity of a warm-core Gulf Stream ring. *Deep-Sea Res.* 32: 1287–1300.

Hitchcock, G.L., Langdon, C. and Smayda, T.J. (1987) Short-term changes in the biology of a Gulf Stream warm-core ring: phytoplankton biomass and productivity. *Limnol. Oceanogr.* 32: 919–928.

Hogg, N.G., Pickart, R.S., Hendry, R.M. and Smethie, W.J. (1986) The northern recirculation gyre of the Gulf Stream. *Deep-Sea Res.* 33: 1139–1165.

Holbrook, N.M., Denny, M.W. and Koehl, M.A.R. (1991) Intertidal 'trees': consequences of aggregation on the mechanical and photosynthetic properties of sea-palms, *Postelia palmaeformis* Ruprecht. *J. Exp. Mar. Biol. Ecol.* 146: 39–67.

Holligan, P.M. (1981) Biological implications of fronts on the northwest European continental shelf. *Phil. Trans. Roy. Soc. A* 302:547–562.

Holligan, P.M., Harris, R.P., Newell, R.C., Harbour, D.S., Head, R.N., Linley, E.A.S., Lucas, M.I., Tranter, P.R.G. and Weekly, C.M. (1984a) Vertical distribution and partitioning of organic carbon in mixed, frontal and stratified waters of the English Channel. *Mar. Ecol. Prog. Ser.* 14: 111–127.

Holligan, P.M., Williams, P.J. LeB., Purdie, D. and Harris R.P. (1984b) Photosynthesis, respiration and nitrogen supply of plankton populations in stratified, frontal and tidally mixed shelf waters. *Mar. Ecol. Prog. Ser.* 17: 201–213.

Holligan, P.M., Pingree, R.D. and Mardell, G.T. (1985) Oceanic solitons, nutrient pulses and phytoplankton growth. *Nature* 314: 348–350.

Holm-Hansen, O. (1985) Nutrient cycles in Antarctic marine ecosystems, pp. 6–10. In: W.R. Siegfried, P.R. Condy and R.M. Laws (Eds) *Antarctic Nutrient Cycles and Food Webs*. Springer-Verlag, Berlin. 700 pp.

Holm-Hansen, O., El-Sayed, S.Z., Franceschini, G.A. and Cuhel, R.L. (1977) Primary production and the factors controlling phytoplankton growth in the Southern Ocean, pp. 11–50. In: G.A. Llano (Ed) *Adaptations Within Antarctic Ecosystems*. Smithsonian Institution, Washington, DC. 1252 pp.

Honjo, S. and Manganini, S.J. (1993) Annual biogenic particle fluxes to the interior of the North Atlantic Ocean; studied at 34° N 21° W and 48° N 21° W. *Deep-Sea Res.* 40: 587–607.

Horel, J.D. and Wallace, J.M. (1981) Planetary-scale atmospheric phenomena associated with the southern oscillation. *Monthly Weather Rev.* 109: 813–829.

Horne, E.P.W. (1978) Physical aspects of the Nova Scotian shelfbreak fronts, pp. 59–86. In: M.J. Bowman and W.E. Esaias (Eds) *Oceanic Fronts in Coastal Processes*. Springer-Verlag, New York.

Horne, E.P.W., Loder, J.W., Harrison, W.G., Mohn, R., Lewis, M.R., Irwin, B. and Platt, T. (1989) Nitrate supply and demand at the Georges Bank tidal front. *Scientia Marina* 53 (2–3): 145–158.

Houghton, J.T., Jenkins, G.J. and Ephraums, J.J. (Eds) (1990) *Climate Change: The IPCC Assessment*. Cambridge University Press. Cambridge, UK. 365 pp.

Houghton, J.T., Callander, B.A. and Varney, S.K. (Eds) (1992) *Climate Change 1992: The Supplementary Report to the IPCC Scientific Assessment*. Cambridge University Press, Cambridge, UK. 97 pp.

Howarth, R.W., Butler, T., Lunde, K., Swaney, D. and Chu, C.R. (1993) Turbulence and planktonic nitrogen fixation: a mesocosm experiment. *Limnol. Oceanogr.* 38: 1696–1711.

Hunter, J.R. (1972) Swimming and feeding behaviour of larval anchovy, *Engraulis mordax. Fish. Bull. US* 73: 453–462.

Huntsman, S.A. and Barber, R.T. (1977) Primary production off northwest Africa: the relationship to wind and nutrient conditions. *Deep-Sea Res.* 24: 25–33.

Hutchings, J.A. and Myers, R.A. (1994) What can be learned from the collapse of a renewable resource? Atlantic cod, *Gadus morhua* of Newfoundland and Labrador. *Can. J. Fish. Aquat. Sci.* 51: 2126–2146.

Hutchinson, G.E. (1967) *A Treatise on Limnology,* Vol. 2. Wiley, New York. 1116 pp.

Huthnance, J.M. (1973) Tidal current asymmetries over the Norfolk sandbanks. *Estuar. Coast. Mar. Sci.* 1: 89–99.

Huyer, A. (1976) A comparison of upwelling events in two locations: Oregon and northwest Africa. *J. Mar. Res.* 34: 531–546.

Huyer, A. (1983) Coastal upwelling in the California system. *Progr. Oceanogr.* 12: 259–284.

Iles, T.D. and Sinclair, M. (1982) Atlantic herring: stock discreetness and abundance. *Science* 215: 627–633.

Imbrie, J. and Kipp, N.G. (1971) A new micropalaeontological method for quantitative palaeoclimatology: application to a late Pleistocene Caribbean core, pp. 77–181. In: K. K. Turekian (Ed) *The Late Cenozoic Glacial Ages.* Yale University Press, New Haven.

Iselin C. O'D. (1936) Preliminary report on long period variations in the transport of the Gulf Stream. *Pap Phys. Oceanogr. Meteorol.* 8: 1–40.

Isemer, H.J. and Hasse, L. (1987) *The Bunker Climate Atlas of the North Atlantic Ocean,* Vol. 2: Air–sea interactions. Springer-Verlag, Berlin.

Ivanoff, A. (1977) Oceanic absorption of solar energy, pp. 47–71. In: E.B. Kraus (Ed) *Modelling and Prediction of the Upper Layers of the Ocean.* Pergamon Press, New York.

Jackson, G.A. (1980) Phytoplankton growth and zooplankton grazing in oligotrophic oceans. *Nature* 284: 439–441.

Jackson, G.A. (1989) Simulation of bacterial attraction and adhesion to falling particles in an aquatic environment. *Limnol. Oceanogr.* 34: 514–530.

Jamart, B.M., Winter, D.F., Banse, K., Anderson, G.C. and Lam, R.K. (1977) A theoretical study of phytoplankton growth and nutrient distribution in the Pacific Ocean of the northwest U.S. coast. *Deep-Sea Res.* 24: 753–773.

Jamart, B.M., Winter, D.F. and Banse, K. (1979) Sensitivity analysis of a mathematical model of phytoplankton growth and nutrient distribution in the Pacific Ocean off the northwest US coast. *J. Plankt. Res.* 1: 267–290.

Jenkins, W.J. (1982) Oxygen utilization rates in North Atlantic subtropical gyre and primary production in oligotrophic systems. *Nature* 300: 246–248.

Jenkins, W.J. and Goldman, J.C. (1985) Seasonal oxygen cycling and primary production in the Sargasso Sea. *J. Mar. Res.* 43: 465–491.

Jerlov, N.J. (1976) *Marine Optics.* Elsevier Oceanography Series 14. Elsevier, Amsterdam.

Jones, B.H. and Halpern, D. (1981) Biological and physical aspects of a coastal upwelling event observed during March–April 1974 off northwest Africa. *Deep-Sea Res.* 28A: 71–81.

Jones, P.D., Wigley, T.M.L., Folland, C.K., Parker, D.E., Angell, J.K., Lebedeff, S. and Hansen, J.E. (1988) Evidence for global warming in the past decade. *Nature* 332: 790.

Jørgensen, B.B. and Revsbech, N.P. (1985) Diffusive boundary layers and the oxygen uptake of sediments and detritus. *Limnol. Oceanogr.* 30: 111–122.

Jumars, P.A., Deming, J.W., Hill, P.S., Karp-Boss, L., Yager, P.L. and Dade, W.B. (1993) Physical constraints on marine osmotrophy in an optimal foraging context. *Mar. Microb. Food Webs* 7: 121–159.

Jury, M.R. and Brundrit, G.B. (1992) Temporal organization of upwelling in the Southern Benguela ecosystem by resonant coastal trapped waves in the ocean and atmosphere. *S. Afr. J. Mar. Sci.* 12: 219–224.

Kamykowsky, D., McCollum, S.A. and Kirkpatrick, G.J. (1988) Observations and a model concerning the translational velocity of a photosynthetic marine dinoflagellate under variable environmental conditions. *Limnol. Oceanogr.* 33: 57–65.

Karl, D.M., Knauer, G.A. and Martin, J.H. (1988) Downward flux of particulate organic matter in the ocean: a particle decomposition paradox. *Nature* 332: 438–441.

Karl, D.M., Leteller, R., Hebel, D., Tupas, L., Dore, J., Christian, J. and Winn, C. (1995) Ecosystem changes in the North Pacific subtropical gyre attributed to the 1991–92 El Niño. *Nature* 373: 230–234.

Kawasaki, T. (1992) Mechanisms governing fluctuations in pelagic fish populations, pp. 873–879. In: A.I.L. Payne, K.H. Brink, K.H. Mann and R. Hilborn (Eds) Benguela trophic functioning. *S. Afr. J. Mar. Sci.* 12.

Kepkay, P.E. (1994) Particle aggregation and the biological reactivity of colloids. *Mar. Ecol. Prog. Ser.* 109: 293–304.

Kerr, R.A. (1983) Orbital variation–ice age link strengthened. *Science* 219: 272–274.

Kerr, R.A. (1985) Long-lived small eddies are criss-crossing the oceans carrying the effects of local mixing hundreds and even thousands of kilometres. *Science* 230: 793.

Kerr, R.A. (1989) Greenhouse skeptic out in the cold. *Science* 246: 1118–1119.

Kessler, W.S. and McPhaden, M.J. (1995) Oceanic equatorial waves and the 1991–93 El Niño. *J. Climate* 8: 1757–1774.

Kidd, R. and Sander, F. (1979) Influence of the Amazon River discharge on the marine production system off Barbados, West Indies. *J. Mar. Res.* 37: 669–681.

Killworth, P.D. (1977) Mixing on the Weddell Sea continental slope. *Deep-Sea Res.* 24: 427–448.

Kinder, T.H. and Bryden, H.L. (1988) Gibraltar experiment. Woods Hole Oceanogr. Inst. Tech. Rept. WHOI-88-30.

King, F.D. and Devol, A.H. (1979) Estimates of vertical eddy diffusion through the thermocline from phytoplankton nitrate uptake rates in the mixed layer of the eastern tropical Pacific. *Limnol. Oceanogr.* 24: 645–651.

King, F.D., Cucci, T.L. and Townsend, D.W. (1987) Microzooplankton and macrozooplankton glutamate dehydrogenase as indices of the relative contribution of these fractions to ammonium regeneration in the Gulf of Maine. *J. Plankt. Res.* 9: 277–289.

Kingsford, M.J. and Choat, J.H. (1986) Influence of surface slicks on the distribution and inshore movement of small fish. *Mar. Biol.* 91: 161–171.

Kjelson, M.A., Raquel, P.F. and Fisher, F.W. (1982) Life history of fall-run juvenile Chinook salmon, *Oncorhynchus tshawytscha,* in the Sacramento–San Joaquin estuary, California, pp. 393–411. In: V.S. Kennedy (Ed) *Estuarine Comparisons.* Academic Press, New York.

Kleckner, R.C. and McCleave, J.D. (1985) Spatial and temporal distribution of American eel larvae in relation to North Atlantic Ocean current systems. *Dana* 4: 67–92.

Kleckner, R.C. and McCleave, J.D. (1988) The northern limit of spawning by Atlantic eels (*Anguilla* spp.) in the Sargasso Sea in relation to thermal fronts and surface water masses. *J. Mar. Res.* 46: 647–667.

Klein, B. and Siedler, G. (1989) On the origin of the Azores current. *J. Geophys. Res.* 94: 6159–6168.

Koehl, M.A.R. (1984) Mechanisms of particle capture by copepods at low Reynolds numbers: possible modes of selective feeding, pp. 135–166. In: D.G. Meyers and J.R. Strickler (Eds) *Trophic Interaction within Aquatic Ecosystems.* American Association for the Advancement of Science. Washington, D.C.

Koehl, M.A.R. (1986) Seaweeds in moving water: form and mechanical function, pp. 603–634. In: T.J. Givnish (Ed) *On the Economy of Plant Form and Function.* Cambridge University Press, Cambridge. 717 pp.

Koehl, M.A.R. and Alberte, R.S. (1988) flow, flapping and photosynthesis of *Nereocystis leutkeana*: a functional comparison of undulate and flat blade morphologies. *Mar. Biol.* 99: 435–444.

Koehl, M.A.R. and Strickler, J.R. (1981) Copepod feeding currents: food capture at low Reynolds number. *Limnol. Oceanogr.* 26: 1062–1073.

Krauss, W. (1986) The North Atlantic current. *J. Geophys. Res.* 91: 5061–5074.

Krauss, W. (1993) Ekman drift in homogeneous water. *J. Geophys. Res.* 98: 20187–20209.

Kremer, J.N. and Nixon, S.W. (1978) *A Coastal Marine Ecosystem: Simulation and Analysis.* Springer-Verlag, New York. 217 pp.

Kumar, N., Anderson, R.F., Mortlock, R.A., Froelich, P.N., Kubik, P., Dittrich-Hannen, B. and Suter, M. (1995) Increased biological productivity and export production in the glacial southern ocean. *Nature* 378: 675–680.

Lamb, H.H. (1969) The new look of climatology. *Nature* 223: 1209–1215.

Lande, R. and Lewis, M.R. (1989) Models of photoadaptation and photosynthesis by algal cells in the turbulent mixed layer. *Deep-Sea Res.* 36: 1161–1175.

Lande, R., Li, W.K., Horne, E.P.W. and Wood, A.M. (1989) Phytoplankton growth rates estimated from depth profiles of cell concentration and turbulent diffusion. *Deep-Sea Res.* 36: 1141–1159.

Lande, R. and Yentsch, C.S. (1988) Internal waves, primary production and the compensation depth of marine phytoplankton. *J. Plankt. Res.* 10: 565–571.

Landry, F.P., Miller, T.J. and Leggett, W.C. (1995) The effects of small-scale turbulence on the ingestion rate of fathead minnow (*Pimephales promelas*) larvae. *Can. J. Fish. Aquat. Sci.* 52: 1714–1719.

Lange, C.B., Burke, S.K. and Berger, W.H. (1990) Biological production off southern California is linked to climate change. *Clim. Change* 16: 319–329.

Lasker, R. (1975) Field criteria for the survival of anchovy larvae: the relation between inshore chlorophyll maximum layers and successful first feeding. *Fish. Bull. U.S.* 73: 847–855.

Lasker, R. (1978) The relation between oceanographic conditions and larval anchovy food in the California current: identification of factors leading to recruitment failure. *Rapp. P.-v. Réun. Cons. Int. Explor. Mer.* 173: 212–230.

Lasker, R. (1988) Food chains and fisheries: an assessment after 20 years, pp. 173–182. In: B.J. Rothschild (Ed) *Towards a Theory on Biological–Physical Interactions in the World Ocean.* Kluwer, Dordrecht.

Laurs, R.M., Yuen, H.S.H. and Johnson, J.H. (1977) Small-scale movements of albacore, *Thunnus alalunga*, in relation to ocean features as indicated by ultrasonic tracking and oceanographic sampling. *Fish. Bull. US* 75: 347–355.

Lawler, J.P.,Weinstein, M.P., Chen, H.Y. and Englert, T.L. (1988) Modeling of physical and behavioral mechanisms influencing recruitment of spot and Atlantic croaker to the Cape Fear estuary. *Am. Fish. Soc. Symp.* 3: 115–131.

Laws, E.A., DiTullio, G.R. and Redalje, D.G. (1987) High phytoplankton growth and production rates in the North Pacific subtropical gyre. *Limnol. Oceanogr.* 32: 905–918.

Lazier, J.R.N. (1980) Oceanographic conditions at ocean weather ship Bravo, 1964–1974. *Atmosph.–Ocean* 18: 227–238.

Lazier, J.R.N. (1988) Temperature and salinity changes in the deep Labrador Sea, 1962–1986. *Deep-Sea Res.* 35: 1247–1253.

Lazier, J.R.N. and Mann, K.H. (1989) Turbulence and diffusive layers around small organisms. *Deep-Sea Res.* 36: 1721–1733.

Le Fèvre, J. (1986) Aspects of the biology of frontal systems. *Adv. Mar. Biol.* 23: 164–299.

Le Fèvre, J. and Grall, J.R. (1970) On the relationships of *Noctiluca* swarming off the western coast of Brittany with hydrological features and plankton characteristics of the environment. *J. Exp. Mar. Biol. Ecol.* 4: 287–306.

Le Fèvre, J. and Frontier, S. (1988) Influence of temporal characteristics of physical phenomena on plankton dynamics, as shown by northwest European marine ecosystems, pp. 245–272. In: B.J. Rothschild (Ed) *Towards a Theory on Biological–Physical Interactions in the World Ocean*. Kluwer, Dordrecht.

LeBlond, P.H., Hickey, B.M. and Thompson, R.E. (1986) Runoff-driven coastal flow off British Columbia, pp. 309–318. In: S. Skreslet (Ed) *The Role of Freshwater Outflow in Coastal Marine Systems*. Springer-Verlag, New York.

Lee, T.N., Atkinson, L.P. and Legeckis, R. (1981) Observations on a Gulf Stream frontal eddy on the Georgia continental shelf, April 1977. *Deep-Sea Res.* 28: 347–378.

Leetma, A., McCreary, J.P. and Moore, D.W. (1981) Equatorial currents: observations and theory, pp. 186–196. In: B.A. Warren and C. Wunch (Eds) *Evolution of Physical Oceanography*. M.I.T. Press, Cambridge, Mass.

Legendre, L. (1981) Hydrodynamic control of marine phytoplankton production: the paradox of stability, pp. 191–207. In: J.C.J. Nihoul (Ed) *Ecohydrodynamics. Proceedings of the 12th International Liège Colloquium on Ocean Hydrodynamics*. Elsevier, Amsterdam.

Legendre, L. and Demers, S. (1984) Towards dynamic biological oceanography and limnology. *Can. J. Fish. Aquat. Sci.* 41: 2–19.

Legendre, L. and Demers, S. (1985) Auxiliary energy, ergoclines and aquatic biological production. *Naturaliste Can. (Rev. écol. syst.)* 112: 5–14.

Legendre, L., Demers, S. and LeFaivre, D. (1986) Biological production at marine ergoclines, pp. 1–54. In: J.C.J. Nihoul (Ed) *Marine Interfaces Ecohydrodynamics*. Elsevier, Amsterdam.

Legrand, M., Feniet-Saigne, C., Saltzman, E.S., Germain, C., Barkov, N.I. and Petrov, V.N. (1991) Ice-core record of oceanic emissions of dimethylsulphide during the last climate cycle. *Nature* 350: 144–146.

Lehman, J.T. and Scavia, D. (1982a) Microscale patchiness of nutrients in plankton communities. *Science* 216: 729–730.

Lehman, J.T. and Scavia, D. (1982b) Microscale nutrient patches produced by zooplankton. *Proc. Natl. Acad. Sci.* 79: 5001–5005.

Lewis, M.R. and Platt, T. (1982) Scales of variability in estuarine ecosystems, pp. 3–20. In: V.S. Kennedy (Ed) *Estuarine Comparisons*. Academic Press, New York.

Lewis, M.R., Cullen, J.J. and Platt, T. (1983) Phytoplankton and thermal structure in the upper ocean: consequences of nonuniformity in chlorophyll profile. *J. Geophys. Res.* 88 (C4): 2565–2570.

Lewis, M.R., Cullen, J.J. and Platt, T. (1984a) Relationships between vertical mixing and photoadaptation of phytoplankton: similarity criteria. *Mar. Ecol. Prog. Ser.* 15: 141–149.

Lewis, M.R., Horne, E.P.W., Cullen, J.J., Oakey, N.S. and Platt, T. (1984b) Turbulent motions may control phytoplankton photosynthesis in the upper ocean. *Nature* 311: 49–50.

Lewis, M.R., Harrison, W.G., Oakey, N.S., Hebert, D. and Platt, T. (1986) Vertical nitrate fluxes in the oligotrophic ocean. *Science* 234: 870–873.

Lindeboom, H., Van Raaphorst, W., Beukema, J., Cadée, G. and Swennen, C. (In press) Sudden changes in the North Sea and Wadden Sea: oceanic influences underestimated? *ICES Symposium*.

Lindstrom, S.S., Qian, X. and Watts, D.R. (Submitted) Vertical motion in the Gulf Stream and its relation to meanders. *J. Geophys. Res.*

Loder, J.W. (1980) Topographic rectification of tidal currents on the sides of George's Bank. *J. Phys. Oceanogr.* 10: 1399–1416.

Loder, J.W., Brickman, D. and Horne, E.P.W. (1992) Detailed structure of currents and hydrography on the northern side of Georges bank. *J. Geophys. Res.* 97: 14331–14342.

Loder, J.W. and Greenberg, D.A. (1986) Predicted positions of tidal fronts in the Gulf of Maine region. *Continental Shelf Res.* 6: 397–414.

Loder, J.W. and Platt, T. (1985) Physical controls on phytoplankton production at tidal fronts, pp. 3–21. In: P.E. Gibbs (Ed) *Proceedings of the 19th European Marine Biology Symposium.* Cambridge University Press, Cambridge, UK.

Loder, J.W. and Wright, D.G. (1985) Tidal rectification and frontal circulation on the sides of George's Bank. *J. Mar. Res.* 43: 581–604.

Long, R.R. (1954) Some aspects of the flow of stratified fluids II. Experiments with a two-fluid system. *Tellus* 6: 97–115.

Longhurst, A.R. (1976) Interactions between zooplankton and phytoplankton profiles in the eastern tropical Pacific Ocean. *Deep-Sea Res.* 23: 729–754.

Longhurst, A.R. (1981) Significance of spatial variability, pp. 415–441. In: A.R. Longhurst (Ed) *Analysis of Marine Ecosystems.* Academic Press, London. 741 pp.

Longhurst, A.R. (1995) Seasonal cycles of pelagic production and consumption. *Progress in Oceanography.* 36: 77–167.

Longhurst, A.R. and Harrison, W.C. (1989) The biological pump: profiles of plankton production and consumption in the upper ocean. *Prog. Oceanogr.* 22: 47–123.

Longhurst, A.R. and Herman, A.W. (1981) Do oceanic zooplankton aggregate at, or near, the deep chlorophyll maximum? *J. Mar. Res.* 39: 353–356.

Longhurst, A., Sathyendranath, S., Platt, T. and Caverhill, C. (1995) An estimate of global primary production in the ocean from satellite radiometer data. *J. Plankt. Res.* 17: 1245–1271.

Longhurst, A.R. and Williams, R. (1979) Materials for plankton modelling: vertical distribution of Atlantic zooplankton in summer. *J. Plankt. Res.* 1: 1–28.

Lorius, C., Barkov, N.I., Jouzel, J., Korotkevich, Y.S., Kolyakov, V.M. and Raynaud, D. (1988) Antarctic ice core: CO_2 and climatic change over the last climatic cycle. EOS. *Trans. Am. Geophys. Union* 69: 681–684.

Lough, R.G., Valentine, P.C., Potter, D.C., Auditore, P.J., Bolz, G.R., Neilson, J.D. and Perry, R.I. (1989) Ecology and distribution of juvenile cod and haddock in relation to sediment type and bottom currents on eastern Georges Bank. *Mar. Ecol. Prog. Ser.* 56: 1–12.

Lovelock, J.E. (1979) *Gaia.* Oxford University Press, Oxford.

MacIntyre, S., Alldredge, A.L. and Gotschalk, C.C. (1995) Accumulation of marine snow at density discontinuities in the water column. *Limnol. Oceanogr.* 40: 449–468.

MacIsaac, J.J., Dugdale, R.C., Barber, R.T., Blasco, D. and Packard, T.T. (1985) Primary production cycle in an upwelling center. *Deep Sea Res.* 32: 503–529.

MacKenzie, B.R. and Leggett, W.C. (1991) Quantifying the contribution of small-scale turbulence to the encounter rates between larval fish and their zooplankton prey: effect of wind and tide. *Mar. Ecol. Prog. Ser.* 73: 149–160.

MacKenzie, B.R., Miller, T.J., Cry, S. and Leggett, W.C. (1994) Evidence for a dome-shaped relationship between turbulence and larval fish ingestion rates. *Limnol. Oceanogr.* 39: 1790–1799.

Malone, T.C. (1982) Factors influencing the fate of sewage-derived nutrients in the lower Hudson estuary and New York Bight, pp. 301–320. In: G.F. Mayer (Ed) *Ecological Stress in the New York Bight: Science and Management.* Estuarine Research Foundation. Columbia, South Carolina.

Malone, T.C. (1984) Anthropogenic nitrogen loading and assimilation capacity of the Hudson River estuarine system, USA, pp. 291–311. In: V.S. Kennedy (Ed) *The Estuary as a Filter.* Academic Press, Orlando, Florida.

Mann, C.R. (1967) The termination of the Gulf Stream and the beginning of the North Atlantic current. *J. Geophys. Res.* 14: 337–359.

Mann, K.H. (1984) Fish production in open ocean ecosystems, pp. 435–458. In: M.J.R. Fasham (Ed) *Flows of Energy and Materials in Marine Ecosystems: Theory and Practice.* Plenum Press, New York.

Mann, K.H. (1988) Towards predictive models for coastal marine ecosystems, pp. 291–316. In: L.R. Pomeroy and J.J. Alberts (Eds) *Concepts of Ecosystem Ecology.* Springer-Verlag, New York.

Mann, K.H. and Drinkwater, K.F. (1994) Environmental influences on fish and shellfish production in the Northwest Atlantic. *Environ. Rev.* 2: 16–32.

Marcogliese, D.J. and Cone, D.K. (1993) What metazoan parasites tell us about the evolution of American and European eels. *Evolution* 47: 1632–1635.

Margalef, R. (1974) Asociación o exclusion en lo distribucion des especies del mimo género en algas unicelulares. *Mem. R. Acad. Cienc. Artes Barcelona* 42: 353–372.

Margalef, R. (1978a) Life-forms of phytoplankton as survival alternatives in an unstable environment. *Oceanologica Acta* 1: 493–509.

Margalef, R. (1978b) What is an upwelling system? pp. 12–14. In: R. Boje and M. Tomczak (Eds) *Upwelling Ecosystems.* Springer-Verlag, Berlin.

Markle, D.F., Scott, W.B. and Kohler, A.C. (1980) New and rare records of Canadian fishes and the influence of hydrography on resident and nonresident Scotian Shelf ichthyofauna. *Can. J. Fish. Aquat. Sci.* 37: 49–65.

Marra, J. (1978) Effect of short-term variations in light intensity on photosynthesis of a marine phytoplankter: a laboratory simulation study. *Mar. Biol.* 46: 191–202.

Marshall, S.M. and Orr, A.P. (1928) The photosynthesis of diatom cultures in the sea. *J. Mar. Biol. Assoc.* 15: 321–364.

Martin, J.H., Knauer, G.A., Karl, D.M. and Broenkow, W.W. (1987) VERTEX: carbon cycling in the northeast Pacific. *Deep-Sea Res.* 34: 267–285.

Martin, J.H. *et al.* (1994) Testing the iron hypothesis in ecosystems of the equatorial Pacific ocean. *Nature* 371: 123–129.

Mathisen, O.A., Thorne, R.E., Trumble, R.J. and Blackburn, M. (1978) Food consumpion of pelagic fish in an upwelling area, pp. 111–123. In: R. Boje and M. Tomczak (Eds) *Upwelling Ecosystems*. Springer-Verlag, Berlin. 303 pp.

May, R.M. and Oster, G.F. (1976) Bifurcations and dynamic complexity in simple ecological models. *Am. Nat.* 110: 573–599.

Mazé, R. (1983) Formation d'ondes internes induits dans un golfe par le passage d'un dépression et par la marée. Application au Golfe de Gascogne. *Ann. Hydrograph.* 8: 45–58.

Mazé, R., Camus, Y. and Le Tareau, J.Y. (1986) Formation de gradients thermiques à la surface de l'ocean, au-dessus d'un talus, par interaction entre les ondes internes et le mélange dû au vent. *J. Cons. Perm. Int. Explor. Mer.* 42: 221–240.

McCarthy, J.J. and Goldman, J.C. (1979) Nitrogenous nutrition of marine phytoplankton in nutrient-depleted waters. *Science* 203: 670–672.

McCarthy, J.J. and Nevins, J.L. (1986) Sources of nitrogen for primary production in warm-core rings 79-E and 81-D. *Limnol. Oceanogr.* 31: 690–700.

McCleave, J.D., Kleckner, R.C. and Castonguay, M. (1987) Reproductive sympatry of American and European eels and the implications for migration and taxonomy, pp. 286–297. In: M.J. Dadswell, R.J. Klauda, C.M. Moffitt, R.L. Saunders, R.A. Rulifson and J.E. Cooper (Eds) Common strategies of anadromous and catadromous fish. *Am. Fish. Soc.* Bethesda, Md.

McClimans, T.A. (1986) Laboratory modelling of dynamic processes in fjords and shelf waters, pp. 67–84. In: S. Skreslet (Ed) *The Role of Freshwater Outflow in Coastal Marine Ecosystems.* Springer-Verlag, New York.

McCreary, J.P. and Anderson, D.L.T. (1991) An overview of coupled ocean–atmosphere models of El Niño and the Southern Oscillation. *J. Geophys. Res.* 96: 3125–3150.

McGillicuddy, D.J. (1995) One-dimensional numerical simulation of primary production: Lagrangian and Eulerian formulations. *J. Plankt. Res.* 17: 405–412.

McGowan, J.A. (1985) El Niño 1983 in the southern California bight, pp. 166–184. In: W.S. Wooster and D.L. Fluharty (Eds) *El Niño North: Niño Effects in the Eastern Subarctic Pacific Ocean.* Washington Sea Grant Program, University of Washington, Seattle.

McGowan, J.A. and Hayward, T.L. (1978) Mixing and oceanic productivity. *Deep-Sea Res.* 25: 771–793.

McNider, R.T., Schroeder, W.W., Goodrick, S.L. and Davis, A.M. (Submitted) On the relationship between the katabatic wind system of the Antarctic and biological productivity in the Southern Ocean. *Science.*

McPhaden, M.J. (1993) TOGA-TAO and the 1991–93 El Niño–Southern Oscillation event. *Oceanography* 6: 36–44.

Menzel, D.W. and Ryther, J.H. (1960) The annual cycle in primary production in the Sargasso Sea off Bermuda. *Deep-Sea Res.* 6: 351–367.

Mertz, G. and Myers, R.A. (1994) The ecological impact of the Great Salinity Anomaly in the northern North–West Atlantic. *Fish. Oceanogr.* 3: 1–14.

Micklin, P.P. (1988) Desiccation of the Aral Sea: a water management disaster in the Soviet Union. *Science* 241: 1170–1176.

Miller, A.J., Cayan, D.R., Barnett, T.P., Graham, N.E. and Oberhuber, J.M. (1994) The 1976–77 climate shift of the Pacific Ocean. *Oceanography* 7: 21–26.

Miller, C.B., Batchelder, H.P., Brodeur, R.D. and Pearcy, W.G. (1985) Response of zooplankton and ichthyoplankton off Oregon to the El Niño event of 1983, pp. 185–187. In: W.S. Wooster and D.L. Fluharty (Eds) *El Niño North: Niño Effects in the Eastern Subarctic Pacific Ocean.* University of Washington, Seattle.

Miller, G.R. (1966) The flux of tidal energy out of the deep oceans. *J. Geophys. Res.* 71: 2485–2489.

Milliman, J.D. and Boyle, E. (1975) Biological uptake of dissolved silica in the Amazon River estuary. *Science* 189: 995–997.

Mills, E.L. (1989) *Biological Oceanography: An Early History, 1870–1960.* Cornell University Press, Ithaca, NY. 378 pp.

Minas, H.J., Codispoti, L.A. and Dugdale, R.C. (1982) Nutrients and primary production in the upwelling region off northwest Africa. *Rapp. P.-v. Réun. Cons. Int. Explor. Mer.* 180: 148–183.

Minas, H.J., Minas, M. and Packard, T.T. (1986) Productivity in upwelling areas deduced from hydrographic and chemical fields. *Limnol. Oceanogr.* 31: 1182–1206.

Mitchell, J.F.B. (1989) The 'greenhouse' effect and climate change. *Rev. Geophys.* 27: 115–139.

Mitchell, J.F.B., Johns, T.C., Gregory, J.M. and Tett, S.F.B. (1995) Climate response to increasing levels of greenhouse gases and sulphate aerosols. *Nature* 376: 501–504.

Mix, A.C. (1989) Influence of productivity variations on long-term atmospheric CO_2. *Nature* 337: 541–544.

Mommaerts, J.P., Pichot, G., Ozer, J., Adam, Y. and Baeyens, W. (1984) Nitrogen cycling and budget in Belgian coastal waters: North Sea areas with and without river inputs. *Rapp. P.-v. Réun. Cons. Int. Explor. Mer.* 183: 57–69.

Mooers, C.N.K., Flagg, C.N. and Boicourt, W.C. (1978) Prograde and retrograde fronts, pp. 43–58. In: M.J. Bowman and W.E. Esaias (Eds) *Oceanic Fronts in Coastal Processes.* Springer-Verlag, New York.

Morin, P., Wafar, M.V.M. and Le Corré, P. (1993) Estimation of nitrate flux in a tidal front from satellite-derived temperature data. *J. Geophys. Res.* 98 (C3): 4689–4695.

Muck, P., Rojas de Mendiola, B. and Antonietti, E. (1989) Comparative studies on feeding in the larval anchoveta (*Engraulis ringens*) and sardine (*Sardinops sagax*), pp. 86–96. In: D. Pauly, P. Muck, J. Mendo and I. Tsukayama (Eds) *The Peruvian Upwelling Ecosystem: Dynamics and Interactions.* International Center for Living Resources, Manila, the Philippines.

Muck, P. and Sanchez, G. (1987) The importance of mackerel and horse mackerel predation for the Peruvian anchoveta stock (a population and feeding model), pp. 279–293. In: D. Pauly and I. Tsukayama (Eds) *The Peruvian Anchoveta and Its Upwelling System: Three Decades of Change.* International Center for Living Resources Management, Manila, the Philippines.

Muelbert, J.H., Luis, M.R. and Kelley, D.E. (1994) The importance of small-scale turbulence in the feeding of herring larvae. *J. Plankt. Res.* 16: 927–944.

Muench, R.D. and Heggie, D.T. (1978) Deep water exchange in Alaskan subarctic fjords, pp. 239–267. In: B. Kjerve (Ed) *Estuarine Transport Processes.* University of South Carolina Press.

Müller-Navara, S. and Mittelstaedt, E. (1985) *Schadstoffbereitung und Schadstoffbelastung in der Nordsee. Eine Medellstudie.* Deutsches Hydrographisches Institut, Hamburg, 1–50.

Munk, P. (1993) Differential growth of larval sprat *Sprattus sprattus* across a tidal front in the North Sea. *Mar. Ecol. Prog. Ser.* 99: 17–27.

Munk, P. and Christiansen, V. (1990) Larval growth and drift pattern and the separation of spawning groups in the North Sea. *J. Fish. Biol.* 37: 135–148.

Munk, W. and Riley, G.A. (1952) Absorption of nutrients by aquatic plants. *J. Mar. Res.* 11: 215–240.

Muschenheim, D.K. (1987a) The dynamics of near-bed seston flux and suspension-feeding benthos. *J. Mar. Res.* 45: 473–496.

Muschenheim, D.K. (1987b) The role of hydrodynamic sorting of seston in the nutrition of a benthic suspension feeder, *Spio setosa* (Polychaeta: Spionidae). *Biol. Oceanogr.* 4: 265–288.

Myers, R.A. and Drinkwater, K. (1989) The influence of Gulf Stream warm core rings on recruitment of fish in the northwest Atlantic. *J. Mar. Res.* 47: 635–656.

Myers, R.A., Drinkwater, K.F., Barrowman, N.J. and Baird, J.W. (1993a) Salinity and recruitment of Atlantic cod (*Gadus morhua*) in the Newfoundland region. *Can. J. Fish. Aquat. Sci.* 50: 1599–1609.

Myers, R.A., Mertz, G. and Bishop. C.A. (1993b) Cod spawning in relation to physical and biological cycles of the northern Northwest Atlantic. *Fish. Oceanogr.* 2: 154–165.

Mysak, L.A. (1986) El Niño, interannual variability and fisheries in the northeast Pacific Ocean. *Can. J. Fish. Aquat. Sci.* 43: 464–497.

Namias, J. and Cayan, D.R. (1984) El Niño: Implications for forecasting. *Oceanus* 27 (2): 41–47.

Narimousa, S. and Maxworthy, T. (1985) Two-layer model of shear-driven coastal upwelling in the presence of bottom topography. *J. Fluid. Mech.* 159: 503–531.

Narimousa, S. and Maxworthy, T. (1987) Coastal upwelling on a sloping bottom: the formation of plumes, jets and pinched-off cyclones. *J. Fluid. Mech.* 176: 169–190.

Nelson, D.M., McCarthy, J.J., Joyce, T.M. and Ducklow, H.W. (1989) Enhanced near-surface nutrient availability and new production resulting from the frictional decay of a Gulf Stream warm-core ring. *Deep-Sea Res.* 36: 705–714.

Nelson, G. (1992) Equatorward wind and atmospheric pressure spectra as metrics for primary productivity in the Benguela system. *S. Afr. J. Mar. Sci.* 12: 19–28.

New, A.L. (1988) Internal tidal mixing in the Bay of Biscay. *Deep-Sea Res.* 35: 691–697.

Nichols, F.H. (1985) Increased benthic grazing: an alternative explanation for low phytoplankton biomass in northern San Francisco Bay during the 1976–77 drought. *Estuar. Coast. Shelf Sci.* 21: 379–388.

Nichols, F.H., Cloern, J.E., Luoma, S.N. and Peterson, D.H. (1986) The modification of an estuary. *Science* 231: 567–573.

Norcross, J.J. and Stanley, E.M. (1967) Inferred surface and bottom drift, June 1963 through October 1964. Circulation of Shelf waters off the Chesapeake bight. *Prof. Pap. Environ. Sci. Serv. Admin.* 3: 11–42.

Nunes Vas, R.A. (1994) Turbulence closure model of estuarine stratification. *J. Geophys. Res.* 99: 16143–16160.

O'Boyle, R.M., Sinclair, M., Conover, R.J., Mann, K.H. and Kohler, A.C. (1984) Temporal and spatial distribution of ichthyoplankton communities of the Scotian Shelf in relation to biological, hydrological and physiographic features. *Rapp. P.-v. Reun. Cons. Int. Explor. Mer.* 183: 27–40.

O'Reilly, J.E., Evans-Zetlin, C. and Thomas, J.P. (1981) The relationship between surface and average water column concentrations of chlorophyll-a in northwestern Atlantic shelf water. ICES CM 1981/L: 17, 19 pp. (mimeo).

O'Reilly, J.E. and Busch, D.A. (1984) Phytoplankton primary production on the northwest Atlantic shelf. *Rapp. P.-v. Reun. Cons. Int. Explor. Mer.* 183: 255–268.

Oakey, N.S. and Elliott, J.A. (1980) Dissipation in the mixed layer near Emerald Basin, pp. 123–133. In: J.C.L. Nihoul (Ed) *Marine Turbulence.* Elsevier, Amsterdam, 270 pp.

Odum, H.T. (1967a) Biological circuits and marine systems of Texas, pp. 99–157. In: T.A. Olson and F.J. Burgess (Eds) *Pollution and Marine Ecology.* Wiley, New York.

Odum, H.T. (1967b) Energetics of world food production, pp. 55–94. In: I.L. Bennett (chairman). *The World Food Problem,* a report of the President's Science Advisory Committee, Vol. 3. The White House, Washington, D.C.

Okubo, A. (1971) Horizontal and vertical mixing in the sea, pp. 89–168. In: D.W. Hood (Ed) *Impingement of Man on the Oceans.* Wiley-Interscience, New York.

Okubo, A. (1987) Fantastic voyage into the deep: marine biofluid mechanics, pp. 32–47. In: E. Teromoto and M. Yamaguti (Eds) *Mathematical Topics in Population Biology, Morphogenesis and Neurosciences.* Springer-Verlag, New York.

Oliver, J.K. and Willis, B.L. (1987) Coral-spawn slicks in the Great Barrier Reef: preliminary observations. *Mar. Biol.* 94: 521–529.

Olson, D.B. (1986) Lateral exchange with Gulf Stream warm-core ring surface layers. *Deep-Sea Res.* 33: 1691–1704.

Olson, D.B., Schott, F.A., Zantopp, R.J. and Leaman, K.D. (1984) The mean circulation east of the Bahamas as determined from a recent measurement program and historical XBT data. *J. Phys. Oceanogr.* 14: 1470–1487.

Ortner, P.B., Wiebe, P.H. and Cox, J.L. (1980) Relationships between oceanic epizooplankton distributions and the seasonal deep chlorophyll maximum in the northwest Atlantic Ocean. *J. Mar. Res.* 38: 507–531.

Ortner, P.B., Wiebe, P.H. and Cox, J.L. (1981) Reply to 'Do oceanic zooplankton aggregate at, or near, the deep chlorophyll maximum?' *J. Mar. Res.* 39: 357–359.

Osborn, T.R. (1978) Measurements of energy dissipation adjacent to an island. *J. Geophys. Res.* 83: 2939–2957.

Osborn, T.R. (1980) Estimates of the local rate of vertical diffusion from dissipation measurements. *J. Phys. Oceanogr.* 10: 83–89.

Owen, R.W. (1968) Oceanographic conditions in the northeast Pacific and their relation to the albacore fishery. *Fish. Bull. U.S.* 66: 503–526.

Packard, T.T., Blasco, D. and Barber, R.T. (1978) *Mesodinium rubrum* in the Baja California upwelling system, pp. 73–89. In: R. Boje and M. Tomczak (Eds) *Upwelling Ecosystems.* Springer-Verlag, Berlin.

Paine, A.I.S., Brink, K.H. and Hilborn, R. (Eds) *Benguela Trophic Functioning.* S. Afr. J. Mar. Sci. 12. Sea Fisheries Institute, Cape Town. 1108 pp.

Palomera, I. and Rubies, P. (1982) Kinds and distribution of fish eggs and larvae off northwest Africa in April/May 1973. *Rapp. P.-v. Réun. Cons. Int. Explor. Mer.* 180: 356–358.

Pape, E.H. and Garvine, R.W. (1982) The subtidal circulation in Delaware Bay and adjacent shelf waters. *J. Geophys. Res.* 87: 7955–7970.

Parker, C.E. (1971) Gulf Stream rings in the Sargasso Sea. *Deep-Sea Res.* 18: 981–993.

Parrish, R.H., Bakun, A., Husby, D.M. and Nelson, C.S. (1983) Comparative climatology of selected environmental processes in relation to eastern boundary current pelagic fish production, pp. 731–777. In: G.D. Sharp and J. Csirke (Eds) *Proceedings of the Expert Consultation to Examine Changes in Abundance and Species Composition of Neritic Fish Resources.* F.A.O. Fisheries Report 291: (3). Food and Agriculture Organization of the United Nations, Rome.

Parsons, T.R. and Lalli, C.M. (1988) Comparative oceanic ecology of the plankton communities of the subarctic Atlantic and Pacific Oceans. *Oceanogr. Mar. Biol. Ann. Rev.* 26: 317–359.

Pasciak, W.J. and Gavis, J. (1974) Transport limitation of nutrient uptake in phytoplankton. *Limnol. Oceanogr.* 19: 881–888.

Pasciak, W.J. and Gavis, J. (1975) Transport limited nutrient uptake rates in *Dytilum brightwellii. Limnol. Oceanogr.* 20: 604–617.

Patterson, M.R., Sebens, K.P. and Olson, R.R. (1991) *In situ* measurements of flow effects on primary production and dark respiration in reef corals. *Limnol. Oceanogr.* 36: 936–948.

Pauly, D. and Tsukayama, I. (1987) On the implementation of management-oriented fishery research: the case of the Peruvian anchoveta, pp. 1–13. In: D. Pauly and I. Tsukayama (Eds) *The Peruvian Anchoveta and its Upwelling Ecosystem: Three Decades of Change*. ICLARM Stud. Rev. 1987.

Pawlik, J.R. and Butman, C.A. (1993) Settlement of a marine tube worm as a function of current velocity: interacting effects of hydrodynamics and behavior. *Limnol. Oceanogr.* 38: 1730–1740.

Pawlik, J.R., Butman, C.A. and Starzak, V.R. (1991) Hydrodynamic facilitation of gregarious settlement of a reef-building worm. *Science* 251: 421–424.

Payne, A.I.L., Gulland, J.A. and Brink, K.H. (Eds) (1987) The Benguela and comparable ecosystems. *S. Afr. J. Mar. Sci.* 5: 1–956.

Pearcy, W. (1983) Abiotic variations in regional environments, pp. 30–34. In: W.S. Wooster (Ed) *From Year to Year*. Washington Sea Grant Publication, University of Washington, Seattle.

Pearcy, W., Fisher, J., Brodeur, R. and Johnson, S. (1985) Effects of the 1983 El Niño on coastal nekton off Oregon and Washington, pp. 188–204. In: W. S. Wooster and D. L. Fluharty (Eds) *El Niño North: Niño Effects in the Eastern Subarctic Pacific Ocean*. Washington Sea Grant Program, University of Washington, Seattle.

Pearcy, W.G. and Keene, D.F. (1974) Remote sensing of water colour and sea surface temperatures off the Oregon coast. *Limnol. Oceanogr.* 19: 573–583.

Pedlosky, J. (1990) The dynamics of the oceanic subtropical gyres. *Science* 248: 316–322.

Peffley, M.B. and O'Brien, J.J. (1976) A three-dimensional simulation of coastal upwelling off Oregon. *J. Phys. Oceanogr.* 6: 164–180.

Peinert, R. (1986) Production, grazing and sedimentation in the Norwegian coastal current, pp. 361–374. In: S. Skreslet (Ed) *The Role of Freshwater Outflow in Coastal Marine Ecosystems*. Springer-Verlag, Berlin.

Perry, R.I., Harding, G.C., Loder, J.W., Drinkwater, K.F., Tremblay, J., Taggart, C. and Sinclair, M. (1990) The Georges Bank frontal system: mechanisms of plankton retention or dispersal. International Council for the Exploration of the Sea. Council Meeting, 1990, collected papers. L8.

Peterson, W.T., Miller, C.B. and Hutchinson, A. (1979) Zonation and maintenance of copepod populations in the Oregon upwelling zone. *Deep-Sea Res.* 26: 467–494.

Petrie, B., Topliss, B.J. and Wright, D.W. (1987) Coastal upwelling and eddy development off Nova Scotia. *J. Geophys. Res.* 29 (C12): 12979–12991.

Petterssen, S. (1969) *Introduction to Meteorology*. McGraw-Hill, New York.

Phillips, O.M. (1977) *The Dynamics of the Upper Ocean*, 2nd Ed. Cambridge University Press, Cambridge, UK.

Pineda, J. (1991) Predictable upwelling and shoreward transport of planktonic larvae by internal tidal bores. *Science* 253: 548–550.

Pineda, J. (1994) Internal tidal bores in the nearshore: warm-water fronts, seaward gravity currents and the onshore transport of neustonic larvae. *J. Mar. Res.* 52: 427–458.

Pingree, R.D. (1978) Cyclonic eddies and cross-frontal mixing. *J. Mar. Biol. Ass. U.K.* 58: 955–963.

Pingree, R.D. (1979) Baroclinic eddies bordering the Celtic Sea in late summer. *J. Mar. Biol. Ass. U.K.* 59: 689–698.

Pingree, R. D., Bowman, M. J. and Esaias, W. E. (1978a) Headland fronts, pp. 78–86. In: M. J. Bowman and W. E. Esaias (Eds) *Oceanic Fronts in Coastal Processes*. Springer-Verlag, Berlin.

Pingree, R. D., Forster, G. R. and Morrison, G. K. (1974) Turbulent convergent tidal fronts. *J. Mar. Biol. Ass. U.K.* 54: 469–479.

Pingree, R.D. and Griffiths, D.K. (1980) A numerical model of the M2 tide in the Gulf of St. Lawrence. *Oceanologica Acta* 3: 221–225.

Pingree, R.D., Holligan, P.M., Mardell, G.T. and Head, R.N. (1976) The influence of physical stability on spring, summer and autumn phytoplankton blooms in the Celtic Sea. *J. Mar. Biol. Ass. U.K.* 56: 845–873.

Pingree, R.D., Holligan, P.M. and Mardell, G.T. (1978) The effects of vertical stability on phytoplankton distributions in the summer on the northwest European shelf. *Deep-Sea Res.* 25: 1011–1028.

Pingree, R.D. and Mardell, G.T. (1981) Slope turbulence, internal waves and phytoplankton growth at the Celtic Sea shelf-break. *Phil. Trans. Roy. Soc. Lond.* A 302: 663–682.

Pingree, R.D. and Mardell, G.T. (1985) Solitary internal waves in the Celtic Sea. *Prog. Oceanogr.* 14: 431–444.

Pingree, R.D., Mardell, G.T. and New, A.L. (1986) Propagation of the internal tides from the upper slopes of the Bay of Biscay. *Nature* 321: 154–158.

Pingree, R.D. and Pennycuik, L. (1975) Transfer of heat, fresh water and nutrients through the seasonal thermocline. *J. Mar. Biol. Ass. U.K.* 55: 261–274.

Pingree, R.D., Pugh, P.R., Holligan, P.M. and Forster, G.R. (1975) Summer phytoplankton blooms and red tides along tidal fronts in the approaches to the English Channel. *Nature* 258: 672–677.

Platt, T. (1984) Primary productivity in the central North Pacific: comparison of oxygen and carbon fluxes. *Deep-Sea Res.* 31: 1311–1319.

Platt, T. and Harrison, W.G. (1986) Reconciliation of carbon and oxygen fluxes in the upper ocean. *Deep-Sea Res.* 33: 55–58.

Platt, T., Harrison, W.G., Lewis, M.R., Li, W.K.W., Sathyendranath, S., Smith, R.E. and Vezina, A.F. (1989) Biological production of the oceans: the case for consensus. *Mar. Ecol. Prog. Ser.* 52: 77–88.

Platt, T., Prakash, A. and Irwin, B. (1972) Phytoplankton, nutrients and flushing of inlets on the coast of Nova Scotia. *Nat. Can.* 99: 253–261.

Platt, T. and Subba Rao, D.V. (1975) Primary production of marine microphytes, pp. 249–280. In: J.P. Cooper (Ed) *Photosynthesis and Productivity in Different Environments.* Cambridge University Press, Cambridge, UK.

Platt, T. and Sathyendranath, S. (1988) Oceanic primary production: estimation by remote sensing at local and regional scales. *Science* 241: 1613–1620.

Platt, T., Woods, J.D., Sathyendranath, S. and Barkmann, W. (1994) Net primary production and stratification in the ocean, pp. 247–254. In: D.M. Johannesen, R.D. Muench and D.E. Overland (Eds) *The Polar Oceans and Their Role in Shaping the Global Environment.* Geophysical Monograph 85. American Geophysical Union. Washington, DC. 525 pp.

Polovina, J.J., Mitchum, G.T., Graham, N.E., Craig, M.P., Demartini, E.E. and Flint, E.N. (1994) Physical and biological consequences of a climate event in the central north Pacific. *Fish. Oceanogr.* 3: 15–21.

Pond, S. and Pickard, G.L. (1983) *Introductory Dynamical Oceanography*, 2nd Ed. Pergamon Press, Oxford.

Prandtl, L. (1969) *The Essentials of Fluid Dynamics.* Blackie, London. 451 pp.

Price, H.J. and Paffenhofer, G.-A. (1980) Kinematographic analyses of the feeding of calanoid copepods over a range of algal sizes and concentrations. Abstr. Papers A.S.L.O. 3rd Winter Meeting.

Price, J.F., Weller, R.A. and Schudlich, R.R. (1987) Wind-driven ocean currents and Ekman transport. *Science* 238: 1534–1538.

Pritchard, D.W. (1967) What is an estuary: physical viewpoint, pp. 52–63. In: G.H. Lauff (Ed) *Estuaries.* American Association for the Advancement of Science, Washington, DC.

Prospero, J.M. and Savoie, D.L. (1989) Comparison of oceanic and continental sources of non-sea-salt sulphate over the Pacific Ocean. *Nature* 339: 685–687.

Provenzano, A.J., McConaugh, J.R., Philips, K.B., Johnson, D.F. and Clark, J. (1983) Vertical distribution of first stage larvae of the blue crab *Callinectes sapidus* at the mouth of Chesapeake Bay. *Estuar. Coast. Shelf Sci.* 16: 489–499.

Purcell, E.M. (1977) Life at low Reynolds number. *Am. J. Physics* 45: 3–11.

Purcell, E.M. (1978) The effect of fluid motions on the absorption of molecules by suspended particles. *J. Fluid Mech.* 84: 551–559.

Rasmusson, E.M. and Carpenter, T.H. (1982) Variations in tropical sea surface temperature and surface wind fields associated with the southern oscillation/El Niño. *Monthly Weather Rev.* 110: 354–384.

Rasmusson, E.M. and Wallace, J.M. (1983) Meteorological aspects of the El Niño/southern oscillation. *Science* 222: 1195–1202.

Rattray, M. and Hansen, D.V. (1962) A similarity solution for circulation in an estuary. *J. Mar. Res.* 20: 121–133.

Reid, J.L. and Shulenberger, E. (1986) Oxygen saturation and carbon uptake near 28 N, 155 W. *Deep-Sea Res.* 33: 267–271.

Revelante, N. and Gilmartin, M. (1976) The effects of Po River discharge on phytoplankton dynamics in the northern Adriatic Sea. *Mar. Biol.* 34: 259–271.

Revelle, R. (1982) Carbon dioxide and world climate. *Sci. Am.* 247: 35–43.

Revelle, R. (1983) The oceans and the carbon dioxide problem. *Oceanus* 26: 3–9.

Rhines, P. (1986) Vorticity dynamics of the oceanic general circulation. *Ann. Rev. Fluid Mech.* 18: 433–497.

Richards, F.A. (Ed) (1981) *Coastal Upwelling.* American Geophysical Union. Washington, DC. 529 pp.

Richardson, C.A., Crisp, D.J. and Runham, N.W. (1980) Factors influencing shell growth in *Cerastoderma edule.* *Proc. R. Soc. Lond. B* 210: 513–521.

Richardson, P.L. (1983) Gulf Stream rings, pp. 19–45. In: A.R. Robinson (Ed) *Eddies in Marine Science.* Springer-Verlag, New York.

Richardson, P.L., Cheney, R.E. and Worthington, L.V. (1978) A census of Gulf Stream rings, spring 1985. *J. Geophys. Res.* 83: 6136–6145.

Richman, J.G., de Szoeke, R.A. and Davis, R.E. (1987) Measurements of near-surface shear in the ocean. *J. Geophys. Res.* 92: (C3): 2851–2858.

Riley, G.A. (1937) The significance of the Mississippi River drainage for biological conditions in the northern Gulf of Mexico. *J. Mar. Res.* 1: 60–74.

Riley, G.A. (1941) Plankton studies IV: George's Bank. *Bull. Bingham Oceanogr. Coll.* 7: 1–73.

Riley, G.A. (1942) The relationship of vertical turbulence and spring diatom flowering. *J. Mar. Res.* 5: 67–87.

Riley, G.A. (1970) Particulate organic matter in sea water. *Adv. Mar. Biol.* 8: 1–118.

Riley, G.A., Stommel, H. and Bumpus, D.F. (1949) Quantitative ecology of the plankton of the western North Atlantic. *Bull. Bingham Oceanogr. Coll.* 12: 1–169.

The Ring Group. (1981) Gulf Stream cold-core rings: their physics, chemistry and biology. *Science* 212: 1091–1100.

Roberts, J. (1975) *Internal Gravity Waves in the Ocean.* Marcell Dekker, New York.

Roberts, L. (1989) Global warming: blaming the sun. *Science* 246: 992–993.

Robinson, A.R. (1983) Overview and summary of eddy science, pp. 3–15. In: A.R. Robinson (Ed) *Eddies in Marine Science.* Springer-Verlag, New York.

Robinson, C.K.L. (1994) The influence of ocean climate on coastal plankton and fish production. *Fish. Oceanogr.* 3: 159–171.

Robinson, G.A., Colebrook, J.M. and Cooper, G.A. (1975) The continuous plankton recorder survey: plankton in the ICNAF area in 1972. ICNAF Spec. Pub. No. 10: 95–103.

Roden, G.I. (1961) On the wind-driven circulation in the Gulf of Tehuantepec and its effects on sea-surface temperature. *Geofis. Int.* 1: 55–76.

Roemmich, D. and McGowan, J. (1995) Climatic warming and the decline of zooplankton in the California current. *Science* 267: 1324–1326.

Rogers, J.C. (1984) The association between the North Atlantic oscillation and the southern oscillation in the northern hemisphere. *Monthly Weather Rev.* 112: 1999–2015.

Roman, M.R., Gauzens, A.L. and Cowles, T.J. (1985) Temporal and spatial changes in epipelagic microplankton and mesoplankton biomass in warm-core Gulf Stream Ring 82-B. *Deep-Sea Res.* 32: 1007–1022.

Rothschild, B.J. (1988) Biodynamics of the sea: the ecology of high dimensionality systems, pp. 527–548. In: B.J. Rothschild (Ed) *Toward a Theory on Biological–Physical Interactions in the World Ocean.* Kluwer, Dordrecht. 650 pp.

Rothschild, B.J. and Osborn, T.R. (1988) Small-scale turbulence and plankton contact rates. *J. Plankt. Res.* 10: 465–474.

Rowe, G.T., Clifford, C.H. and Smith, K.L., Jr. (1977) Nitrogen regeneration in sediments off Cap Blanc, Spanish Sahara. *Deep-Sea Res.* 24: 57–63.

Rowe, P.M. and Epifanio, C.E. (1994) Flux and transport of larval weakfish in Delaware Bay, USA. *Mar. Ecol. Prog. Ser.* 110: 115–120.

Rowell, T.W. and Trites, R.W. (1985) Distribution of larval and juvenile *Illex* (Mollusca: Cephalopoda) in the Blake Plateau region (Northwest Atlantic). *Vie Milieu Ser. C* 35: 149–161.

Royce, W.F., Smith, L.S. and Hartt, A.C. (1968) Models of oceanic migrations of Pacific salmon and comments on guidance mechanisms. *Fish. Bull.* 66: 441–462.

Royer, T.C. (1989) Upper ocean temperature variability in the northeast Pacific: is it an indicator of global warming? *J. Geophys. Res.* 94: 18175–18183.

Rozengurt, M.A. and Haydock, I. (1991) Effects of freshwater development and water pollution policies on the world's river–delta–estuary–coastal zone ecosystems, pp. 85–99. In: H.S. Bolton (Ed) *Coastal Wetlands.* American Society of Civil Engineers, New York.

Rozengurt, M.A., Josselyn, M. and Hertz, M. (1985) The impact of water diversions on the river–delta–estuary–sea ecosystems of San Francisco Bay and the Sea of Azov, pp. 35–62. In: D.M. Goodrich (Ed) *San Francisco Bay: Issues, Resources, Status and Management.* NOAA Estuary-of-the-month Seminar Series No. 6. US Dept. Commerce, NOAA Estuarine Programs Office, Washington, DC.

Rubenstein, D.I. and Koehl, M.A.R. (1977) The mechanism of filter feeding: some theoretical considerations. *Am. Nat.* 111: 981–994.

Russell, F.S. (1973) A summary of the observations of the occurrence of planktonic stages of fish off Plymouth 1924–72. *J. Mar. Biol. Ass. U.K.* 53: 347–355.

Ryther, J.H. (1967) Occurrence of red water off Peru. *Nature* 214: 1318–1319.

Ryther, J.H. (1969) Photosynthesis and fish production in the sea. *Science* 166: 72–80.

Saiz, E. and Alcaraz, M. (1991) Effects of small-scale turbulence on development time and growth of *Acartia grani* (Copepoda: Calanoida). *J. Plankt. Res.* 13: 873–883.

Saiz, E. and Alcaraz, M. (1992) Free-swimming behaviour of *Acartia clausi* under turbulent water movement. *Mar. Ecol. Prog. Ser.* 80: 229–236.

Saiz, E., Alcaraz, M. and Paffenhöfer, G.-A. (1992) Effect of small-scale turbulence on feeding rate and gross growth efficiency of three *Acartia* species. *J. Plankt. Res.* 14: 1085–1097.

Sakshaug, E. and Slagstad, D. (1992) Sea ice and wind: effects on primary productivity in the Barents Sea. *Atmosphere–Ocean* 30: 579–591.

Sakshaug, E., Slagstad, D. and Holm-Hansen, O. (1991) Factors controlling the development of phytoplankton blooms in the Antarctic Ocean—A mathematical model. *Mar. Chem.* 35: 259–271.

Sambrotto, R.N. *et al.* (1993) Elevated consumption of carbon relative to nitrogen in the surface ocean. *Nature* 363: 248–250.

Sander, F. (1981) A preliminary assessment of the main causative mechanisms of the 'Island Mass' effect of Barbados. *Mar. Biol.* 64: 199–205.

Sandstrom, H. and Elliott, J.A. (1984) Internal tides and solitons on the Scotian Shelf: a nutrient pump at work. *J. Geophys. Res.* 89: 6415–6426.

Sarnthein, M., Winn, K., Duplessy, J.-C. and Fontugne, M.R. (1988) Global variations of surface ocean productivity in low and mid-latitudes: influence on CO_2 reservoirs of the deep ocean and the atmosphere during the last 21,000 years. *Paleoceanography* 3: 361–399.

Saunders, P.M. (1982) Circulation in the eastern North Atlantic. *J. Mar. Res.* 40: (suppl.) 641–657.

Savidge, G. (1976) A preliminary study of the distribution of chlorophyll-a in the vicinity of fronts in the Celtic and western Irish Seas. *Estuar. Coast. Mar. Sci.* 4: 617–625.

Savidge, G. (1981) Studies of the effects of small-scale turbulence on phytoplankton. *J. Mar. Biol. Ass. U.K.* 61: 477–488.

Savidge, G. and Foster, P. (1978) Phytoplankton biology of a thermal front in the Celtic Sea. *Nature* 271: 155–157.

Scheltema, R.S. (1966) Evidence for transatlantic transport of gastropod larvae belonging to the genus *Cymatium*. *Deep-Sea Res.* 13: 83–95.

Schmidt, J. (1922) The breeding places of the eel. *Phil. Trans. Roy. Soc. B* 211: 178–208.

Schneider, S.H. (1989) The greenhouse effect: science and policy. *Science* 243: 771–781.

Schumacher, G.J. and Whitford, L.A. (1965) Respiration and ^{32}P uptake in various species of freshwater algae as affected by current. *J. Phycol.* 1: 78–80.

Schwartz, S.E. (1988) Are global cloud albedo and climate controlled by marine phytoplankton? *Nature* 336: 441–445.

Shanks, A.L. (1983) Surface slicks associated with tidally forced internal waves may transport pelagic larvae of benthic invertebrates and fishes shoreward. *Mar. Ecol. Prog. Ser.* 13: 311–315.

Shanks, A.L. (1985) Behavioural basis of internal-wave induced shoreward transport of megalopae of the crab *Pachygrapsus crassipes*. *Mar. Ecol. Prog. Ser.* 24: 289–295.

Shanks, A.L. and Trent, J.D. (1979) Marine snow: microscale nutrient patches. *Limnol. Oceanogr.* 24: 850–854.

Shanks, A.L. and Wright, W.G. (1987) Internal-wave-mediated shoreward transport of cyprids, megalopae, and gammarids and correlated longshore differences in the settling rate of intertidal barnacles. *J. Exp. Mar. Biol. Ecol.* 114: 1–13.

Shannon, L.V. (1985) The Benguela ecosystem Part 1. Evolution of the Benguela, physical features and processes. *Oceanogr. Mar. Biol. Ann. Rev.* 23: 105–182.

Shea, R.E. and Broenkow, W.W. (1982) The role of internal tides in the nutrient enrichment of Monterey Bay, California. *Estuar. Coastal Shelf Sci.* 15: 57–66.

Sheldon, R.A, Prakash, A. and Sutcliffe, W. (1972) The size distribution of particles in the ocean. *Limnol. Oceanogr.* 17: 327–340.

Shelton, P.A. and Hutchings, L. (1982) Transport of anchovy, *Engraulis capensis* Gilchrist, eggs and early larvae by a frontal jet current. *J. Cons. Int. Explor. Mer.* 40: 185–198.

Shulenberger, E. and Reid, J.L. (1981) The Pacific shallow oxygen maximum, deep chlorophyll maximum and primary productivity reconsidered. *Deep-Sea Res.* 28: 901–919.

Shushkina, E.A. (1985) Production of principal ecological groups of plankton in the epipelagic zone of the ocean. *Oceanology* 25: 653–658.

Silver, M.W. and Alldredge, A.L. (1981) Bathypelagic marine snow: deep-sea algal and detrital community. *J. Mar. Res.* 39: 501–530.

Silver, M.W., Shanks, A.L. and Trent, J.D.F. (1978) Marine snow: microplankton habitat and source of small-scale patchiness in pelagic populations. *Science* 201: 371–373.

Simpson, J.H. (1971) Density stratification and microstructure in the western Irish Sea. *Deep-Sea Res.* 18: 309–319.

Simpson, J.H. (1981) The shelf-sea fronts: implications of their existence and behaviour. *Phil. Trans. R. Soc. Lond. A* 302: 531–546.

Simpson, J.H. and Bowers, D. (1981) Models of stratification and frontal movement in shelf seas. *Deep-Sea Res.* 28: 727–738.

Simpson, J.H. and Hunter, J.R. (1974) Fronts in the Irish Sea. *Nature* 250: 404–406.

Simpson, J.H. and Pingree, R.D. (1978) Shallow sea fronts produced by tidal stirring, pp. 29–42. In: M.J. Bowman and W.E. Esaias (Eds) *Oceanic Fronts in Coastal Processes.* Springer-Verlag, New York.

Simpson, J.H. and Tett, P.B. (1986) Island stirring effects on phytoplankton growth, pp. 41–76. In: M.J. Bowman, C.M. Yentsch and W.T. Peterson (Eds) *Tidal Mixing and Plankton Dynamics.* Springer-Verlag, New York.

Simpson, J.J. (1992) Response of the Southern California current system to the mid-latitude North Pacific warming events of 1982–1983 and 1940–1941. *Fish. Oceanogr.* 1: 57–79.

Sinclair, M. (1988) *Marine Populations: An Essay on Population Regulation and Speciation.* University of Washington Press, Seattle and London.

Sinclair, M., Bugden, G.L., Tang, C.L., Therriault, J.-C. and Yeats, P.A. (1986) Assessment of effects of freshwater runoff variability on fisheries production in coastal waters, pp. 139–160. In: S. Skreslet (Ed) *The Role of Freshwater Outflow in Coastal Marine Ecosystems.* Springer-Verlag, Heidelberg.

Sinclair, M., Tremblay, M.J. and Bernal, P. (1985) El Niño events and variability in a Pacific Mackerel (*Scomber japonicus*) survival index: support for Hjort's second hypothesis. *Can. J. Fish. Aquat. Sci.* 42: 602–608.

Sissenwine, M.P., Cohen, E.B. and Grosslein, M.D. (1984) Structure of the George's Bank ecosystem. *Rapp. P.-v. Réun. Cons. Int. Explor. Mer.* 183: 243–254.

Skreslet, S. (1986) *The Role of Freshwater Outflow in Coastal Marine Ecosystems.* Springer-Verlag, Berlin. 453 pp.

Smayda, T.J. (1970) The suspension and sinking of phytoplankton in the sea. *Oceanogr. Mar. Biol. Ann. Rev.* 8: 353–414.

Smith, P.C. (1989) Circulation and dispersal on Brown's Bank. *Can. J. Fish. Aquat. Sci.* 46: 539–559.

Smith, P.E. and Eppley, R.W. (1982) Primary production and the anchovy population in the southern California bight: comparison of time series. *Limnol. Oceanogr.* 27: 1–17.

Smith, P.E. and Lasker, R. (1978) Position of larval fish in an ecosystem. *Rapp. P.-v. Réun. Cons. Int. Explor. Mer.* 173: 77 84.

Smith, S.D. (1980) Wind stress and heat flux over the ocean in gale force winds. *J. Phys. Oceanogr.* 10: 709–726.

Smith, S.D. and Dobson, F.W. (1984) The heat budget at ocean weather ship Bravo. *Atmosphere–Ocean* 22: 1–22.

Smith, S.L. and Whitledge, T.E. (1977) The role of zooplankton in the regeneration of nitrogen in a coastal upwelling system off NW Africa. *Deep-Sea Res.* 24: 49–56.

Sommer, U. (1988) Some size relationships in phytoplankton motility. *Hydrobiologia* 161: 125–131.

Sorokin, Y.I., Sukhanova, I.N., Konolova, G.V. and Pavelyeva, E.B. (1975) Primary production and phytoplankton in the area of equatorial divergence in the equatorial Pacific. *Trans. Inst. Oceanol.* 102: 108–122.

Soutar, A. and Isaacs, J.D. (1969) History of fish populations inferred from fish scales in anaerobic sediments off California. *Calif. Coop. Ocean. Fish. Invest. Rep.* 13: 63–70.

Soutar, A. and Isaacs, J.D. (1974) Abundance of pelagic fish during the 19th and 20th centuries as recorded in anaerobic sediment off California. *Fish Bull.* 72: 257–273.

Southward, A.J., Boalch, G.T. and Maddock, L. (1988) Fluctuations in the herring and pilchard fisheries of Devon and Cornwall linked to changes in climate since the 16th century. *J. Mar. Biol. Ass. U.K.* 68: 423–445.

Stacey, M.W., Pond, S. and LeBlond, P.H. (1986) A wind-forced Ekman spiral as a good statistical fit to low-frequency currents in a coastal strait. *Science* 233: 470–472.

St. John, M.A., Harrison, P.J. and Parsons, T.R. (1992a) Tidal wake mixing: localized effects on primary production and zooplankton distribution in the Strait of Georgia, British Columbia. *J. Exp. Mar. Biol. Ecol.* 164: 261–274.

St. John, M.A., MacDonald, J.S., Beamish, R.J. and Choromanski, E. (1992b) The Fraser River plume: some preliminary observations on the distribution of juvenile salmon, herring and their prey. *Fish. Oceanogr.* 1: 153–162.

St. John, M.A. and Pond, S. (1992) Tidal plume generation around a promontory: effects on nutrient concentrations and primary productivity. *Continental Shelf Res.* 12: 339–354.

Steele, J.H. (1981) Some varieties of biological oceanography. In: B.A. Warren and C. Wunsch (Eds) *Evolution of Physical Oceanography.* M.I.T. Press, Cambridge, Mass.

Steele, J.H. (1988) Scale selection for biodynamic theories, pp. 513–526. In: B. J. Rothschild (Ed) *Towards a Theory on Biological-Physical Interactions in the World Ocean.* Kluwer, Amsterdam.

Steele, J.H. (1989) Discussion: scale and coupling in ecological systems, pp. 177–180. In: J. Roughgarden, R.M. May and S.A. Levin (Eds) *Perspectives in Ecological Theory.* Princeton University Press, Princeton, New Jersey.

Steele, J.H. and Henderson, E.W. (1984) Modeling long-term fluctuations in fish stocks. *Science* 224: 985–987.

Steele, J.H. and Yentsch, C.S. (1960) The vertical distribution of chlorophyll. *J. Mar. Biol. Ass. U.K.* 39: 217–226.

Stevens, D.E., Kohlhorst, D.W., Miller, L.W. and Kelly, D.W. (1985) The decline of striped bass in the Sacramento–San Joaquin Estuary, California. *Trans. Am. Fish Soc.* 114: 12–30.

Stommel, H. (1948) The westward intensification of wind-driven ocean currents. *Trans. Am. Geophys. Union.* 29: 202–206.

Stommel, H. (1958) The abyssal circulation. *Deep-Sea Res.* 5: 80–82.

Stommel, H. (1965) *The Gulf Stream—A Physical Description,* 2nd Ed. Cambridge University Press, Cambridge, UK.

Stommel, H., Niiler, P. and Anati, D. (1978) Dynamic topography and recirculation of the North Atlantic. *J. Mar. Res.* 36: 449 468.

Stramska, M. and Dickey, T.D. (1993) Phytoplankton blooms and the vertical structure of the upper ocean. *J. Mar. Res.* 51: 819–842.

Strass, V. and Woods, J.D. (1988) Horizontal and seasonal variation of density and chlorophyll profiles between the Azores and Greenland, pp. 113–136. In: B.J. Rothschild (Ed) *Towards a Theory on Biological–Physical Interactions in the World Ocean.* Kluwer, Dordrecht.

Strickler, J.R. (1984) Sticky water: a selective force in copepod evolution, pp. 187–242. In: D.G. Meyers and J.R. Strickler (Eds) *Trophic Interactions within Aquatic Ecosystems.* American Association for the Advancement of Science. Washington, DC.

Sundby, S., Ellertsen, B. and Fossum, P. (1994) Wind effects on the vertical distribution of first-feeding cod larvae and their prey, and on the encounter rates between them. Paper no. 26. In: J. Jacobsson *et al.* (Eds) *ICES Symposium on Cod and Climate Change.* International Council for the Exploration of the Sea, Copenhagen. 693 pp.

Sundby, S. and Fossum, P. (1990) Feeding conditions of Arcto-Norwegian cod larvae compared with the Rothschild-Osborne theory on small-scale turbulence and plankton contact rates. *J. Plankt. Res.* 12: 1153–1162.

Sundquist, E.T. (1993) The global carbon dioxide budget. *Science* 259: 934–941.

Sutcliffe, W.H. (1972) Some relations of land drainage, nutrients, particulate material and fish catches in two eastern Canadian bays. *J. Fish. Res. Bd. Can.* 29: 357–362.

Sutcliffe, W.H. (1973) Correlations between seasonal river discharge and local landings of the American lobster (*Homarus americanus*) and Atlantic halibut (*Hippoglossus hippoglossus*) in the Gulf of St. Lawrence. *J. Fish. Res. Bd. Can.* 30: 856–859.

Sutcliffe, W.H., Jr., Loucks, R.H., Drinkwater, K.F. and Coote, A.R. (1983) Nutrient flux onto the Labrador Shelf from Hudson Strait and its biological consequences. *Can. J. Fish. Aquat. Sci.* 40: 1692–1701.

Suthers, I.M. and Frank, K.T. (1989) Inter-annual distributions of larval and pelagic juvenile cod (*Gadus morhua*) in southwestern Nova Scotia determined with two different gear types. *Can. J. Fish. Aquat. Sci.* 46: 591–602.

Sverdrup, H.U. (1953) On conditions for the vernal blooming of phytoplankton. *J. Cons. Perm. Int. Exp. Mer.* 18: 287–295.

Sverdrup, H.U., Johnson, M.W. and Fleming, R.H. (1942) *The Oceans: Their Physics, Chemistry and General Biology.* Prentice-Hall, Englewood Cliffs, New Jersey. 1060 pp.

Swallow, J.C. (1976) Variable currents in mid-ocean. *Oceanus* 19: 18–25.

Swallow, J.C. and Hamon, B.V. (1960) Some measurements of deep-sea currents in the eastern North Atlantic. *Deep-Sea Res.* 6: 155–168.

Swift, J.H., Aagaard, K. and Malmberg, S.-A. (1980) The contribution of the Denmark Strait overflow to the deep North Atlantic. *Deep-Sea Res.* 27: 29–42.

Sy, A. (1988) Investigation of large-scale circulation patterns in the central North Atlantic: the North Atlantic current, the Azores current and the Mediterranean water plume in the area of the mid-Atlantic ridge. *Deep-Sea Res.* 35: 383–413.

Takahashi, T. (1989) The carbon dioxide puzzle. *Oceanus* 32 (2): 22–29.

Tanimoto, Y., Iwasaka, N., Hanawa, K. and Toba, Y. (1993) Characteristic variations of sea surface temperature with multiple time scales in the North Pacific. *J. Climate* 6: 1153–1160.

Taunton-Clark, J. and Shannon, L.V. (1988) Annual and inter-annual variability in the southeast Atlantic during the 20th century. *S. Afr. J. Mar. Sci.* 6: 97–106.

Taylor, A.H., Harris, J.R.W. and Aiken, J. (1986) The interaction of physical and biological processes in a model of the vertical distribution of phytoplankton under stratification, pp. 313–330. In: J.C. Nihoul (Ed) *Marine Interfaces Ecohydrodynamics.* Elsevier, New York.

Tee, K.T., Smith, P.C. and LeFaivre, D. (1993) Topographic upwelling off Southwest Nova Scotia. *J. Phys. Oceanogr.* 23: 1703–1726.

Templeman, W. (1972) Year class success in some North Atlantic stocks of cod and haddock. *Spec. Publ. Int. Comm. NW Atl. Fish.* 8: 223–239.

Tenore, K.R. *et al.* (1982) Coastal upwelling in the Rias Bajas, N.W. Spain: contrasting the benthic regimes of the Rias de Arosa and de Duras. *J. Mar. Res.* 40: 701–772.

Tett, P. (1981) Modelling phytoplankton production at shelf-sea fronts. *Phils. Trans. Roy. Soc. Lond.* A 302: 605–615.

Thomas, W.H. and Gibson, C.H. (1990) Quantified small-scale turbulence inhibits a red-tide dinoflagellate, *Gonyaulax polyedra* Stein. *Deep-Sea Res.* 37: 1583–1593.

Thomas, W.H. and Gibson, C.H. (1992) Effects of small-scale turbulence on the dinoflagellate *Gymnodinium sanguineum* (*splendens*): contrasts with *Gonyaulax* (*Lingulodinium*) *polyedra*, and the fishery implication. *Deep-Sea Res.* 39: 1429–1437.

Thompson, J.D. (1977) Ocean deserts and ocean oases. In: M. N. Glantz (Ed) *Desertification.* Westview Press, Boulder, Colorado.

Thompson, J.D. (1978) Role of mixing in the dynamics of upwelling systems, pp. 203–222. In: R. Boje and M. Tomczak (Eds) *Upwelling Ecosystems.* Springer-Verlag, Berlin.

Thomson, K.A., Ingraham, W. J., Healey, M.C., LeBlond, P.H., Groot, C. and Healey, C.G. (1992) The influence of ocean currents on latitude of landfall and migration speed of sockeye salmon returning to the Fraser River. *Fish. Oceanogr.* 1: 163–179.

Thordardottir, T. (1986) Timing and duration of spring blooming south and southwest of Iceland, pp. 345–360. In: S. Skreslet (Ed) *The Role of Freshwater Outflow in Coastal Marine Ecosystems*. Springer-Verlag, Berlin.

Tolmazin, D. (1985) Changing coastal oceanography of the Black Sea. 1: Northwest shelf. *Prog. Oceanogr.* 15: 217–276.

Townsend, D.W., Graham, J.J. and Stevenson, D.K. (1986) Dynamics of larval herring (*Clupea harengus*) production in tidally mixed waters of the eastern coastal Gulf of Maine, pp. 253–277. In: M.J. Bowman, C.M. Yentsch and W.D. Peterson (Eds) *Tidal Mixing and Plankton Dynamics*. Springer-Verlag, New York.

Traganza, E.D., Conrad, J.C. and Breaker, L.C. (1981) Satellite observations of a cyclonic upwelling system and a giant plume in the California current, pp. 228–241. In: F.A. Richards (Ed) *Coastal Upwelling*. Am. Geophys. Union, Washington, DC.

Traganza, E.D., Redalje, D.G. and Garwood, R.W. (1987) Chemical flux, mixed layer entrainment and phytoplankton blooms at upwelling fronts in the California coastal zone. *Continental Shelf Res.* 7: 89–105.

Traganza, E.D., Silva, V.M., Austin, D.M., Hanson, W.L. and Bronsink, S.H. (1983) Nutrient mapping and recurrence of coastal upwelling centers by satellite remote sensing: its implication to primary production and the sediment record, pp. 61–83. In: E. Suess and J. Thiede (Eds) *Coastal Upwelling: Its Sediment Record*. Part A. Plenum Press, New York.

Trager, G.C., Hwang, J.S. and Strickler, J.R. (1990) Barnacle suspension feeding in variable flow. *Mar. Biol.* 105: 117–127.

Tranter, D.J., Carpenter, D.J. and Leech, G.S. (1986) The coastal enrichment effect of the east Australian current eddy field. *Deep-Sea Res.* 33: 1705–1728.

Tranter, D.J., Parker, R.R. and Creswell, G.R. (1980) Are warm-core eddies unproductive? *Nature* 284: 540–542.

Tremblay, M. J. and Sinclair, M. (1992) Planktonic sea scallop larvae (*Placopecten megallanicus*) in the Georges Bank region: broadscale distribution in relation to physical oceanography. *Can. J. Fish. Aquat. Sci.* 49: 1597–1615.

Trenberth, K.E. (1990) Recent observed interdecadal climate changes in the northern hemisphere. *Bull. Amer. Meteor. Soc.* 71: 988–993.

Trites, R.W. (1983) Physical oceanographic features and processes relevant to *Illex illecebrosus* spawning in the western North Atlantic and subsequent larval distribution. *NAFO Sci. Coun. Studies* 6: 39–55.

Trumble, R.J., Mathisen, O.A. and Stuart, D.W. (1981) Seasonal food production and consumption by nekton in the northwest African upwelling system, pp. 458–463. In: F. A. Richards (Ed) *Coastal Upwelling*. Am. Geophys. Union, Washington, DC.

Tsujita, T. (1957) The fisheries oceanography of the east China Sea and Tsuchima Strait. 1. The structure and ecological character of the fishing grounds. *Bull. Seikai Reg. Fish. Lab.* 13: 1–47.

Tsukamoto, K. (1992) Discovery of the spawning area for the Japanese eel. *Nature* 356: 789–791.

Turner, J.S. (1973) *Buoyancy Effects in Fluids*. University Press, Cambridge, UK.

Tyler, A.V. and Westerheim, S.J. (1986) Effect of transport, temperature and stock size on recruitment of Pacific cod. *Int. N. Pac. Fish. Comm. Bull.* 47: 175–189.

Tyler, M.A. and Seliger, H.H. (1978) Annual subsurface transport of a red tide dinoflagellate to its bloom area: water circulation patterns and organism distributions in the Chesapeake Bay. *Limnol. Oceanogr.* 23: 227–246.

Tyler, M.A. and Seliger, H.H. (1981) Selection for a red tide organism: Physiological responses to the physical environment. *Limnol. Oceanogr.* 26: 310–324.

Uye, S-I., Yamaoka, T. and Fujisawa, T. (1992) Are tidal fronts good recruitment areas for herbivorous copepods? *Fish. Oceanogr.* 1: 216–232.

Valdivia, J.E. (1978) The anchoveta and El Niño. *Rapp. P.-v. Reun. Cons. Int. Explor. Mer.* 173: 196–202.

Valentine, P.C. and Lough, R.G. (1991) The sea-floor environment and the fishery of eastern Georges Bank. Open-File Report 91-439. *US Geol. Survey.* NMFS Woods Hole, Mass. 25 pp.

van Loon, H. and Rogers, J.C. (1978) The seesaw in winter temperatures between Greenland and northern Europe. Part I. General description. *Monthly Weather Rev.* 106: 296–310.

Vandevelde, T., Legendre, L., Theriault, J.-C., Demers, S. and Bah, A. (1987) Surface chlorophyll maximum and hydrodynamics of the water column. *J. Mar. Res.* 45: 377–396.

Vincent, A. and Kurc, G. (1969) Hydrologie: variations saisonnierres de la situation thermique du Golfe de Gascogne en 1967. *Rév. Trav. Inst. Peches Maritimes.* 33: 203–212.

Vinogradov, M.E. (1981) Ecosystems of equatorial upwellings, pp. 69–93. In: A.R. Longhurst (Ed) *Analysis of Marine Ecosystems*. Academic Press, London. 741 pp.

Vogel, S. (1981) *Life in Moving Fluids: The Physical Biology of Flow*. Willard Grant Press, Boston. 352 pp.

Von Schwind, J.J. (1980) *Geophysical Fluid Dynamics for Oceanographers*. Prentice-Hall, New York. 307 pp.

Waldron, H.N. and Probyn, T.A. (1992) Nitrate supply and potential new production in the Benguela system. *S. Afr. J. Mar. Sci.* 12: 29–39.

Walsh, J.J. (1983) Death in the sea: enigmatic phytoplankton losses. *Prog. Oceanogr.* 12: 1–86.

Walsh, J.J. and Dieterle, D.A. (1988) Use of satellite ocean colour observations to refine understanding of global geochemical cycles, pp. 287–317. In: T. Rosswall, R.G. Woodmansee and P.G. Risser (Eds). *Scales and Global Change.* Wiley, New York.

Walsh, J.J., Rowe, G.T., Iverson, R.L. and McRoy, C.P. (1981) Biological export of shelf carbon is a sink of the global CO_2 cycle. *Nature* 291: 196–201.

Walsh, J.J., Whitledge, T.E., Esaias, W., Smith, R.L., Huntsman, S.A., Santander, H. and de Mendiola, B.R. (1980) The spawning habitat of the Peruvian anchovy, *Engraulis ringens. Deep-Sea Res.* 27: 1–27.

Ware, D.M. and Thomson, R.E. (1991) Link between long-term variability in upwelling and fish production in the northeast Pacific Ocean. *Can. J. Fish. Aquat. Sci.* 48: 2296–2306.

Warren, B. (1983) Why is no deep water formed in the North Pacific? *J. Mar. Res.* 41: 327–347.

Warren, B.A. (1981) Deep circulation of the world ocean. In: B.A. Warren and C. Wunsch (Eds) *Evolution of Physical Oceanography.* Scientific Surveys in Honour of Henry Stommel. M.I.T. Press, Cambridge, Mass. 620 pp.

Watanabe, T. and Mizuno, K. (1994) Decadal changes in the thermal structure in the North Pacific. *International WOCE Newsletter* 15: 10–13.

Webb, K.L. and D'Elia, C.F. (1980) Nutrient and oxygen redistribution during a spring neap tidal cycle in a temperate estuary. *Science* 207: 983–985.

Webster, P.J. (1983) Large scale structure of the tropical atmosphere, pp. 235–276. In: B. Hoskins and R. Pearce (Eds) *Large Scale Dynamical Processes in the Atmosphere.* Academic Press, New York. 397 pp.

Weinstein, M.P., Weiss, S.L., Hodson, R.G. and Gerry, L.R. (1980) Retention of three taxa of postlarval fishes in an intensely flushed tidal estuary, Cape Fear River, North Carolina. *Fish. Bull.* 78: 419–435.

Westlake, D.F. (1967) Some effects of low-velocity currents on the metabolism of aquatic macrophytes. *J. Exp. Bot.* 18: 187–205.

Wheeler, W.N. (1978) Ecophysiological studies on the giant kelp *Macrocystis,* Ph.D. Thesis. University of California, Santa Barbara.

Wheeler, W.N. (1980) Effect of boundary layer transport on the fixation of carbon by the giant kelp *Macrocystis pyrifera. Mar. Biol.* 56: 103–110.

White, F.M. (1988) *Heat and Mass Transfer.* Addison-Wesley, New York.

Whitford, L.A. and Schumacher, G.J. (1961) Effect of current on mineral uptake and respiration by a freshwater alga. *Limnol. Oceanogr.* 6: 423–425.

Wildish, D.J. and Kristmanson, D.D. (1979) Tidal energy and sublittoral macrobenthic animals in estuaries. *J. Fish. Res. Bd. Can.* 36: 1197–1206.

Wildish, D.J. and Peer, D.L. (1983) Tidal current speed and production of benthic macrofauna in the lower Bay of Fundy. *Can. J. Fish. Aquat. Sci.* 40 (suppl. 1.): 309–321.

Wildish, D.J. and Kristmanson, D.D. (1984) Importance to mussels of the benthic boundary layer. *Can. J. Fish. Aquat. Sci.* 41: 1618–1625.

Williams, P.J. Le B. and Muir, L.R. (1981) Diffusion as a constraint on the biological importance of microzones in the sea, pp. 209–218. In: J.C. Nihoul (Ed) *Ecohydrodynamics.* Elsevier, Amsterdam.

Witman, J.D., Leichter, J J., Genovese, S.J. and Brooks, D.A. (1993) Pulsed phytoplankton supply to the rocky subtidal zone: Influence of internal waves. *Proc. Natl. Acad. Sci. USA* 90: 1686–1690.

Wolanski, E. and Hamner, W.M. (1988) Topographically controlled fronts in the ocean and their biological influence. *Science* 241: 177–181.

Wolanski, E., Drew, E., Abel, K.M. and O'Brien, J. (1988) Tidal jets, nutrient upwelling and their influence on the productivity of the alga *Halimeda* in the Ribbon Reefs, Great Barrier Reef. *Estuar. Coast. Shelf Sci.* 26: 169–201.

Wolf, K.U. and Woods, J.D. (1988) Lagrangian simulation of primary production in the physical environment— the deep chlorophyll maximum and nutricline, pp. 51–70. In: B.J. Rothschild (Ed) *Towards a Theory on Biological-Physical Interactions in the World Ocean.* Kluwer, Dordrecht.

Wolff, W.J. (1977) A benthic food budget for the Gravelingen Estuary, the Netherlands, and a consideration of the mechanisms causing high benthic secondary production in estuaries, pp. 267–280. In: B.C. Coull (Ed) *Ecology of Marine Benthos.* Belle W. Baruch Library in Marine Science No. 6. University of South Carolina Press.

Wong, C.K. (1980) Copepod predation and prey defense. M.Sc. Thesis, University of Ottawa, Ontario.

Wood, L. and Hargis, W.J. (1971) Transport of bivalve larvae in a tidal estuary, pp. 21–44. In: D.J. Crisp (Ed) *4th European Marine Biology Symposium.* Cambridge University Press, Cambridge.

Woods, J.D. (1968) Wave-induced shear instability in summer thermocline. *J. Fluid Mech.* 32: 791–800.

Woods, J.D. (1977) Parameterization of unresolved motions, pp. 118–140. In: E.B. Kraus (Ed) *Modelling and Prediction of the Upper Layers of the Ocean.* Pergamon Press, New York.

Woods, J.D. (1988) Scale upwelling and primary production, pp. 7–38. In: B.J. Rothschild (Ed) *Towards a Theory on Biological-Physical Interactions in the World Ocean.* Kluwer, Dordrecht.

Woods, J.D. and Barkmann, W. (1986) The response of the upper ocean to solar heating. I. The mixed layer. *Quart. J. R. Met. Soc.* 112: 1–27.

Woods, J.D. and Onken, R. (1982) Diurnal variation and primary production in the ocean—preliminary results of a Lagrangian ensemble model. *J. Plankton Res.* 4: 735–756.

Wooster, W.S., Bakun, A. and McLain, D.R. (1976) The seasonal upwelling cycle along the eastern boundary of the North Atlantic. *J. Mar. Res.* 34: 131–141.

Wooster, W.S., and Fluharty, D.L. (Eds) (1985) *El Niño North: Niño Effects in the Eastern Subarctic Pacific Ocean.* Washington Sea Grant, Seattle. 361 pp.

Worthington, L.V. (1976) On the North Atlantic circulation. *Johns Hopkins Studies in Oceanography*, No. 6, 110 pp. Johns Hopkins Marine Laboratory, Monterey, Calif.

Wright, D.G., Hendry, R.M., Loder, J.W. and Dobson, F.W. (1986) Oceanic changes associated with global increase in atmospheric carbon dioxide: a preliminary report for the Atlantic coast of Canada. *Can. Tech. Rep. Fish. Aquat. Sci.* 1426. 78 pp.

Wroblewski, J.S. (1982) Interaction of currents and vertical migration in maintaining *Calanus marshallae* in the Oregon upwelling zone—a simulation. *Deep-Sea Res.* 29: 665–686.

Wroblewski, J.S. and Cheney, J. (1984) Ichthyoplankton associated with a warm-core ring off the Scotian Shelf. *Can. J. Fish. Aquat. Sci.* 41: 294–303.

Wyatt, T. and Jenkinson, I.R. (1993) The North Atlantic turbine: views of production processes from a mainly North Atlantic perspective. *Fish. Oceanogr.* 2: 231–243.

Wyrtki, K. (1962) The upwelling in the region between Java and Australia during the south-east monsoon. *Aust. J. Mar. Freshwater Res.* 13: 217–225.

Wyrtki, K. (1964) Upwelling in the Costa Rica dome. *Fish. Bull. U.S.* 63: 335–372.

Xie, L. and Hsieh, W.W. (1989) Predicting the return migration routes of the Fraser River sockeye salmon (*Onchorhynchus nerka*). *Can. J. Fish. Aquat. Sci.* 46: 1287–1292.

Yamanaka, H. (1985) Effect of El Niño on fish migration and yield in the western Pacific Ocean, pp. VI-16 to VI-22. In: *International Conference on the TOGA Scientific Programme.* WMO/TD No. 65. World Climate Research Programme Publication Series No. 4.

Yamazaki, H., Osborn, T.R. and Squires, K.D. (1991) Direct numerical simulation of planktonic contact in turbulent flow. *J. Plankt. Res.* 13: 629–644.

Yen, J., Sanderson, B., Strickler, J.R. and Okubo, A. (1991) Feeding currents and energy dissipation by *Eucheta rimana*, a subtropical pelagic copepod. *Limnol. Oceanogr.* 36: 362–369.

Yentsch, C.S. and Phinney, D.A. (1985) Rotary motions and convection as a means of regulating primary production in warm core rings. *J. Geophys. Res.* 90: (C2) 3237–3248.

Yoder, J.A., Atkinson, L.P., Bishop, S.S., Hofmann, E.E. and Lee, T.N. (1983) Effect of upwelling on phytoplankton productivity of the outer southeastern United States continental shelf. *Continental Shelf Res.* 4: 385–404.

Yoder, J.A., Atkinson, L.P., Lee, T.N., Kim, H.H. and McLain, C.R. (1981) Role of Gulf Stream frontal eddies in forming phytoplankton patches on the outer southeast shelf. *Limnol. Oceanogr.* 26: 1103–1110.

Zebiak, S.E. and Cane, M.A. (1987) A model El Niño–Southern Oscillation. *Monthly Weather Rev.* 115: 2262–2278.

Zeldis, J.R. and Jillett, J.B. (1982) Aggregation of pelagic *Munida gregaria* (Fabricius) (Decapoda, Anomura) by coastal fronts and internal waves. *J. Plankt. Res.* 4: 839–857.

Index